Jean Cleymans
A Life for Physics

Jean Cleymans
A Life for Physics

Editors

Raghunath Sahoo
Dinesh Kumar Srivastava
Edward Sarkisyan-Grinbaum
Airton Deppman

Basel • Beijing • Wuhan • Barcelona • Belgrade • Novi Sad • Cluj • Manchester

Editors

Raghunath Sahoo
Indian Institute of
Technology Indore
Indore
India

Dinesh Kumar Srivastava
National Institute of
Advanced Studies
Bengaluru
India

Edward Sarkisyan-Grinbaum
Experimental Physics
Department, CERN
Geneva
Switzerland

Airton Deppman
Instituto de Física,
Universidade de Sao Paulo
Sao Paulo
Brazil

Editorial Office
MDPI AG
Grosspeteranlage 5
4052 Basel, Switzerland

This is a reprint of articles from the Special Issue published online in the open access journal *Physics* (ISSN 2624-8174) (available at: https://www.mdpi.com/journal/physics/special_issues/JeanCleymans).

For citation purposes, cite each article independently as indicated on the article page online and as indicated below:

Lastname, A.A.; Lastname, B.B. Article Title. *Journal Name* **Year**, *Volume Number*, Page Range.

ISBN 978-3-7258-1167-0 (Hbk)
ISBN 978-3-7258-1168-7 (PDF)
doi.org/10.3390/books978-3-7258-1168-7

Cover image courtesy of Raghunath Sahoo
Prof. Jean Cleymans during 2015 in Indore, India.

Contents

Preface

This Special Issue is dedicated to the memory of our friend and colleague, Professor Jean Cleymans, Emeritus Professor of Physics at the University of Cape Town, South Africa, who unexpectedly passed away on the 22nd of February 2021. We were, and still are, shocked. Professor Cleymans was known as an active person, a great organizer, and a brilliant collaborator, always ready to help. This is a big loss for science; he made fundamental contributions to our understanding of particle production in high-energy physics, new matter formation in heavy ion collisions, theoretical and experimental studies of quantum chromodynamics, and statistical approaches in particle physics, such as the statistical thermal model and non-extensivity, on which he was a leading figure.

His participation in the STAR experiment at the Relativistic Heavy Ion Collider at the Brookhaven National Laboratory, USA, and in the ALICE experiment at the Large Hadron Collider at CERN, Geneva, Switzerland, was of high importance and lent the results a great visibility and world-wide recognition. Prof. Cleymans conceived, planned, organized, and was successfully leading the South Africa–CERN Program within the UCT-CERN Research Centre in the University of Cape Town.

We are honored to organize this Special Issue, which is a call for contributions mixed with personal experiences, scientific studies, and reviews—a real tribute to the contribution of Professor Cleymans to high-energy physics.

**Raghunath Sahoo, Dinesh Kumar Srivastava, Edward Sarkisyan-Grinbaum, and
Airton Deppman**
Editors

physics

Editorial

Jean Cleymans: Scientist, Mentor, and Friend Extraordinaire

Dinesh Kumar Srivastava

National Institute of Advanced Studies, Bengaluru 560012, India; dinesh.srivastava@nias.res.in

Citation: Srivastava, D.K. Jean Cleymans: Scientist, Mentor, and Friend Extraordinaire. *Physics* **2022**, *4*, 690–696. https://doi.org/10.3390/physics4030047

Received: 5 February 2022
Accepted: 17 June 2022
Published: 22 June 2022

Publisher's Note: MDPI stays neutral with regard to jurisdictional claims in published maps and institutional affiliations.

1. Evolution of a Friendship

When in 1988, I decided to start working on the physics of quark-gluon plasma and relativistic heavy-ion collisions, I was reasonably well-entrenched in my chosen field of low-energy nuclear reactions and break-up of light nuclei, having worked for over 17 years in that field. I had to first brush up on the details of my high-energy physics concepts and then become acquainted with the field in general, before starting my work on electromagnetic probes of quark-gluon plasma, for which my experimental colleagues were preparing to build detectors. The most useful introductory sources included a review by J. Cleymans et al. [1], the crisp and brief yet exhaustive book by B. Müller [2], and *Applications of Perturbative QCD* by R.D. Field [3], to which I often returned and sent students, whenever I had doubts. "QCD" stands here for quantum chromodynamics.

After these introductory readings, I started working on photons and dileptons and read several papers by Jean on low-mass dileptons from the bremsstrahlung of quarks and gluons. One important point of these papers was that the provided details allowed me to derive every step and reproduce all the results, without any difficulty, and then proceed to use those treatments for my own studies. During the summer of 1993, I attended an extended workshop at Santa Barbara, where Jean was also present. Unfortunately, I had only a very brief overlap with his stay. However, during his talk on the topic of low-mass dileptons, I asked him a question.

We were still in the days of overhead projectors and transparencies. He just went to the whiteboard and in his extremely clean board work and elegant handwriting, worked out the entire derivation. I was stunned. I can hardly take a class without holding notes in my hands, and I still write down all my points on the slides for my talks. After the talk, I went to him, introduced myself, and told him that I was reading his papers and conducting some follow-up work as a result, adding, "Sir, you must be a very popular teacher!" He blushed—as only he could—and which I was to witness again and again during years of association, and it started a friendship and collaboration with a scientist extraordinaire, who became my mentor and a close friend and confidant for years to come. A student, Dipali Pal, wrote several papers on low-mass dilepton production, and we benefitted from more discussions with him on these.

Our discussions continued over e-mails. It became intense and very fruitful when Helmut Satz and Xin-Nian Wang planned a collaborative project on hard probes, inviting several of us to CERN (European Organization for Nuclear Research, Geneva, Switzerland), LBNL (Lawrence Berekeley, National Laboratory, Berkeley, CA, USA), Trento, Jyvaskyla, and Lisbon, in turn, to work on our predictions for many hard probes using perturbative QCD calculations [4], which provided a strong platform for a lasting association for many practitioners of hard probes of quark-gluon plasma. Many of these associations were further cemented during various conferences and workshops. Jean came to most of the International Conferences on Physics and Astrophysics of Quark-Gluon Plasma, which we organised in India, and we benefitted from his valuable advice in organising these events.

In early 1996, Jean invited me to Cape Town (Figure 1) for a visit lasting about a month. Without me knowing, he had also invited Krzysztof Redlich for a part of that period. I knew Krzysztof from his famous paper [5] but had no previous acquaintance with him.

Both were excited about our paper on single photons in S + Au collisions [6], and very intense discussions started about all the inputs of that work.

Figure 1. Jean Cleymans in the balcony of his office in Cape Town.

However, let me pause here to recall the hospitality of Jean during this period. Almost every day, he took us out for excellent dining, and on weekends, he drove us down to some of the most beautiful places, of which there is no scarcity in Cape Town. His wife, Ria, and daughters, Silvie and Silke, extended a very warm welcome to us and won our deep affection and admiration by their patience with three grown-ups arguing most vociferously about the applicability of hydrodynamics, formation time, and production of photons, and none of them yielding an inch of space! I wish to put on record the extreme patience shown by Ria during these discussions, which often lasted till late in the evening, and I would like to express my gratitude to the family for sharing Jean with us.

2. More on Our Scientific Collaborations

These discussions resulted in incorporating a new equation of state for a hot hadronic matter, which included all the hadrons in the particle data book in chemical equilibrium and led to two publications [6,7]. The first one [6] was later to form the basis for an explanation of single photons in Pb + Pb collisions at CERN SPS (Super Proton Synchrotron), BNL RHIC (Brookhaven National Laboratory, Relativistic Heavy Ion Collider, Upton, NY, USA), and CERN LHC (Large Hadron Collider) in terms of the formation of quark-gluon plasma, and the second [7] provided a prediction for the variation of the number of photons produced as a function of charged particle multiplicity in such collisions [8], experimentally confirmed some time ago.

Jean invited me to Cape Town, twice again—once for the excellent Strange Quark Matter meeting in Cape Town in 2004, held inside a pre-apartheid era prison (!), and then for a longer visit—both under an Indo-South Africa Collaboration program that we had established. Rekha, my wife, joined me during my later visit and considers it the most enjoyable and satisfying foreign trip she ever made with me (Figure 2).

Figure 2. Rekha, the author, and Jean: Table Mountain, Cape Town.

I was to meet some of his brilliant collaborators—Duncan Elliot, David Hislop, Bruce Becker, Sarah-Louise Blyth, Mark Horner, Spencer Wheaton, Azwinndini Muronga, and Zeblon Vilakazi—many of whom became our friends for life. Zeblon and Azwinndini were also to visit India later.

Alas, a plan to watch a Cricket Test Match at the famed Eden Gardens Stadium in Kolkata with Duncan could not materialise due to his very untimely death in a mountaineering accident in the Andes. Jean also organised that I deliver lectures at iThemba Labs, the University of Stellenbosch, and the University of Western Cape, which were very valuable experiences for me and won me many new friends.

Jean was extremely popular with our students, and several of them worked with him as post-docs. Some of these collaborations started over discussions during lunch or dinner. Thus, Raghunath Sahoo, while still a PhD student, had made a very interesting observation that transverse energy deposited in nucleus–nucleus collisions divided by the number of charged particles produced at SPS energies was independent of the centrality of the collisions. The value was slightly larger but also near-constant at RHIC energies. We were having lunch at an unusual restaurant designed similar to a truck driver's dining place when Raghunath used paper napkins to explain his experimental findings. Soon, it was realised that since most of the particles were pions, and the measurements were at a given rapidity (near zero at RHIC), this could possibly be understood in terms of the famous paper of Jean with Krzysztof, which showed that $\langle E \rangle / \langle N \rangle$ (where E denotes the total energy and N denotes the number of particles produced) for nuclear collisions at all energies from SIS (Schwer-Ionen-Synchrotron, Darmstadt, Germany) energies to LHC was about 1 GeV/nucleon [9]—thus started a collaboration, which expanded to include the Tsallis statistics to describe all the features of particle production in nucleus–nucleus, proton–nucleus, and proton–proton collisions at relativistic energies, in a long series of papers. Jean visited the Indian Institute of Technology, Indore, where Raghunath had joined and established a very vibrant group to meet his young collaborators there, who remember his inspirational visit with awe and affection (Figure 3).

Figure 3. Jean Cleymans at Rani Roopmati's Palace, Mandu (near Indore).

The list of students from India who worked with him is large—Danish Azmi (Figure 4) and Trambak Bhattacharya were post-docs with him, and several others worked closely with him and made valuable contributions. Many of his collaborations with students from India started when they approached him with some questions. This happened with Natasha Sharma and many others. His friendly and generous nature made him extremely approachable, and he revelled in the achievements of the students, which further endeared him to them.

Figure 4. Jean Cleymans with Danish Azmi and the young Azmi, Cape Town.

3. A Glimpse of Jean's Scientific Contributions

Several others will write more extensively and eloquently about his scientific contributions. I, for one, consider many of his studies to be among the most valuable and having a lasting effect on the field—namely, his pioneering and exhaustive work on the application of Tsallis statistics for a quantitative and detailed description of particle production in relativistic collisions of nucleons and nuclei [10]; his observation that particle ratios in such collisions provided a robust measure of chemical equilibration between them [11]; his paper [9] in which he established the boundary of (chemical) freeze-out in such collisions for all centre of mass energies; and his study with Jorgen Randrup [12], which used the large body of his studies on ratios of particle production and their momentum distributions to establish that the maximum freeze-out density has a baryonic chemical potential of 400–500 MeV, which is above the critical value, and that it is reached for a fixed-target bombarding energy of 20–30 GeV/A; this provided a most convincing justification for FAIR (Darmstadt, Germany) and NICA (Dubna, Russia) facilities as well as a hope for a bountiful harvest of exciting results for the nuclear equation of state to study neutron stars.

His contribution to establishing a theoretical and experimental school for the study of relativistic heavy-ion collisions and particle physics in South Africa is too well-known. He trained his students well and thoroughly, and they were welcomed at all laboratories across the world; they have surely by now established a collaborative network of their own. Jean established the University of Cape Town—CERN Centre, initiated a collaboration with the ALICE Collaboration (CERN), and set up a grid computing facility there. Zeblon Vilakazi went on to head the South Africa—ATLAS Collaboration (CERN). Jean further initiated an extensive collaboration with the Joint Institute of Nuclear Research, Dubna, towards the building of NICA (Nuclotron-based Ion Collider fAcility) and its future utilisation.

His generous hospitality and great standing in the field (Figure 5), so evident from numerous awards and honours, also brought scientists from across the world to the many prestigious conferences he organised, which opened the world to the students of South Africa.

Figure 5. Ria and Jean Cleymans at CERN.

I worked for about a year after my retirement on a book entitled *Climate Change and Energy Options for a Sustainable Future*, which drew extensively on years of my efforts at science outreach and efforts at dispelling fears about nuclear energy. I shared a draft copy of it with him, well before it was published. He liked it immensely and praised it whole-heartedly when I had a video chat with him around mid-February 2021, when the book was about to be released. He ordered several copies of it to be given as a present to friends and relatives. He wanted to make a present of one of these to Ria on her birthday. It arrived on her birthday, but Ria and I will always regret that he was not there to present it to her in person (Figure 6).

Figure 6. Ria Cleymans.

4. Epilogue

I have only talked about the scientific contributions of Jean Cleymans (Figure 7). His sense of humour was legendary. His patience while training students was phenomenal. He was extremely well-read, had a vast and deep knowledge of history, literature, music, and culture, and spoke several languages. He introduced me to several excellent authors and historians. This opened another facet of his and his family to me. As I occasionally indulged in writing short stories in English, I would give them to him and Ria and invariably receive warm and affectionate comments. Ria remains one of the closest friends and a confidant of my wife, Rekha.

It is also extremely tragic that Ria lost her elder daughter, Sylvie, soon after Jean's passing. My family and I hope that she and her daughter, Silke, find the courage and strength to bear this double tragedy and loss.

Figure 7. Jean Willy André Cleymans (5 August 1944–22 February 2021).

Funding: This research received no external funding.

Acknowledgments: This is a very personal homage to Jean Cleymans, a dear friend, a brilliant scientist, a generous and caring host, and a great human being. His scientific contributions have been discussed here very briefly. The honours that Jean received from across the world, and in South Africa, are all too well-known. I have used photographs taken by different friends at different times, from my collection, and I do not even remember who took these pictures. I thank them all in advance and apologise for not seeking their permission before using these. I would also like to apologise to the friends and colleagues whose names I have forgotten to mention in my grief.

Conflicts of Interest: The author declares no conflict of interest.

References

1. Cleymans, J.; Gavai, R.; Suhonen, E. Quarks and gluons at high temperatures and densities. *Phys. Rep.* **1986**, *130*, 217–292. [CrossRef]
2. Müller, B. *The Physics of Quark Gluon Plasma*; Springer: Berlin/Heidelberg, Germany, 1985.
3. Field, R.D. *Applications of Perturbative QCD*; Addison-Wesley Publishing Company, Inc.: Redwood City, CA, USA, 1989. Available online: https://www.desy.de/~jung/qcd_and_mc_2009-2010/R.Field-Applications-of-pQCD.pdf (accessed on 10 June 2022).
4. Satz, H.; Wang, X.N. Hard processes in p + A and A + A collisions. Introduction. *Int. J. Mod. Phys. E* **2003**, *12*, 147–148. [CrossRef]
5. Baier, R.; Nakkagawa, H.; Niégawa, A.; Redlich, K. Production rate of hard thermal photons and screening of quark mass singularity. *Z. Phys. C* **1992**, *53*, 433–438. [CrossRef]
6. Srivastava, D.K.; Sinha, B. Single photons from S + Au collisions at the CERN Super Proton Synchrotron and the quark-hadron phase transition. *Phys. Rev. Lett.* **1994**, *73*, 2421–2424. [CrossRef] [PubMed]
7. Cleymans, J.; Redlich, K.; Srivastava, D.K. Thermal particle and photon production in Pb+ Pb collisions with transverse flow. *Phys. Rev. C* **1997**, *55*, 1431–1442. [CrossRef]
8. Cleymans, J.; Redlich, K.; Srivastava, D.K. Equation of state of hadronic matter and electromagnetic radiation from relativistic heavy ion collisions. *Phys. Lett. B* **1998**, *420*, 261–266. [CrossRef]
9. Cleymans, J.; Redlich, K. Unified description of freeze-out parameters in relativistic heavy ion collisions. *Phys. Rev. Lett.* **1998**, *81*, 5284–5286. [CrossRef]
10. Cleymans, J. The Tsallis Distribution for p–p collisions at the LHC. *J. Phys. Conf. Ser.* **2013**, *455*, 012049. [CrossRef]
11. Cleymans, J.; Satz, H. Thermal hadron production in high energy heavy ion collisions. *Z. Phys. C* **1993**, *57*, 135–147. [CrossRef]
12. Cleymans, J.; Randrup, J. Maximum freeze-out baryon density in nuclear collisions. *Phys. Rev. C* **2006**, *74*, 047901. [CrossRef]

Article

Jet Transport Coefficient at the Large Hadron Collider Energies in a Color String Percolation Approach [†]

Aditya Nath Mishra [1], Dushmanta Sahu [2] and Raghunath Sahoo [2,3,*]

[1] Wigner Research Centre for Physics, 29-33 Konkoly-Thege Miklós Str., 1121 Budapest, Hungary; aditya.nath.mishra@cern.ch

[2] Department of Physics, Indian Institute of Technology Indore, Simrol, Indore 453552, India; dushmanta.sahu@cern.ch

[3] European Organization for Nuclear Research (CERN), CH 1211 Geneva, Switzerland

* Correspondence: raghunath.sahoo@cern.ch

[†] This paper is dedicated to the loving memory of Professor Jean Cleymans, a teacher, collaborator, an excellent human being and a great source of inspiration.

Abstract: Within the color string percolation model (CSPM), jet transport coefficient, $\hat{q}$, is calculated for various multiplicity classes in proton-proton and centrality classes in nucleus-nucleus collisions at the Large Hadron Collider energies for a better understanding of the matter formed in ultra-relativistic collisions. $\hat{q}$ is studied as a function of final state charged particle multiplicity (pseudorapidity density at midrapidity), initial state percolation temperature and energy density. The CSPM results are then compared with different theoretical calculations from the JET Collaboration those incorporate particle energy loss in the medium.

Keywords: jet quenching; color string percolation; quark-gluon plasma

Citation: Mishra, A.N.; Sahu, D.; Sahoo, R. Jet Transport Coefficient at the Large Hadron Collider Energies in a Color String Percolation Approach. *Physics* **2022**, *4*, 315–328. https://doi.org/10.3390/physics4010022

Received: 12 January 2022
Accepted: 23 February 2022
Published: 16 March 2022

Publisher's Note: MDPI stays neutral with regard to jurisdictional claims in published maps and institutional affiliations.

1. Introduction

The main objective of tera-electron volt energy heavy-ion collisions is to form a quark–gluon plasma (QGP)—the deconfined state of quarks and gluons, by creating extreme conditions of temperature and/or energy density [1,2], a scenario that might have been the case after a few microseconds of the creation of the universe. Jets, collimated emission of a multitude of hadrons originating from the hard partonic scatterings, play an important role as hard probes of QGP. These hard jets lose their energy through medium-induced gluon radiation and collisional energy loss, as a consequence of which one observes suppression of high transverse momentum particles and the phenomenon is known as jet quenching [3–9]. This is a direct signature of a highly dense partonic medium, usually formed in high energy heavy-ion collisions. The first evidence of the jet quenching phenomenon has been observed at the Relativistic Heavy-Ion Collider (RHIC) [10–23] via the measurement of inclusive hadron and jet production at high transverse momentum (p_T), γ-hadron correlation, di-hadron angular correlations and the dijet energy imbalance. The jet quenching phenomena are also widely studied in heavy-ion collisions at the Large Hadron Collider (LHC) [24–38]. All the measured observables are found to be strongly modified in central heavy-ion collisions relative to minimum bias proton-proton collisions, when compared to expectations based on treating heavy-ion collisions as an incoherent superposition of independent nucleon-nucleon collisions.

A number of theoretical models that incorporate parton energy loss have been proposed to study the observed jet quenching phenomena, namely, Baier–Dokshitzer–Mueller–Peigne–Schiff–Zakharov (BDMPS-Z) [7,39,40], Gyulassy–Levai–Vitev (GLV) [41–43] and its CUJET implementation [44], high-twist (HT) approach (HT-M (Majumder) and HT-BW (Berkeley–Wuhan)) [45–49], Amesto–Salgado–Wiedemann (ASW) [50,51], Arnold–Moore–Yaffe (AMY) model [52,53], MARTINI model [54], BAMPS model [55], and linear Boltzmann transport (LBT) model [56]. Most of the theoretical models assumed a static potential for jet-medium interactions, which result in a factorized dependence of parton energy loss on the jet transport coefficient, $\hat{q}$. The coefficient, which describes the average transverse momentum square transferred from the traversing parton per unit mean free path, is a common parameter that modulates the energy loss of jets in a strongly-interacting quantum chromodynamics (QCD) medium [7,9]. $\hat{q}$ is also related to the gluon distribution density of the medium and therefore characterizes the medium property as probed by an energetic jet [7,57]. Thus the collision energy and system size dependence study of jet transport coefficient will not only improve our understanding of experimental results on jet quenching but also can directly provide some information about the internal structure of the hot and dense QCD matter [57,58].

In the present paper, $\hat{q}$ and its relation with various thermodynamic properties of the QCD matter are studied in the framework of the color string percolation model (CSPM) [59–64] which is inspired by QCD. This can be used as an alternative approach to color glass condensate (CGC) [64] and is related to the Glasma approach [65]. In CSPM, it is assumed that color strings are stretched between the projectile and the target, which may decay into new strings via $q\bar{q}$ pair production and subsequently hadronize to produce observed hadrons [66]. These color strings may be viewed as small discs in the transverse plane filled with color field created by colliding partons. The final state particles are produced by the Schwinger mechanism, emitting $q\bar{q}$ pairs in this field [67]. With the increasing collision energy and size of the colliding nuclei, the number of strings grows and they start interacting to form clusters in the transverse plane. This process is very much similar to discs in the 2-dimensional percolation theory [60,62,68,69]. At a certain critical density, called critical percolation density ($\xi_c \geq 1.2$), a macroscopic cluster appears that marks the percolation phase transition [60,62,68–71]. The combination of the string density dependent cluster formation and the 2-dimensional percolation clustering phase transition are the basic elements of the non-perturbative CSPM. In CSPM, the Schwinger barrier penetration mechanism for particle production and the fluctuations in the associated string tension due to the strong string interactions make it possible to define a temperature. The critical density of percolation is related to the effective critical temperature and thus percolation may provide information on deconfinement in the high-energy collisions [63,64]. The CSPM approach has been successfully used to describe the initial stages in the soft region in high-energy collisions [59,64,68,72–77]. In addition to this, CSPM has also been quite successful in estimating various thermodynamic and transport properties of the matter formed in ultra-relativistic energies [78–84].

The paper runs as follows. Section 2 presents the formulation and methodology of the CSPM approach. Section 3 presents the results and discussions. Finally, the important findings of this study are summarized in Section 4.

2. Formulation and Methodology

In the CSPM, the charged hadron multiplicity, μ_n, where n stands for the number of strings in a cluster, reduces with the increase of string interactions while the mean of the squared transverse momentum, $\langle p_T^2 \rangle_n$, of these charged hadrons increases, to conserve the total transverse momentum. The μ_n and the $\langle p_T^2 \rangle_n$ of the particles produced by a cluster are proportional to the color charge and color field, respectively [62,64], and can be defined as

$$\mu_n = \sqrt{\frac{nS_n}{S_1}}\mu_1; \quad \langle p_t^2 \rangle_n = \sqrt{\frac{nS_1}{S_n}}\langle p_T^2 \rangle_1, \tag{1}$$

where S_n denotes the transverse overlap area of a cluster of n-strings and the subscript '1' refers to a single string with a transverse overlap area $S_1 = \pi r_0^2$ with the string radius, $r_0 = 0.2$ fm [64], respectively. For the case when strings are just touching each other $S_n = nS_1$, and $\mu_n = n\mu_1$, $\langle p_T^2 \rangle_n = \langle p_T^2 \rangle_1$. When strings fully overlap $S_n = S_1$ and therefore $\mu_n = \sqrt{n}\mu_1$ and $\langle p_T^2 \rangle_n = \sqrt{n}\langle p_T^2 \rangle_1$, so that the multiplicity is maximally suppressed and the $\langle p_T^2 \rangle_n$ is maximally enhanced. This implies a simple relation between the multiplicity and transverse momentum $\mu_n \langle p_T^2 \rangle_n = n\mu_1 \langle p_T^2 \rangle_1$, which denotes the conservation of the total transverse momentum. In the thermodynamic limit, one can obtain the average value of nS_1/S_n for all the clusters [60,62] as

$$\left\langle n\frac{S_1}{S_n} \right\rangle = \frac{\xi}{1-e^{-\xi}} \equiv \frac{1}{F(\xi)^2}. \tag{2}$$

Here, $F(\xi)$ is the color suppression factor by which the overlapping strings reduce the net-color charge of the strings. With $F(\xi) \to 1$ as $\xi \to 0$ and $F(\xi) \to 0$ as $\xi \to \infty$, where $\xi = N_s S_1/S_N$ is the percolation density parameter. Equation (1) can be written as $\mu_n = nF(\xi)\mu_1$ and $\langle p_T^2 \rangle_n = \langle p_T^2 \rangle_1/F(\xi)$. It is worth noting that CSPM is a saturation model, similar to the CGC, where $\langle p_T^2 \rangle_1/F(\xi)$ plays the same role as the saturation momentum scale Q_s^2 in the CGC model [65,85].

In the present study, $F(\xi)$ in proton-proton (pp) collisions at the center-of-mass energies $\sqrt{s}$ = 5.02 and 13 TeV is extracted for various multiplicity classes using ALICE experiment results on transverse momentum spectra of charged particles [86]. In case of Pb-Pb collisions at the nucleon-nucleon center-of-mass energies $\sqrt{s_{NN}}$ = 2.76 and 5.02 TeV [87] and Xe-Xe collisions at $\sqrt{s_{NN}}$ = 5.44 TeV [88], $F(\xi)$ values are obtained from the centrality-dependent transverse-momentum spectra of charged particles measured by ALICE. To evaluate the initial value of $F(\xi)$ from data, a parameterization [68] of the experimental data of p_T distribution in low-energy pp collisions at $\sqrt{s}$ = 200 GeV (minimum bias), where strings have very low overlap probability, is used. The p_T spectrum of charged particles can be described by a power-law [64]:

$$\frac{d^2N_{ch}}{dp_T^2} = \frac{a}{(p_0 + p_T)^\alpha}, \tag{3}$$

where a is the normalisation factor and p_0, α are fitting parameters given as, $p_0 = 1.98$ and $\alpha = 12.87$ [64]. This parameterization is used in high-multiplicity pp and centrality-dependent heavy-ion (AA) collisions to take into account the interactions of the strings [64]. The parameter p_0 in Equation (3) is for independent strings and gets modified to

$$p_0 \to p_0 \left(\frac{\langle nS_1/S_n \rangle^{\mathrm{mod}}}{\langle nS_1/S_n \rangle_{pp}} \right)^{1/4}. \tag{4}$$

Using Equations (4) and (2) in Equation (3), one gets:

$$\frac{d^2N_{ch}}{dp_T^2} = \frac{a}{(p_0\sqrt{F(\xi)_{pp}/F(\xi)^{\mathrm{mod}}} + p_T)^\alpha}, \tag{5}$$

where $F(\xi)^{\mathrm{mod}}$ is the modified color suppression factor and is used in extracting $F(\xi)$ both in pp and AA collisions. The spectra were fitted using Equation (5) in the softer sector with p_T in the range 0.15–1.0 GeV/c, where c is the speed of light. In pp collisions at low energies, only two strings are considered to exchange with low probability of interactions, so that $\langle nS_1/S_n \rangle_{pp} \approx 1$, which transforms Equation (5) into

$$\frac{d^2N_{ch}}{dp_T^2} = \frac{a}{(p_0\sqrt{1/F(\xi)^{\mathrm{mod}}} + p_T)^\alpha}. \tag{6}$$

In the thermodynamic limit, the color suppression factor $F(\xi)$ is related to the percolation density parameter, ξ, as

$$F(\xi) = \sqrt{\frac{1-e^{-\xi}}{\xi}}. \tag{7}$$

3. Results and Discussion

In the present study, $F(\xi)$ is extracted in the multiplicity-dependent pp collisions at $\sqrt{s}$ = 5.02 and 13 TeV [86] and centrality-dependent Pb-Pb collisions at $\sqrt{s_{NN}}$ = 2.76 and 5.02 TeV [87], and Xe-Xe collisions at $\sqrt{s_{NN}}$ = 5.44 TeV [88] from the charged particles p_T spectra, measured by the ALICE experiment at the LHC.

Figure 1 shows ξ and $F(\xi)$ as functions of final charged particle scaled pseudorapidity density at midrapidity (hereafter, multiplicity, for brevity) for pp, Xe-Xe and Pb-Pb collisions. The error in $F(\xi)$ is obtained by changing the fitting ranges of the transverse momentum spectra and is found within ~3%. For a better comparison of pp and AA collisions, $\langle dN_{ch}/d\eta \rangle$, where N_{ch} is the measured charged particle multiplicity, is scaled by the transverse overlap area, $S_{\perp}$, for both pp and AA collisions. For pp collisions, multiplicity-dependent $S_{\perp}$ is calculated using the IP-Glasma model [89]. In the case of heavy-ion collisions, the transverse overlap area is obtained using the Glauber model calculations [90]. It is observed that $F(\xi)$ falls onto a universal scaling curve for proton-proton and nucleus-nucleus collisions. Particularly, in the most central heavy-ion collisions (high number of tracks) and high-multiplicity pp collisions, $F(\xi)$ values fall in a line. This suggests that the color suppression factor is independent of collision energies and collision systems in the domain of high final state multiplicity. Further, what decides the color suppression factor is the final state multiplicity density of the system, which turns out to be the initial parton density in a system for the case of an isentropic expansion.

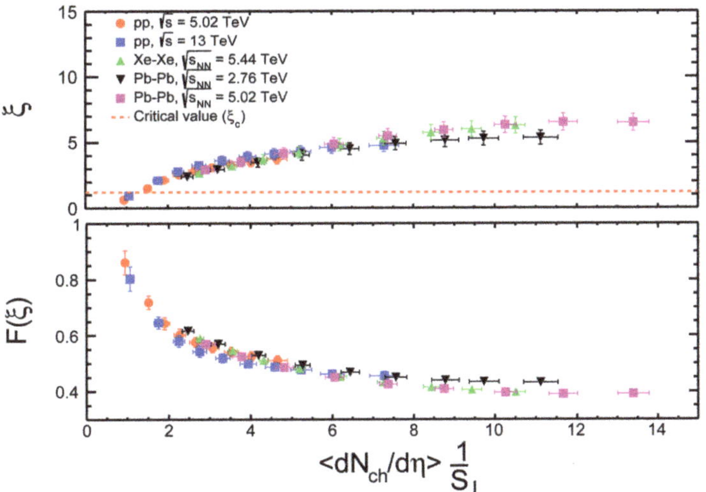

Figure 1. Percolation density parameter, ξ (**upper panel**), and color suppression factor, $F(\xi)$ (**bottom panel**), as functions of charged particle multiplicity (the pseudorapidity density at midrapidity) within $|\eta| < 0.8$ scaled with the transverse overlap area $S_{\perp}$ in pp, Xe-Xe and Pb-Pb collisions. For pp collisions, multiplicity-dependent $S_{\perp}$ is obtained from IP-Glasma model [89]. In case of Xe-Xe and Pb-Pb collisions, $S_{\perp}$ values are obtained using the Glauber model [90]. See text for details.

3.1. Temperature

The connection between $F(\xi)$ and the initial percolation temperature $T(\xi)$ involves the Schwinger mechanism for particle production [63,64,67] and can be expressed as [63,68]

$$T(\xi) = \sqrt{\frac{\langle p_T^2 \rangle_1}{2F(\xi)}}. \tag{8}$$

Here, one adopts the point of view that the universal hadronization temperature, T_h, is a good measure of the upper end of the cross-over phase transition temperature [91]. The single string average transverse momentum $\langle p_T^2 \rangle_1$ is calculated at the critical percolation density parameter $\xi_\text{c} = 1.2$ with the universal hadronization temperature $T_\text{h} = 167.7 \pm 2.6$ MeV [91]. This gives $\sqrt{\langle p_T^2 \rangle_1} = 207.2 \pm 3.3$ MeV.

In this way, at $\xi_\text{c} = 1.2$, the connectivity percolation transition at $T(\xi_\text{c})$ models the thermal deconfinement transition. The temperature obtained for most central Pb-Pb collisions at $\sqrt{s_\text{NN}} = 2.76$ TeV found to be of $\sim$223 MeV, whereas the direct photon measurement up to $p_T < 10$ GeV/c gives the initial temperature $T_\text{i} = 297 \pm 12(\text{stat}) \pm 41(\text{syst})$ much clearer. MeV for 0–20% central Pb-Pb collisions at $\sqrt{s_\text{NN}} = 2.76$ TeV measured by the ALICE Collaboration [92]. The measured temperature shows that the temperature obtained using Equation (8) can be termed as the temperature of the percolation cluster.

Figure 2 shows a plot of initial temperature from CSPM as a function of $\langle dN_\text{ch}/d\eta \rangle$ scaled by $S_\perp$. Temperatures from both pp and AA collisions fall on a universal curve when multiplicity is scaled by the transverse overlap area. The horizontal line at $\sim$167.7 MeV is the universal hadronization temperature obtained from the systematic comparison of the statistical thermal model parametrization of hadron abundances measured in high energy e^+e^-, pp and AA collisions [91]. One can see that temperature for higher multiplicity classes in pp collisions at $\sqrt{s} = 5.02$ and 13 TeV, are higher than the hadronization temperature and similar to those observed in Xe-Xe at $\sqrt{s_\text{NN}} = 5.44$ TeV and Pb-Pb collisions at $\sqrt{s_\text{NN}} = 2.76$ and 5.02 TeV.

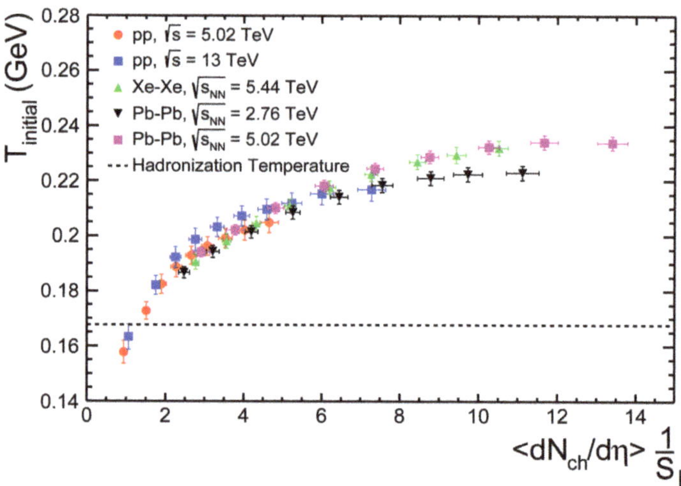

Figure 2. Initial percolation temperature vs. $\langle dN_\text{ch}/d\eta \rangle$, scaled by $S_\perp$, from pp, Pb-Pb and Xe-Xe collisions. The line $\sim$167.7 MeV is the universal hadronization temperature [91].

3.2. Energy Density

The calculation of the bulk properties of hot QCD matter and characterization of the nature of the QCD phase transition is one of the most important and fundamental problems in finite-temperature QCD. The QGP, according to CSPM, is born in local thermal equilibrium because the temperature is determined at the string level. Beyond the initial temperature, $T > T_\text{c}$ the CSPM perfect fluid may expand according to Bjorken boost invariant 1-dimension hydrodynamics [93]. In this framework, the initial energy density is given by:

$$\varepsilon = \frac{3}{2} \frac{\frac{dN_\text{ch}}{dy} \langle m_T \rangle}{S_\text{N} \tau_\text{pro}}, \tag{9}$$

where ε is the energy density, S_N is the transverse overlap area and τ_{pro}, the production time for a boson (gluon), is described by [94]

$$\tau_{\text{pro}} = \frac{2.405\hbar}{\langle m_T \rangle}. \tag{10}$$

Here, $m_T = \sqrt{m^2 + p_T^2}$ is the transverse mass and $\hbar$ is the reduced Planck constant. For evaluating ε, the charged particle multiplicity (rapidity density) dN_{ch}/dy at midrapidity is used and m is taken as the pion mass (pions being the most abundant particles in the multiparticle production process such as that discussed here), which gives the lower bound of the energy density. For the estimation of $\langle m_T \rangle$, the p_T spectra of pions at different collision energies and collision species in the p_T range 0.15 GeV/c $< p_T < 1$ GeV/c are used.

The purpose of estimating the initial percolation temperature and the initial energy density in the framework of CSPM is to study the jet transport coefficient as a function of these global observables for different collision species and collision energies at the LHC. Let us now proceed to estimate $\hat{q}$ in the CSPM framework.

3.3. Jet Transport Coefficient

The final state hadrons, produced in ultra-relativistic collisions at large transverse momenta, are strongly suppressed in central collisions compared to peripheral collisions. This suppression of hadrons at high p_T, which is usually referred to as jet quenching, is believed to be the result of the parton energy loss induced by multiple collisions in the strongly interacting medium. Thus, we are encouraged to study the jet transport coefficient, $\hat{q}$, which encodes the parton energy loss in the medium. It is also related to the p_T broadening of the energetic partons propagating inside the medium. In kinetic theory framework, $\hat{q}$ can be estimated by the formula [95],

$$\hat{q} = \rho \int d^2q_\perp \, q_\perp^2 \frac{d\sigma}{d^2q_\perp}, \tag{11}$$

where ρ is the number density of the constituents of the medium, $q_\perp$ is the transverse momentum exchange between the jet and the medium, and $d\sigma/d^2q_\perp$ denotes the differential scattering cross-section of the particles inside the medium.

The jet transport coefficient, $\hat{q}$, and the shear viscosity-to-entropy density ratio, η/s, transport parameters describing the exchange of energy and momentum between fast partons and medium, are directly related to each other as [57,96–98]

$$\frac{\eta}{s} \approx \frac{3}{2} \frac{T^3}{\hat{q}}. \tag{12}$$

Within the CSPM approach, η/s can be expressed as [64,83]

$$\frac{\eta}{s} = \frac{TL}{5(1 - e^{-\xi})}, \tag{13}$$

here L is the longitudinal extension of the string ~ 1 fm [63]. One can get final expression for jet transport coefficient from Equation (12) as:

$$\hat{q} \approx \frac{3}{2} \frac{T^3}{\eta/s} \approx \frac{15}{2} \frac{T^2(1 - e^{-\xi})}{L}. \tag{14}$$

The jet quenching parameter $\hat{q}$ is plotted as a function of initial percolation temperature in Figure 3. Interestingly, one observes a linear increase in $\hat{q}$, with the increase in temperature for both pp and AA collisions. At low temperatures, the value of jet quenching parameter is around 0.02 GeV^3. This value increases gradually and at high temperatures, it reaches the value around 0.08 GeV^3.

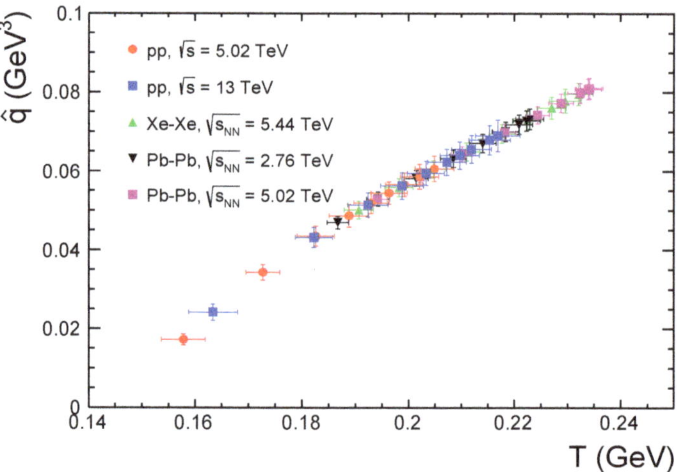

Figure 3. Jet quenching parameter, $\hat{q}$, as a function of temperature within the color string percolation model (CSPM) for *pp* collisions at $\sqrt{s}$ = 5.02 and 13 TeV, Xe-Xe collisions at $\sqrt{s_{\mathrm{NN}}}$ = 5.44 TeV and Pb-Pb collisions at $\sqrt{s_{\mathrm{NN}}}$ = 2.76 and 5.02 TeV.

The JET Collaboration has also extracted $\hat{q}$ values from five different hydrodynamic approaches with the initial temperatures of 346–373 MeV and 447–486 MeV for the most central Au-Au collisions at $\sqrt{s} = 200$ GeV at RHIC and Pb-Pb collisions at $\sqrt{s} = 2.76$ TeV at LHC, respectively [58]. The variation of $\hat{q}$ values between different hydrodynamic models is considered as theoretical uncertainties. The scaled jet quenching parameter $\hat{q}/T^3$ at the highest temperatures reached in the most central Au-Au and Pb-Pb collisions are [58].

$$\frac{\hat{q}}{T^3} \approx \begin{cases} 4.6 \pm 1.2 \text{ at RHIC}, \\ 3.7 \pm 1.4 \text{ at LHC}. \end{cases}$$

The corresponding absolute values for $\hat{q}$ for a 10 GeV quark jet are,

$$\hat{q} \approx \begin{cases} 0.23 \pm 0.05 \\ 0.37 \pm 0.13 \end{cases} \text{GeV}^3 \text{ at } \begin{array}{l} T = 346\text{–}373 \text{ MeV (RHIC)}, \\ T = 447\text{–}486 \text{ MeV (LHC)}, \end{array}$$

at an initial time $\tau_0 = 0.6$ fm/c. In this study, charged particle spectra are used to calculate $\hat{q}$ within the CSPM approach, so one cannot reach the initial temperature published by the JET Collaboration. Therefore, the $\hat{q}$ obtained is significantly smaller than the value reported by the JET Collaboration for the most central Pb-Pb collisions at $\sqrt{s} = 2.76$ TeV at the LHC.

In Figure 4, $\hat{q}$ is plotted as a function of charged particle multiplicity scaled with transverse overlap area for *pp* collisions at $\sqrt{s}$ = 5.02 and 13 TeV, Xe-Xe collisions at $\sqrt{s_{\mathrm{NN}}}$ = 5.44 TeV and Pb-Pb collisions at $\sqrt{s_{\mathrm{NN}}}$ = 2.76 and 5.02 TeV. One can see that $\hat{q}$ shows a steep increase at lower charged particle multiplicities in *pp* collisions and gets saturated at very high multiplicity for all studied energies. This behaviour suggests that at lower multiplicities, the system is not dense enough to highly quench the partonic jets, whereas with the increase of multiplicity, the quenching of jets becomes more prominent.

The dimensionless parameter, T^3-scaled $\hat{q}$ is shown in Figure 5 as a function of charged particle multiplicity scaled with transverse overlap area. In the low multiplicity regime, one can see a steep increase in $\hat{q}/T^3$, and after reaching a maximum at $\langle dN_{\mathrm{ch}}/d\eta \rangle / S_{\perp} \sim 2$, it starts decreasing regardless of the collision system or collision energy. The decrease in $\hat{q}/T^3$ is faster in Pb-Pb and Xe-Xe as compared to the *pp* collisions.

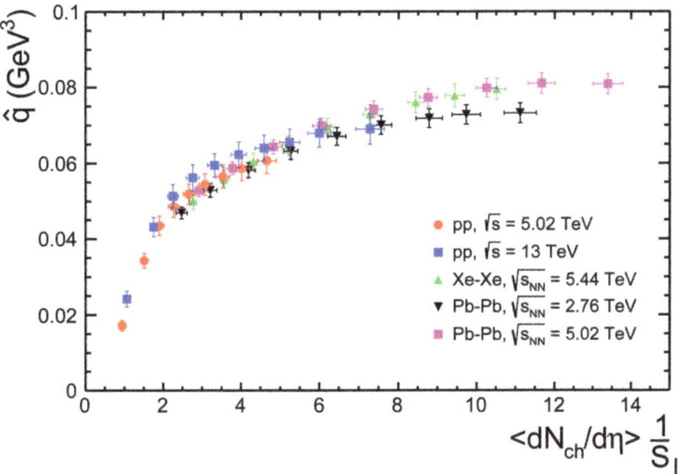

Figure 4. Jet quenching parameter, $\hat{q}$, as a function charged particle multiplicity scaled with transverse overlap area ($S_\perp$) within the CSPM for pp collisions at $\sqrt{s}$ = 5.02 and 13 TeV, Xe-Xe collisions at $\sqrt{s_{NN}}$ = 5.44 TeV and Pb-Pb collisions at $\sqrt{s_{NN}}$ = 2.76 and 5.02 TeV.

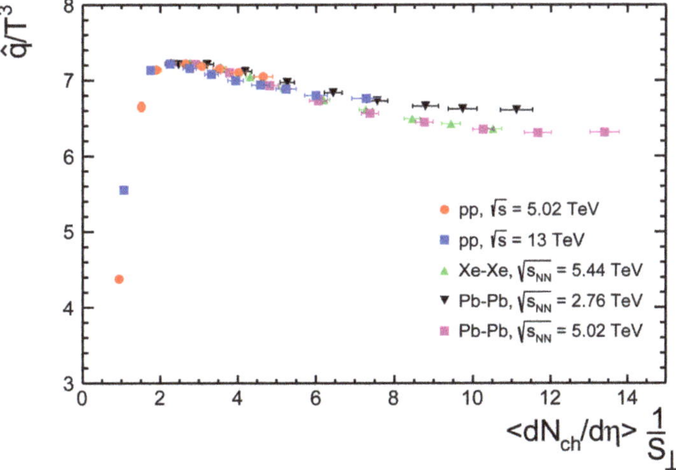

Figure 5. $\hat{q}/T^3$ vs. charged particle multiplicity, scaled by $S_\perp$, for pp collisions at $\sqrt{s}$ = 5.02 and 13 TeV, Xe-Xe collisions at $\sqrt{s_{NN}}$ = 5.44 TeV and Pb-Pb collisions at $\sqrt{s_{NN}}$ = 2.76 and 5.02 TeV.

The variation of $\hat{q}$ as a function of initial energy density is shown in Figure 6. To have a better understanding, the results obtained here are compared with that of cold nuclear matter, massless hot pion gas and ideal QGP calculations [99]. The comparison shows that the CSPM result obtained here is closer to the massless hot pion gas at low energy density. As initial energy density increases, $\hat{q}$ values increase and then show a saturation towards heavy-ion collisions, which produce a denser medium. The saturation behaviour, observed at high energy densities suggests that $\hat{q}$ remains unaffected after a certain energy density. Similar behaviour is observed when $\hat{q}$ is studied as a function of multiplicity (see Figure 4). The jet energy loss inside a denser QCD medium goes towards saturation after a threshold in the final state multiplicity is reached. If one compares the behaviour of η/s as a function of T/T_c for $T > T_c$ (the domain of validity of CSPM), an increasing trend is observed, which is expected to be reflected in a reverse way in the observable $\hat{q}/T^3$.

However, the interplay of higher temperature and lower η/s decides the high temperature behavior of $\hat{q}$ as shown in Figure 6. Further, one observes the CSPM based estimations of $\hat{q}$ showing a deviation from the ideal QGP behaviour for energy densities higher than $1\,\text{GeV}/\text{fm}^3$. This is because the ideal QGP calculations of Ref. [99], assumes ϵ/T^4 a constant value, whereas the CSPM-based estimations show an increasing trend of ϵ/T^4 towards high temperature(energy density or final state multiplicity) [84].

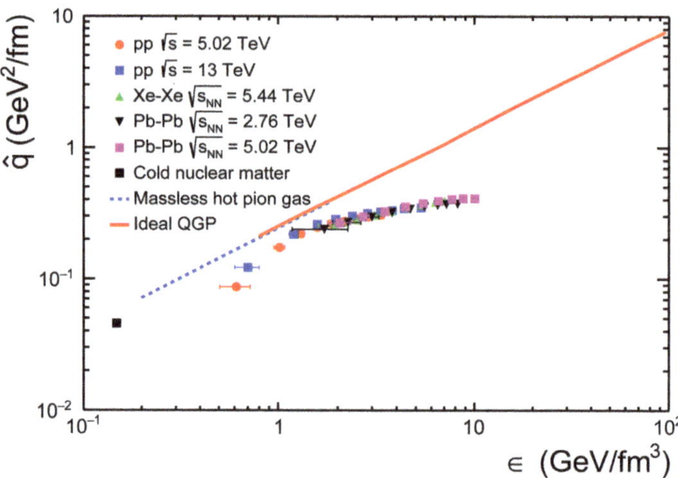

Figure 6. Jet quenching parameter, $\hat{q}$, as a function of initial energy density for *pp* collisions at $\sqrt{s}$ = 5.02 and 13 TeV, Xe-Xe collisions at $\sqrt{s_{\text{NN}}}$ = 5.44 TeV and Pb-Pb collisions at $\sqrt{s_{\text{NN}}}$ = 2.76 and 5.02 TeV. The blue dotted line is for massless pion gas, the solid red curve is for ideal quark-gluon plasma (QGP) and the black square is for cold nuclear matter [99].

In Figure 7, $\hat{q}/T^3$ is plotted as a function of initial temperature. For comparison, the results obtained by the JET Collaboration are also plotted using five different theoretical models that incorporate particle energy loss in the medium. The GLV model [41–43] predicted the general form of the evolution of center-of-mass energy of the high transverse momentum pion nuclear modification factor from Super Proton Synchrotron (SPS) and RHIC to LHC energies. CUJET 1.0 explained the similarity between R_{AA} at RHIC and LHC, despite the fact that the initial QGP density in LHC almost doubles that of RHIC, by taking the effects due to multi-scale running of the QCD coupling $\alpha(Q^2)$ into account [44]. In CUJET 2.0, the CUJET 1.0 is coupled with the 2+1D (2+1 dimensional) viscous hydro fields. By taking GLV-CUJET, the JET Collaboration has estimated the scaled $\hat{q}$, shown by the dashed black line. The HT-BW model uses a 3+1D ideal hydrodynamics to provide the space-time evolution of the local temperature and the flow velocity in the medium along the jet propagation path in heavy-ion collisions. The result obtained from HT-BW model is represented by the blue line. The HT-M model (red line with filled circles) uses a 2+1D viscous hydrodynamic model to provide the space-time evolution of the entropy density [45–49]. The nuclear initial parton scatterings for jet production are carried out by using PYTHIA8 Monte Carlo generator in the MARTINI model [54]. This model describes the suppression of hadron spectra in heavy-ion collisions at RHIC rather well with a fixed value of the strong coupling constant. In the McGill-AMY model [52,53], the scattering and radiation processes are described by thermal QCD and hard thermal loop (HTL) effects [100] and Landau–Pomeranchuck–Migdal (LPM) interference [101]. In this approach, a set of rate equations for their momentum distributions are solved to obtain the evolution of hard jets (quarks and gluons) in the hot QCD medium. One observes that $\hat{q}/T^3$ obtained from the CSPM approach has a similar kind of behaviour as observed by JET Collaboration.

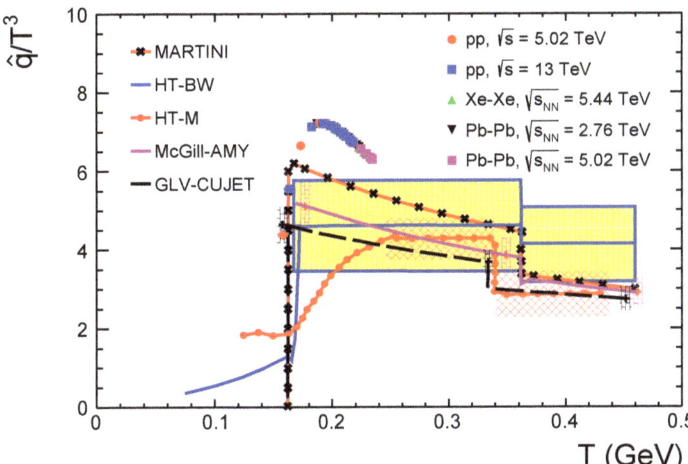

Figure 7. Scaled jet quenching parameter, $\hat{q}/T^3$, as a function of initial temperature for pp collisions at $\sqrt{s}$ = 5.02 and 13 TeV, Xe-Xe collisions at $\sqrt{s_{NN}}$ = 5.44 TeV and Pb-Pb collisions at $\sqrt{s_{NN}}$ = 2.76 and 5.02 TeV. The yellow band shows the estimated uncertainties under the high-twist high-twist Berkeley–Wuhan (HT-BW) model, whereas the red shaded region shows the corresponding uncertainty in the high-twist Majumder (HT-M) model. See text for details.

4. Conclusions

In this paper, the thermodynamic and transport properties of the matter, formed in proton-proton (pp) and heavy-ion (AA) collisions at Large Hadron Collider (LHC) energies, are studied within the framework of the color string percolation model (CSPM). The percolation density parameter is extracted by fitting transverse momentum spectra within CSPM and then initial percolation temperature (T), energy density (ε) and the jet transport coefficient ($\hat{q}$) are estimated. In the present paper, for the first time, the jet transport coefficient of produced hot QCD matter is studied within the color string percolation approach as a function of final state charged particle multiplicity (pseudorapidity density at midrapidity) at the LHC energies. It is shown that $\hat{q}$ increases linearly with initial temperature regardless of the collision system or collision energy.

At very low multiplicity, $\hat{q}$ shows a sharp increase and this dependence becomes weak at high multiplicity (energy density). This behaviour suggests that at lower multiplicity, the system is not dense enough to highly quench the partonic jets, whereas with the increase of multiplicity the quenching of jets becomes more prominent. At very high multiplicity (energy density), $\hat{q}$ saturates with multiplicity (energy density). This allows us to conclude that at very high multiplicity (high energy density), $\hat{q}$ becomes independent of final state multiplicity when scaled by the transverse overlap area of the produced fireball. Interestingly, it is found that for $\hat{q}$ in the low energy density regime, the system behaves almost like a massless hot pion gas. The $\hat{q}/T^3$, obtained from the CSPM approach as a function of temperature, T, has a similar kind of behaviour as observed by the JET Collaboration using five different theoretical models that incorporate particle energy loss in the medium.

In view of the heavy-ion-like signatures seen in TeV high-multiplicity pp collisions at the LHC energies, it would be of high interest to see the jet quenching results in such collisions to infer about the possible quark-gluon plasma droplet formation. The present study of jet transport coefficient as a function of final state multiplicity, initial temperature and energy density will pave the way for such an experimental exploration making LHC pp collisions unique.

Author Contributions: A.N.M., D.S. and R.S. have contributed to the study equally, starting from the cenceptualization of the problem, methodology, paper writing and review. All authors have read and agreed to the published version of the manuscript.

Funding: This research was funded by the Hungarian National Research, Development and Innovation Office (NKFIH) under the contract numbers OTKA K135515, K123815 and NKFIH 2019-2.1.11-TET-2019-00078, 2019-2.1.11-TET-2019-00050 and DAE-BRNS Project No. 58/14/29/2019-BRNS of the Government of India.

Acknowledgments: Aditya Nath Mishra thanks the Hungarian National Research, Development and Innovation Office (NKFIH) and Wigner Scientific Computing Laboratory (WSCLAB, the former Wigner GPU Laboratory). Raghunath Sahoo acknowledges the financial supports under the CERN Scientific Associateship and the financial grants of Dept. of Atomic Energy DAE, the Board of Research in Nuclear Sciences BRNS of the Government of India.

Data Availability Statement: Not applicable.

Conflicts of Interest: The authors declare no conflict of interest.

References

1. Bjorken, J.D. *Energy Loss of Energetic Partons in Quark–Gluon Plasma: Possible Extinction of High p_T Jets in Hadron-Hadron Collisions*; Report FERMILAB-Pub-82-059-THY; Fermi National Accelerator Laboratory: Batavia, IL, USA, 1982. Available online: https://lss.fnal.gov/archive/preprint/fermilab-pub-82-059-t.shtml (accessed on 20 December 2021).
2. dÉnterria, D. Jet quenching. In *Landolt-Börnstein-Group. Elementary Particles, Nuclei and Atoms, 23: Relativistic Heavy Ion Physics*; Stock, R., Ed.; Springer: Berlin/Heidelberg, Germany, 2010. [CrossRef]
3. Blaizot, J.P.; McLerran, L.D. Jets in Expanding quark–gluon plasmas. *Phys. Rev. D* **1986**, *34*, 2739. [CrossRef] [PubMed]
4. Gyulassy, M.; Plumer, M. Jet quenching in dense matter. *Phys. Lett. B* **1990**, *243*, 432. [CrossRef]
5. Wang, X.N.; Gyulassy, M. Gluon shadowing and jet quenching in A + A collisions at $\sqrt{s_{NN}}$ = 200 GeV. *Phys. Rev. Lett.* **1992**, *68*, 1480. [CrossRef]
6. Baier, R.; Dokshitzer, Y.L.; Peigne, S.; Schiff, D. Induced gluon radiation in a QCD medium. *Phys. Lett. B* **1995**, *345*, 277. [CrossRef]
7. Baier, R.; Dokshitzer, Y.L.; Mueller, A.H.; Peigne, S.; Schiff, D. Radiative energy loss and p_T broadening of high-energy partons in nuclei. *Nucl. Phys. B* **1997**, *484*, 265. [CrossRef]
8. Qin, G.Y.; Wang, X.N. Jet quenching in high-energy heavy-ion collisions. *Int. J. Mod. Phys. E* **2015**, *24*, 1530014. [CrossRef]
9. Gyulassy, M.; Vitev, I.; Wang, X.-N.; Zhang, B.-W. Jet quenching and radiative energy loss in dense nuclear matter. In *Quark-Gluon Plasma 3*; Hwa, R.C., Wang, X.-N., Eds.; World Scientific: Singapore, 2004; pp. 123–191. [CrossRef]
10. Adcox, K.; Adler, S.S.; Afanasiev, S.; Aidala, C.; Ajitanand, N.N.; Akiba, Y.; AI-Jamel, A.; Alexander, J.; Amirikas, R.; Aoki, K.; et al. Suppression of hadrons with large transverse momentum in central Au+Au collisions at $\sqrt{s_{NN}}$ = 130 GeV. *Phys. Rev. Lett.* **2002**, *88*, 022301. [CrossRef]
11. Adler, C.; Ahammed, Z.; Allgower, C.; Amonett, J.; Anderson, B.D.; Anderson, M.; Averichev, G.S.; Balewski, J.; Barannikova, O.; Barnby, L.S; et al. Disappearance of back-to-back high p_T hadron correlations in central Au+Au collisions at $\sqrt{s_{NN}}$ = 200 GeV. *Phys. Rev. Lett.* **2003**, *90*, 082302. [CrossRef]
12. Adcox, K.; Adler, S.S.; Afanasiev, S.; Aidala, C.; Ajitanand, N.N.; Akiba, Y.; AI-Jamel, A.; Alexander, J.; Amirikas, R.; Aoki, K.; et al. Centrality dependence of the high p_T charged hadron suppression in Au+Au collisions at $\sqrt{s_{NN}}$ = 130 GeV. *Phys. Lett. B* **2003**, *561*, 82. [CrossRef]
13. Adcox, K.; Adler, S.S.; Afanasiev, S.; Aidala, C.; Ajitanand, N.N.; Akiba, Y.; AI-Jamel, A.; Alexander, J.; Amirikas, R.; Aoki, K.; et al. Suppressed π^0 production at large transverse momentum in central Au+ Au collisions at $\sqrt{s_{NN}}$ = 200 GeV. *Phys. Rev. Lett.* **2003**, *91*, 072301.
14. Adams, J.; Adler, C.; Aggarwal, M.M.; Ahammed, Z.; Amonett, J.; Anderson, B.D.; Anderson, M.; Arkhipkin, D.; Averichev, G.S.; Badyal, S.K.; et al. Transverse momentum and collision energy dependence of high-p_T hadron suppression in Au+Au collisions at ultrarelativistic energies. *Phys. Rev. Lett.* **2003**, *91*, 172302. [CrossRef] [PubMed]
15. Adams, J.; Adler, C.; Aggarwal, M.M.; Ahammed, Z.; Amonett, J.; Anderson, B.D.; Anderson, M.; Arkhipkin, D.; Averichev, G.S.; Badyal, S.K.; et al. Evidence from d+Au measurements for final state suppression of high-p_T hadrons in Au+Au collisions in RHIC. *Phys. Rev. Lett.* **2003**, *91*, 072304. [CrossRef] [PubMed]
16. Roland, C. Charged hadron transverse momentum distributions in Au + Au collisions at $\sqrt{s_{NN}}$ = 200 GeV. *Phys. Lett. B* **2004**, *578*, 297.
17. Arsene, I.; Bearden, I.G.; Beavis, D.; Besliu, C.; Budick, B.; Boggild, H.; Chasman, C.; Christensen, C.H.; Christiansen, P.; Cibor, J.; et al. Transverse momentum spectra in Au+Au and d+Au collisions at $\sqrt{s_{NN}}$ = 200 GeV and the pseudorapidity dependence of high-p_T suppression. *Phys. Rev. Lett.* **2003**, *91*, 072305. [CrossRef] [PubMed]
18. Adare, A.; Afanasiev, S.; Aidala, C.A.; Ajitanand, N.N.; Akiba, Y.; Al-Bataineh, H.; Alexander, J.; Al-Jamel, A.F.; Aoki, K.; Aphecetche, L.; et al. System size and energy dependence of jet-induced hadron pair correlation shapes in Cu+Cu and Au+Au collisions at $\sqrt{s_{NN}}$ = 200 and 62.4 GeV. *Phys. Rev. Lett.* **2007**, *98*, 232302. [CrossRef] [PubMed]

19. Adcox, K.; Adler, S.S.; Afanasiev, S.; Aidala, C.; Ajitanand, N.N.; Akiba, Y.; AI-Jamel, A.; Alexander, J.; Amirikas, R.; Aoki, K.; et al. Formation of dense partonic matter in relativistic nucleus-nucleus collisions at RHIC: Experimental evaluation by the PHENIX collaboration. *Nucl. Phys. A* **2005**, *757*, 184. [CrossRef]
20. Adams, J.; Adler, C.; Aggarwal, M.M.; Ahammed, Z.; Amonett, J.; Anderson, B.D.; Anderson, M.; Arkhipkin, D.; Averichev, G.S.; Badyal, S.K.; et al. Experimental and theoretical challenges in the search for the quark gluon plasma: The STAR Collaboration's critical assessment of the evidence from RHIC collisions. *Nucl. Phys. A* **2005**, *757*, 102. [CrossRef]
21. Back, B.; Baker, M.D.; Ballintijn, M.; Barton, D.S.; Becker, B.; Betts, R.R.; Bickley, A.A.; Bindel, R.; Budzanowski, A.; Busza, W.; et al. The PHOBOS perspective on discoveries at RHIC. *Nucl. Phys. A* **2005**, *757*, 28. [CrossRef]
22. Arsene, I.; Bearden, I.G.; Beavis, D.; Besliu, C.; Budick, B.; Boggild, H.; Chasman, C.; Christensen, C.H.; Christiansen, P.; Cibor, J.; et al. Quark gluon plasma and color glass condensate at RHIC? The Perspective from the BRAHMS experiment. *Nucl. Phys. A* **2005**, *757*, 1. [CrossRef]
23. Adare, A.; Afanasiev, S.; Aidala, C.A.; Ajitanand, N.N.; Akiba, Y.; Al-Bataineh, H.; Alexander, J.; Al-Jamel, A.F.; Aoki, K.; Aphecetche, L.; et al. Quantitative constraints on the opacity of hot partonic matter from semi-inclusive single high transverse momentum pion suppression in Au+Au collisions at $\sqrt{s_{NN}} = 200$ GeV. *Phys. Rev. C* **2008**, *77*, 064907. [CrossRef]
24. Aamodt, K.; Wikne, J.; Mlynarz, J.; Lazzeroni, C.; Bruno, G.; Radomski, S.; Gheata, M.; Stefanek, G.; Piccotti, A.; Cuveland, J.; et al. Suppression of charged particlepProduction at large transverse momentum in central Pb-Pb collisions at $\sqrt{s_{NN}} = 2.76$ TeV. *Phys. Lett. B* **2011**, *696*, 30. [CrossRef]
25. Aad, G.; Abbott, B.; Abdallah, J.; Abdelalim, A.; Abdesselam, A.; Abdinov, O.; Abi, B.; Abolins, M.; Abramowicz, H.; Abreu, H.; et al. Observation of a centrality-dependent dijet asymmetry in lead-lead collisions at $\sqrt{s_{NN}} = 2.77$ TeV with the ATLAS detector at the LHC. *Phys. Rev. Lett.* **2010**, *105*, 252303. [CrossRef] [PubMed]
26. Chatrchyan, S.; Khachatryan, V.; Sirunyan, A.M.; Tumasyan, A.; Adam, W.; Bergauer, T.; Dragicevic, M.; Erö, J.; Fabjan, C.; Friedl, M.; et al. Observation and studies of jet quenching in PbPb collisions at nucleon-nucleon center-of-mass energy = 2.76 TeV. *Phys. Rev. C* **2011**, *84*, 024906. [CrossRef]
27. Chatrchyan, S.; Khachatryan, V.; Sirunyan, A.M.; Tumasyan, A.; Adam, W.; Bergauer, T.; Dragicevic, M.; Erö, J.; Fabjan, C.; Friedl, M.; et al. Dependence on pseudorapidity and centrality of charged hadron production in PbPb collisions at a nucleon-nucleon centre-of-mass energy of 2.76 TeV. *J. High Energy Phys.* **2011**, *1108*, 141. [CrossRef]
28. Aamodt, K.; Wikne, J.; Mlynarz, J.; Lazzeroni, C.; Bruno, G.; Radomski, S.; Gheata, M.; Stefanek, G.; Piccotti, A.; Cuveland, J.; et al. Particle-yield modification in jetlike azimuthal dihadron correlations in Pb-Pb Collisions at $\sqrt{s_{NN}} = 2.76$ TeV. *Phys. Rev. Lett.* **2012**, *108*, 092301. [CrossRef]
29. Chatrchyan, S.; Khachatryan, V.; Sirunyan, A.M.; Tumasyan, A.; Adam, W.; Bergauer, T.; Dragicevic, M.; Erö, J.; Fabjan, C.; Friedl, M.; et al. Study of high-p_T charged particle suppression in PbPb compared to pp collisions at $\sqrt{s_{NN}} = 2.76$ TeV. *Eur. Phys. J. C* **2012**, *72*, 1945. [CrossRef]
30. Chatrchyan, S.; Khachatryan, V.; Sirunyan, A.M.; Tumasyan, A.; Adam, W.; Bergauer, T.; Dragicevic, M.; Erö, J.; Fabjan, C.; Friedl, M.; et al. Jet momentum dependence of jet quenching in PbPb collisions at $\sqrt{s_{NN}} = 2.76$ TeV. *Phys. Lett. B* **2012**, *712*, 176. [CrossRef]
31. Chatrchyan, S.; Khachatryan, V.; Sirunyan, A.M.; Tumasyan, A.; Adam, W.; Bergauer, T.; Dragicevic, M.; Erö, J.; Fabjan, C.; Friedl, M.; et al. Measurement of jet fragmentation into charged particles in pp and PbPb collisions at $\sqrt{s_{NN}} = 2.76$ TeV. *J. High Energy Phys.* **2012**, *1210*, 087.
32. Chatrchyan, S.; Khachatryan, V.; Sirunyan, A.M.; Tumasyan, A.; Adam, W.; Bergauer, T.; Dragicevic, M.; Erö, J.; Fabjan, C.; Friedl, M.; et al. Studies of jet quenching using isolated-photon+jet correlations in PbPb and pp collisions at $\sqrt{s_{NN}} = 2.76$ TeV. *Phys. Lett. B* **2013**, *718*, 773. [CrossRef]
33. Aad, G.; Abbott, B.; Abdallah, J.; Abdelalim, A.; Abdesselam, A.; Abdinov, O.; Abi, B.; Abolins, M.; Abramowicz, H.; Abreu, H.; et al. Measurement of the jet radius and transverse momentum dependence of inclusive jet suppression in lead-lead collisions at $\sqrt{s_{NN}} = 2.76$ TeV with the ATLAS detector. *Phys. Lett. B* **2013**, *719*, 220. [CrossRef]
34. Chatrchyan, S.; Khachatryan, V.; Sirunyan, A.M.; Tumasyan, A.; Adam, W.; Bergauer, T.; Dragicevic, M.; Erö, J.; Fabjan, C.; Friedl, M.; et al. Evidence of b-jet quenching in PbPb collisions at $\sqrt{s_{NN}} = 2.76$ TeV. *Phys. Rev. Lett.* **2014**, *113*, 132301. [CrossRef] [PubMed]
35. CMS Collaboration. Modification of jet shapes in PbPb collisions at $\sqrt{s_{NN}} = 2.76$ TeV. *Phys. Lett. B* **2014**, *730*, 243. [CrossRef]
36. CMS Collaboration. Measurement of jet fragmentation in PbPb and pp collisions at $\sqrt{s_{NN}} = 2.76$ TeV. *Phys. Rev. C* **2014**, *90*, 024908. [CrossRef]
37. Aad, G.; Abbott, B.; Abdallah, J.; Abdelalim, A.; Abdesselam, A.; Abdinov, O.; Abi, B.; Abolins, M.; Abramowicz, H.; Abreu, H.; et al. Measurement of inclusive jet charged-particle fragmentation functions in Pb+Pb collisions at $\sqrt{s_{NN}} = 2.76$ TeV with the ATLAS detector. *Phys. Lett. B* **2014**, *739*, 320. [CrossRef]
38. Aad, G.; Abbott, B.; Abdallah, J.; Abdelalim, A.; Abdesselam, A.; Abdinov, O.; Abi, B.; Abolins, M.; Abramowicz, H.; Abreu, H.; et al. Measurements of the nuclear modification factor for jets in Pb+Pb collisions at $\sqrt{s_{NN}} = 2.76$ TeV with the ATLAS detector. *Phys. Rev. Lett.* **2015**, *114*, 072302. [CrossRef] [PubMed]
39. Zakharov, B. Fully quantum treatment of the Landau-Pomeranchuk-Migdal effect in QED and QCD. *JETP Lett.* **1996**, *63*, 952. [CrossRef]

40. Baier, R.; Dokshitzer, Y.L.; Mueller, A.H.; Peigne, S.; Schiff, D. Radiative energy loss of high-energy quarks and gluons in a finite volume quark-gluon plasma. *Nucl. Phys. B* **1997**, *483*, 291. [CrossRef]
41. Gyulassy, M.; Levai, P.; Vitev, I. Jet quenching in thin quark gluon plasmas. 1. Formalism. *Nucl. Phys. B* **2000**, *571*, 197. [CrossRef]
42. Gyulassy, M.; Levai, P.; Vitev, I. Non-Abelian energy loss at finite opacity. *Phys. Rev. Lett.* **2000**, *85*, 5535. [CrossRef]
43. Gyulassy, M.; Levai, P.; Vitev, I. Reaction operator approach to nonAbelian energy loss. *Nucl. Phys. B* **2001**, *594*, 371. [CrossRef]
44. Buzzatti, A.; Gyulassy, M. Jet flavor tomography of quark gluon plasmas at RHIC and LHC. *Phys. Rev. Lett.* **2012**, *108*, 022301. [CrossRef] [PubMed]
45. Guo, X.F.; Wang, X.N. Multiple scattering, parton energy loss and modified fragmentation functions in deeply inelastic e A scattering. *Phys. Rev. Lett.* **2000**, *85*, 3591. [CrossRef] [PubMed]
46. Wang, X.-N.; Guo, X.-F. Multiple parton scattering in nuclei: Parton energy loss. *Nucl. Phys. A* **2001**, *696*, 788. [CrossRef]
47. Chen, X.F.; Hirano, T.; Wang, E.; Wang, X.N.; Zhang, H. Suppression of high p_T hadrons in Pb+Pb Collisions at LHC. *Phys. Rev. C* **2011**, *84*, 034902. [CrossRef]
48. Majumder, A. Hard collinear gluon radiation and multiple scattering in a medium. *Phys. Rev. D* **2012**, *85*, 014023. [CrossRef]
49. Vitev, I.; Zhang, B.W. Jet tomography of high-energy nucleus-nucleus collisions at next-to-leading order. *Phys. Rev. Lett.* **2010**, *104*, 132001. [CrossRef] [PubMed]
50. Wiedemann, U.A. Gluon radiation off hard quarks in a nuclear environment: Opacity expansion. *Nucl. Phys. B* **2000**, *588*, 303. [CrossRef]
51. Wiedemann, U.A. Jet quenching versus jet enhancement: A quantitative study of the BDMPS-Z gluon radiation spectrum. *Nucl. Phys.* **2001**, *690*, 731. [CrossRef]
52. Arnold, P.B.; Moore, G.D.; Yaffe, L.G. Photon emission from ultrarelativistic plasmas. *J. High Energy Phys.* **2001**, *11*, 057. [CrossRef]
53. Arnold, P.B.; Moore, G.D.; Yaffe, L.G. Photon and gluon emission in relativistic plasmas. *J. High Energy Phys.* **2002**, *6*, 030. [CrossRef]
54. Schenke, B.; Gale, C.; Jeon, S. MARTINI: An Event generator for relativistic heavy-ion collisions. *Phys. Rev. C* **2009**, *80*, 054913. [CrossRef]
55. Fochler, O.; Xu, Z.; Greiner, C. Energy loss in a partonic transport model including bremsstrahlung processes. *Phys. Rev. C* **2010**, *82*, 024907. [CrossRef]
56. He, Y.; Luo, T.; Wang, X.N.; Zhu, Y. Linear Boltzmann transport for jet propagation in the quark–gluon plasma: Elastic processes and medium recoil. *Phys. Rev. C* **2015**, *91*, 054908; Erratum in *Phys. Rev. C* **2018**, *97*, 019902. [CrossRef]
57. Casalderrey-Solana, J.; Wang, X.N. Energy dependence of jet transport parameter and parton saturation in Quark–Gluon plasma. *Phys. Rev. C* **2008**, *77*, 024902. [CrossRef]
58. Burke, K.M.; Buzzatti, A.; Chang, N.; Gale, C.; Gyulassy, M.; Heinz, U.; Jeon, S.; Majumder, A.; Muller, B.; Qin, G.Y.; et al. Extracting the jet transport coefficient from jet quenching in high-energy heavy-ion collisions. *Phys. Rev. C* **2014**, *90*, 014909. [CrossRef]
59. Armesto, N.; Braun, M.A.; Ferreiro, E.G.; Pajares, C. Percolation approach to quark-gluon plasma and J/ψ suppression. *Phys. Rev. Lett.* **1996**, *77*, 3736. [CrossRef] [PubMed]
60. Braun, M.A.; Pajares, C. Implications of percolation of color strings on multiplicities, correlations and the transverse momentum. *Eur. Phys. J. C* **2000**, *16*, 349. [CrossRef]
61. Braun, M.A.; Pajares, C. Transverse momentum distributions and their forward backward correlations in the percolating color string approach. *Phys. Rev. Lett.* **2000**, *85*, 4864. [CrossRef]
62. Braun, M.A.; del Moral, F.; Pajares, C. Percolation of strings and the first RHIC data on multiplicity and transverse momentum distributions. *Phys. Rev. C* **2002**, *65*, 024907. [CrossRef]
63. Dias de Deus, J.; Pajares, C. Percolation of color sources and critical temperature. *Phys. Lett. B* **2006**, *642*, 455. [CrossRef]
64. Braun, M.A.; Dias de Deus, J.; Hirsch, A.S.; Pajares, C.; Scharenberg, R.P.; Srivastava, B.K. De-confinement and clustering of color sources in nuclear collisions. *Phys. Rep.* **2015**, *599*, 1. [CrossRef]
65. de Deus, J.D.; Pajares, C. String percolation and the glasma. *Phys. Lett. B* **2011**, *695*, 211. [CrossRef]
66. Braun, M.A.; del Moral, F.; Pajares, C. Centrality dependence of the multiplicity and transverse momentum distributions at RHIC and LHC and the percolation of strings. *Nucl. Phys. A* **2003**, *715*, 791. [CrossRef]
67. Schwinger, J. Gauge invariance and mass. II. *Phys. Rev.* **1962**, *128*, 2425. [CrossRef]
68. Scharenberg, R.P.; Srivastava, B.K.; Hirsch, A.S. Percolation of color sources and the determination of the equation of state of the quark–gluon plasma (QGP) produced in central Au-Au collisions at $\sqrt{s_{NN}}$ = 200 GeV. *Eur. Phys. J. C* **2011**, *71*, 1510. [CrossRef]
69. Isichenko, M.B. Percolation, statistical topography, and transport in random media. *Rev. Mod. Phys.* **1992**, *64*, 961. [CrossRef]
70. Satz, H. Color deconfinement in nuclear collisions. *Rep. Prog. Phys.* **2000**, *63*, 1511. [CrossRef]
71. Mishra, A.N.; Caulte, E.; Paić, G.; Pajares, C.; Scharenberg, R.P.; Srivastava, B.K. ALICE data in the framework of the color string percolation model. *PoS* **2019**. [CrossRef]
72. Cunqueiro, L.; Dias de Deus, J.; Pajares, C. Nuclear like effects in proton-proton collisions at high energy. *Eur. Phys. J. C* **2010**, *65*, 423. [CrossRef]
73. Andres, C.; Moscoso, A.; Pajares, C. Onset of the ridge structure in AA, pA, and pp collisions. *Phys. Rev. C* **2014**, *90*, 054902. [CrossRef]

74. Scharenberg, R.P. The QGP equation of state by measuring the color suppression factor at RHIC and LHC energies. *PoS* **2013**. [CrossRef]
75. Dias de Deus, J.; Hirsch, A.S.; Pajares, C.; Scharenberg, R.P.; Srivastava, B.K. Clustering of color sources and the shear viscosity of the QGP in heavy ion collisions at RHIC and LHC energies. *Eur. Phys. J. C* **2012**, *72*, 2123. [CrossRef]
76. Srivastava, B.K. Percolation approach to initial stage effects in high energy collisions. *Nucl. Phys. A* **2014**, *926*, 142. [CrossRef]
77. Dias de Deus, J.; Hirsch, A.S.; Pajares, C.; Scharenberg, R.P.; Srivastava, B.K. Transport coefficient to trace anomaly in the clustering of color sources approach. *Phys. Rev. C* **2016**, *93*, 024915. [CrossRef]
78. Sahoo, P.; De, S.; Tiwari, S.K.; Sahoo, R. Energy and centrality dependent study of deconfinement phase transition in a color string percolation approach at RHIC energies. *Eur. Phys. J. A* **2018**, *54*, 136. [CrossRef]
79. Sahoo, P.; Tiwari, S.K.; De, S.; Sahoo, R.; Scharenberg, R.P.; Srivastava, B.K. Thermodynamic and transport properties in Au + Au collisions at RHIC energies from the clustering of color strings. *Mod. Phys. Lett. A* **2019**, *34*, 1950034. [CrossRef]
80. Sahoo, P.; Sahoo, R.; Tiwari, S.K. Wiedemann-Franz law for hot QCD matter in a color string percolation scenario. *Phys. Rev. D* **2019**, *100*, 051503. [CrossRef]
81. Sahoo, P.; Tiwari, S.K.; Sahoo, R. Electrical conductivity of hot and dense QCD matter created in heavy-ion collisions: A color string percolation approach. *Phys. Rev. D* **2018**, *98*, 054005. [CrossRef]
82. Sahu, D.; Tripathy, S.; Sahoo, R.; Tiwari, S.K. Formation of a perfect fluid in pp p-Pb, Xe-Xe and Pb-Pb collisions at the Large Hadron Collider energies. *arXiv* **2001**, arXiv:2001.01252.
83. Sahu, D.; Sahoo, R. Thermodynamic and transport properties of matter formed in pp, p-Pb, Xe-Xe and Pb-Pb collisions at the Large Hadron Collider using color string percolation model. *J. Phys. G* **2021**, *48*, 125104. [CrossRef]
84. Mishra, A.N.; Paić, G.; Pajares, C.; Scharenberg, R.P.; Srivastava, B.K. Deconfinement and degrees of freedom in pp and A-A collisions at LHC energies. *Eur. Phys. J. A* **2021**, *57*, 245. [CrossRef]
85. McLerran, L.; Venugopalan, R. Computing quark and gluon distribution functions for very large nuclei. *Phys. Rev. D* **1994**, *49*, 2233. [CrossRef] [PubMed]
86. Acharya, S.; Adamová, D.; Adhya, S.P.; Adler, A.; Adolfsson, J.; Aggarwal, M.M.; Aglieri Rinella, G.; Agnello, M.; Agrawal, N.; Ahammed, Z.; et al. Charged-particle production as a function of multiplicity and transverse spherocity in pp collisions at $\sqrt{s}$ = 5.02 and 13 TeV. *Eur. Phys. J. C* **2019**, *79*, 857. [CrossRef]
87. Acharya, S.; Adamová, D.; Adhya, S.P.; Adler, A.; Adolfsson, J.; Aggarwal, M.M.; Aglieri Rinella, G.; Agnello, M.; Agrawal, N.; Ahammed, Z.; et al. Transverse momentum spectra and nuclear modification factors of charged particles in pp, p-Pb and Pb-Pb collisions at the LHC. *J. High Energy Phys.* **2018**, *11*, 013.
88. Acharya, S.; Acosta, F.T.; Adamová, D.; Adolfsson, J.; Aggarwal, M.M.; Aglieri Rinella, G.; Agnello, M.; Agrawal, N.; Ahammed, Z.; Ahn, S.U.; et al. Transverse momentum spectra and nuclear modification factors of charged particles in Xe-Xe collisions at $\sqrt{s_{NN}}$ = 5.44 TeV. *Phys. Lett. B* **2019**, *788*, 166. [CrossRef]
89. McLerran, L.; Praszalowicz, M.; Schenke, B. Transverse momentum of protons, pions and Kaons in high multiplicity pp and pA collisions: Evidence for the color glass condensate? *Nucl. Phys. A* **2013**, *916*, 210. [CrossRef]
90. Loizides, C. Glauber modeling of high-energy nuclear collisions at the subnucleon level. *Phys. Rev. C* **2016**, *94*, 024914. [CrossRef]
91. Becattini, F.; Castorina, P.; Milov, A.; Satz, H. A Comparative analysis of statistical hadron production. *Eur. Phys. J. C* **2010**, *66*, 377. [CrossRef]
92. Acharya, S.; Acosta, F.T.; Adamová, D.; Adolfsson, J.; Aggarwal, M.M.; Aglieri Rinella, G.; Agnello, M.; Agrawal, N.; Ahammed, Z.; Ahn, S.U.; et al. Direct photon production in Pb-Pb collisions at $\sqrt{s_{NN}}$ = 2.76 TeV. *Phys. Lett. B* **2016**, *754*, 235.
93. Bjorken, J.D. Highly relativistic nucleus-nucleus collisions: The central rapidity region. *Phys. Rev. D* **1983**, *27*, 140. [CrossRef]
94. Wong, C.Y. *Introduction to High Energy Heavy Ion Collisions*; World Scientific: Singapore, 1994. [CrossRef]
95. Baier, R.; Mehtar-Tani, Y. Jet quenching and broadening: The Transport coefficient q-hat in an anisotropic plasma. *Phys. Rev. C* **2008**, *78*, 064906. [CrossRef]
96. Liu, H.; Rajagopal, K.; Wiedemann, U.A. Calculating the jet quenching parameter from AdS/CFT. *Phys. Rev. Lett.* **2006**, *97*, 182301. [CrossRef] [PubMed]
97. Majumder, A.; Muller, B.; Wang, X.N. Small shear viscosity of a quark–gluon plasma implies strong jet quenching. *Phys. Rev. Lett.* **2007**, *99*, 192301. [CrossRef] [PubMed]
98. Xu, J. Shear viscosity of nuclear matter. *Nucl. Sci. Tech.* **2013**, *24*, 50514.
99. Baier, R. Jet quenching. *Nucl. Phys. A* **2003**, *715*, 209. [CrossRef]
100. Su, N. A brief overview of hard-thermal-loop perturbation theory. *Commun. Theor. Phys.* **2012**, *57*, 409. [CrossRef]
101. Baier, R.; Dokshitzer, Y.L.; Mueller, A.H.; Peigne, S.; Peigne, S.; Schiff, D. The Landau-Pomeranchuk-Migdal effect in QED. *Nucl. Phys. B* **1996**, *478*, 577. [CrossRef]

physics

Article

QCD Phase Boundary and the Hadrochemical Horizon

Berndt Müller [1,2]

[1] Department of Physics, Duke University, Durham, NC 27708, USA; mueller@phy.duke.edu
[2] Wright Laboratory, Department of Physics, Yale University, New Haven, CT 06511, USA

Abstract: I review the physics of the phase boundary between hadronic matter and quark matter from several different points of view. These include thermodynamics, statistical physics, and chemical kinetics. In particular, the review focuses on the role of the chemical freeze-out line and its relation to the concept of valence-quark percolation. The review ends with some recollections of Jean Cleymans.

Keywords: quark-gluon plasma; QCD phase transition; chemical freeze-out

Citation: Müller, B. QCD Phase Boundary and the Hadrochemical Horizon. *Physics* 2022, 4, 597–608. https://doi.org/10.3390/physics4020040

Received: 14 April 2022
Accepted: 4 May 2022
Published: 23 May 2022

Publisher's Note: MDPI stays neutral with regard to jurisdictional claims in published maps and institutional affiliations.

1. The QCD Phase Boundary

The phase diagram of strongly interacting (QCD) matter contains a low-energy density region, in which the mobile constituents are color singlets (hadrons), and a high-energy density domain, in which the mobile constituents are color non-singlet objects, including quarks, antiquarks, gluons and, at high net baryon number density and modest temperature, diquarks. The low-energy density matter is commonly called "hadronic matter" or "hadron gas"; the matter at high-energy density is called "quark-gluon plasma" or "quark matter". Upon finer inspection the diagram may be subdivided into a multitude of specific phases, including various types of color superconductors [1] and a possible phase that has been named "quarkyonic" matter [2].

The first attempt to determine the boundary between hadronic matter and quark matter was made by Hagedorn and Rafelski in the framework of the statistical bootstrap model, where they found a continuous first-order transition line in the QCD phase diagram [3]. However, as numerical simulations of QCD on a lattice have shown, for physical values of the quark masses the two domains are not everywhere separated by a true thermodynamic phase boundary, i.e., a line characterized by singularities in the thermodynamic partition function [4,5]. Yet it is widely believed that such singular boundary exists in some regions of the phase diagram. The most widely studied example of a thermodynamic singularity is a possible critical endpoint of a first-order line separating hadronic matter and quark matter, corresponding to the spontaneous breaking of chiral symmetry at moderate baryon number density and temperature. The first-order transition line may have a second critical endpoint induced by the axial U(1) anomaly at even lower temperature and larger net baryon density [6].

State-of-the-art lattice-QCD calculations show that the critical endpoint, if one exists, must be located in the region $\mu_B/T > 3.5$ of baryochemical potential, μ_B, and temperature, T. first in main parts of the text: Abstract, main text, Figures, tables, Conclusions [7]. Analytical calculations using the functional renormalization group method [8] indicate that the critical endpoint (CEP) lies at $(T_{\mathrm{CEP}}, \mu_{B,\mathrm{CEP}}) \approx (107\,\mathrm{MeV}, 635\,\mathrm{MeV})$ corresponding to $\mu_{B,\mathrm{CEP}}/T_{\mathrm{CEP}} \approx 5.54$. A holographic model of QCD based on an Einstein–Maxwell-dilaton Lagrangian in the anti-de-Sitter AdS_5 five-dimensional space with parameters tuned to reproduce thermodynamic properties of QCD at low net baryon density [9] predicts a similar position of the critical endpoint at $(T_{\mathrm{CEP}}, \mu_{B,\mathrm{CEP}}) \approx (89\,\mathrm{MeV}, 724\,\mathrm{MeV})$. At smaller values of μ_B/T, where the transition between the two regimes is continuous, one can define a pseudocritical line as the location of the maximum of the chiral susceptibility for a fixed ratio μ_B/T. The value of the pseudocritical temperature at $\mu_B = 0$ is $T_c = 156.5 \pm 1.5\,\mathrm{MeV}$ [10].

The phase boundaries defined by the properties of thermodynamic functions are not easily mapped onto experimentally accessible observables. In principle, the internal energy and the pressure enter into the hydrodynamic equations of motion via the equation of state but, in practice, the contributions of singular terms to the equation of state are small and are hardly visible in the expansion dynamics of a quark-gluon plasma fireball. This would be different if the phase boundary would correspond to a strong first-order phase transition with a large latent heat, but such a scenario is not realized in QCD matter. This motivates the consideration of other criteria for a phase boundary. One can distinguish three types of such criteria:

1. Thermodynamic phase boundary. This is the definition discussed above, where the phase boundary is defined as the locus of singularities in the thermodynamic functions, or as the vicinity of such singularities such as in QCD, where singular behavior only occurs in the two-flavor chiral limit.
2. Statistical phase boundary. In this case, the boundary is defined as the locus where certain statistical properties of the matter exhibit a singular behavior. Examples are critical fluctuations at or near a critical point (which also satisfies the thermodynamic criterion) or a percolation threshold.
3. Kinetic phase boundary. Such boundaries exist in dynamical scenarios when kinetic processes that slow down during a cooling or expansion process become much slower than the characteristic cooling or expansion rate and "freeze out". An example is the recombination of hydrogen atoms in the early universe, when the universe became transparent to the photons of the blackbody radiation. A more apt term for this type of boundary would be the photon horizon or optical horizon of the universe, as it is impossible to observe phenomena before that boundary via photons. In the case of the strongly interacting matter, the relevant reaction is the exchange of flavor quantum numbers, heavier than u and d quarks, among hadrons. Here, this boundary is called the "hadrochemical horizon"; more commonly it is called the "chemical freeze-out line".

2. Valence-Quark Percolation

The naïve concept of the deconfinement transition in QCD is that it separates the "normal" phase at low temperature and net baryon density, where quarks are confined to hadrons, from the high-energy density regions of the QCD phase diagram, in which quarks can exist as isolated excitations. In spite of its intuitive allure, this concept is obscured by the mechanism of quark-pair production, which permits a quark to move around easily by dressing itself with a light antiquark. Quark confinement has a rigorous definition in a world in which only gluons exist, i. e., in the pure SU(3) gauge theory, which can be thought of as the hypothetical limit of QCD with very large quark masses when pair production is energetically disfavored. In that idealized limit, quark confinement can be unambiguously identified by the vanishing of the expectation value of the Polyakov loop. In the presence of light dynamical quarks this measure never vanishes, and no rigorous definition of what one means by "quark confinement" in a world with light quarks has been found.

This insight suggests that it may be more rewarding to look at quark confinement from a kinetic point of view rather than an energetic viewpoint. The transport of an individual quark, i. e., a quark identified by its distinct flavor from one location to another in the confined phase requires the transport of a hadron that contains this quark as a constituent. In the case of a strange quark, the least massive vehicle of transport is a kaon, which weighs approximately 0.5 GeV/c. At low temperature ($T < 100$ MeV) kaons are rare excitations in the hadronic gas; this reduces the rate of exchange of strange quarks between hadrons. This argument obviously does not apply to u and d quarks as they can be exchanged as constituents of pions, which provide for a long-range—on the nuclear scale—exchange mechanism. The exchange of a strange quark between hadrons thus predominantly occurs via direct exchange of two quarks—one from each hadron—when the quark cores of the two hadrons come (nearly) into contact. At low temperature or low net baryon density such close encounters among two hadrons are rare, and they are even rarer for close encounters between three or more hadrons. Long-range quark transport, although possible, is, thus, greatly impeded. In this picture, one can think of quark exchange among hadrons as an effective kinetic mechanism for quark transport.

The rate of close encounters of multiple hadrons obviously grows rapidly with increasing density, either by excitations of more hadrons at higher temperature or owing to increased net baryon density at low temperature. At sufficiently high hadron density, the quark cores of all hadrons will be in contact with another hadron core most of the time. There must exist some critical hadron density at which it is possible to find a chain of pairwise touching hadrons bridging across an arbitrary large distance; this critical density is commonly called the "percolation threshold". In this picture, quark deconfinement corresponds to a percolation transition and the QCD phase boundary is identified as a percolation boundary [11].

This concept of valence-quark percolation is different from the concept of color string percolation [12,13], which has been invoked to describe the formation of a deconfined QCD plasma in the initial stage of a nuclear collision. In the color string percolation model the percolation criterion applies to the area density of color strings or flux tubes that are formed immediately after the collision of two nuclei; the model is closely related to the glasma model [14,15] for the initial collision stage. The concept of valence-quark percolation, in contrast, applies to the late stage of the collision when the quark-gluon plasma disassembles into hadrons.

Since hadrons have no well defined surfaces, the valence-quark percolation picture contains some level of ambiguity. For example, one could model hadrons in the spirit of the (Massachusetts Institute of Technology) MIT-Bag Model [16] either as hard spheres or as fuzzy spheres that can overlap. Alternatively, one could model hadrons as valence-quark bags surrounded by meson clouds along the lines of the Cloudy Bag model [17]. For a thermal mass distribution of hadrons, the excluded volume in the hard sphere model might be assumed to be proportional to the mass of the hadrons as in the MIT-Bag Model: $M = 4\mathcal{B}V$ with the bag constant $\mathcal{B} \approx (145 \text{ MeV})^4$ and V denoting the bag volume. Studies of the percolation threshold for various three-dimensional hard sphere models can be found elsewhere [18–20]. The occupied volume fraction, ζ, where a volume spanning cluster appears, lies typically in the range, $\zeta_{cl} \approx 0.6$–0.75. For reference, the volume fraction occupied by closely packed (CP) uniform hard spheres is: $\zeta_{CP} = \pi/(3\sqrt{2}) \approx 0.74$.

3. Chemical Freeze-Out

The quest for an experimental determination of the hadronic gas–quark-gluon plasma boundary is as old as the search for the quark-gluon plasma itself. It was already early recognized that hadron ratios, especially those of hadrons containing strange quarks, carry information about the temperature and baryochemical potential prevalent in the hadronic gas when it disassembles at the end of a heavy-ion collision [21–23]. If these hadrons freeze out soon after the transition from quark-gluon plasma to the hadronic phase, the parameters deduced from thermal fits to the measured hadron yields would even be proxies for the phase boundary.

The premise that hadrochemical abundances freeze out shortly after hadronization was first substantiated by Koch, Müller, and Rafelski [24,25], who studied the evolution of strange hadron abundances in a hot hadronic gas and concluded that they could not be appreciably changed by hadronic reactions at temperatures below the critical temperature: $T < T_c \approx 160$ MeV, where a hadronic gas exists. A similar study by Braun–Munzinger, Stachel, and Wetterich [26] reached the same conclusion: The experimentally determined chemical freeze-out temperature of strange baryons is a good proxy for the phase transition temperature.

An early version of a QCD phase diagram with a (very sparse) set data points from SPS (Super Proton Synchrotron) and AGS (Alternating Gradient Synchrotron) experiments can be found in the review by Harris and Müller [27] (see Figure 4 there). Using a larger set of data points from the SPS, AGS and SIS (Schwerionensynchrotron, Heavy Ion Synchrotron), Cleymans and Redlich [28] observed that the freeze-out line, which we above called the hadrochemical horizon, corresponds to a fixed energy per emitted particle, $\langle E \rangle / \langle N \rangle \approx 1$ GeV, where the brackets denote sample averaging. This contrasts with the condition for the phase boundary which, in the statistical bootstrap model with excluded volume of Hagedorn and Rafelski [3], corresponds to a fixed critical energy density, $\varepsilon_c = 4B$. The value $\varepsilon_c = 0.23$ GeV/fm^3 is close to the value of the energy density at the minimum of the speed of sound in lattice-QCD calculations, $\varepsilon_{\min} = 0.20$ GeV/fm^3 [29]. This was a surprising result, which has since been confirmed by even richer data sets but still lacks a simple explanation. One aspect that made the result surprising is that the particle content varies greatly along the freeze-out line, being meson-dominated at low μ_B but dominated by baryons at high μ_B. The observation showed, however, that the chemical freeze-out line cannot generally be regarded as a proxy for the phase boundary, even if it closely tracks it at low and moderate baryochemical potential.

An extensive set of hadrochemical freeze-out data was obtained in the RHIC (Relativistic Heavy Ion Collider) beam energy scan (BES) [30], which cover the range $\mu_B < 400$ MeV. These will soon be complemented by the higher statistic data from the BES-II at RHIC, which extends at least up to $\mu_B \approx 700$ MeV. Comprehensive thermal fits were performed by Becattini et al. [31] for data from SPS, AGS, and SIS, and by Andronic et al. [32] for both, midrapidity and 4π data from AGS, SPS, and RHIC. Separate fits to STAR and ALICE experiments data for strange and nonstrange hadrons was recently published by Flor et al. [33]. The results for some of these chemical freeze-out fits are shown in Figure 1.

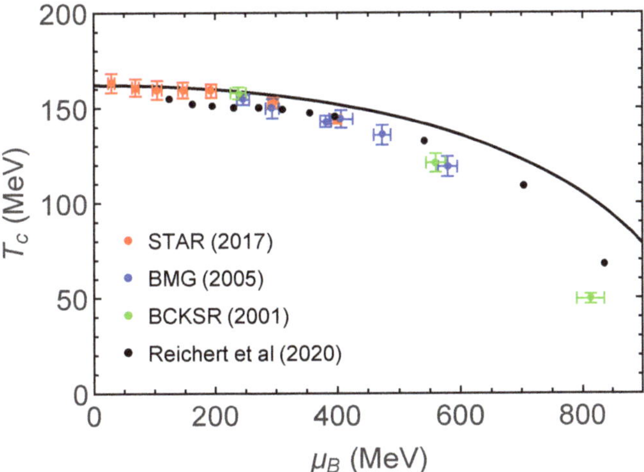

Figure 1. Chemical potential and temperature at chemical freeze-out, determined by thermal model fits to strange hadron yield ratios for data from AGS, RHIC, and LHC experiments [30,32,33]. The STAR experiment data (red) are for 0–10% central collisions. The black dots show the freeze-out parameters obtained from simulations using a hadronic transport model [34]. Blue (BMG) and green (BCKSR) dots are calculations by Becattini et al. from Refs. [31,35], respectively (denoted after the authors' names). The solid line shows the hadron gas phase boundary obtained in the statistical bootstrap model with excluded volume (see Section 4 below).

The analyses compare various measured ("exp") particle yields per unit rapidity, $dN_i^{(\exp)}/dy$, with the thermal predictions ("th") for these yields:

$$\frac{dN_i^{(\text{th})}}{dy} = \frac{dV}{dy} \gamma_s^{|S_i|} g_i \int \frac{d^3p}{(2\pi)^3} n_i(p; T, \mu_B, \mu_s),$$

(1)

where

$$n_i(p; T, \mu_B, \mu_s) = \left(e^{-(\sqrt{p^2 + m_i^2} - \mu_B B_i - \mu_s S_i)/T} \pm 1 \right)^{-1} \tag{2}$$

is the Fermi (or Bose) distribution for given T, μ_B, and strangeness chemical potential, μ_s. Here, p is 3-momentum, m_i is the mass of i-type particle, g_i, B_i and S_i are quantum degeneracy, baryon number and strangeness, respectively, and γ_s is the strange quark fugacity. Some analyses fit the absolute yields, some only ratios of yields, which are independent of the freeze-out volume dV/dy. The fits typically include data for $\pi^+, K^+, p, \Lambda, \Xi$ and their antiparticles; some also include data for Ω and ϕ particles.

The so-called "hadrochemical horizon" is not a sharp line. Hadrons undergo chemical reactions during the early stage of expansion following hadronization of the quark-gluon plasma until they "freeze out" after their last identity changing reaction. This process cannot be observed in action—at least we do not know how to experimentally track it—but it can be studied in detail in microscopic models of the evolution of the fireball after hadronization. Such a study was recently published by Reichert et al. [34] who used the Ultra-relativistic Quantum Molecular Dynamics (UrQMD) model to record the last moment of production of the emitted pions and correlate them with the coarse-grained thermodynamic variables describing the local environment during that moment. These pions continue to rescatter after their final production, which means that their *kinetic* freeze-out occurs later.

The study has two limitations. First, whether the last moment of pion production is a good proxy for the hadrochemical horizon is open to debate. Most pions are emitted by resonance decay without changing the heavier flavor composition of the hadron. When the hadrochemical ratios are analyzed, all states that decay into each other by strong interactions are lumped together, which means that chemical freeze-out as operationally defined by the analysis of stable hadron ratios generally occurs earlier than the last moment of pion emission. Local temperatures recorded in this way therefore represent a lower bound on the temperature of the hadrochemical horizon. The second limitation is that Ref. [34] uses the UrQMD model to describe the collision from beginning to end without explicit creation of a deconfined phase. This means that some of the local temperatures at the last moment of pion production lie above the hadronic phase boundary and introduces an additional upward bias on the recorded temperature. The two countervailing effects both grow in size with increasing μ_B.

With this in mind, it is still worth comparing the results of [34] with the STAR data, as shown in Figure 1. In a hybrid collision model, that includes a hydrodynamic expansion in the quark-gluon plasma phase, the average temperatures would shift to lower values, perhaps by one-third or half the width of the recorded temperature distribution (see Figure 3 in [34]) or 20–30 MeV. Since the first limitation mentioned above generates a bias in the opposite direction (to lower temperatures) the real location of the hadrochemical freeze-out line as defined by stable hadron ratios is likely to lie 5–15 MeV below the black dots. A modified study within a hybrid collision model would be of considerable interest.

A cross-check of the thermodynamic freeze-out parameters at a given collision energy is possible by comparing certain volume-independent ratios of net quantum number fluctuations in the event ensemble with the corresponding ratio of susceptibilities that can be calculated on the lattice. This comparison has been made for net electric charge, Q, and net baryon number, B, [36,37]. The susceptibilities, χ_n, are defined as derivatives of the thermodynamic pressure, P, with respect to the chemical potential, μ_α, that is associated with the conserved quantity $\alpha = Q, B$:

$$\chi_n^{(\alpha)}(T, \mu_\alpha) = \frac{\partial^n (P/T^4)}{\partial(\mu_\alpha/T)^n}. \tag{3}$$

The commonly used fluctuation measures (cumulants) are the mean, M_α, the variance, σ_α^2 and the skewness, S_α. They are related to the susceptibilities as

$$M_\alpha = V\chi_\alpha^{(1)}, \quad \sigma_\alpha^2 = V\chi_\alpha^{(2)}, \tag{4}$$

$$S_\alpha = V^{-1/2}\chi_\alpha^{(3)}/(\chi_\alpha^{(2)})^{3/2}. \tag{5}$$

The unknown volume cancels in the ratios:

$$R_{12} = M/\sigma^2, \quad R_{31} = S\sigma^{3/2}/M, \tag{6}$$

where the index α, indicating which conserved quantum is being considered, is dropped for brevity.

The experimental data for these ratios of cumulants generally agree with those derived from lattice-QCD calculations for the parameters deduced from the chemical freeze-out fits. There are quite a few caveats that come with such comparisons: Experiments measure fluctuations in momentum space, the lattice calculates fluctuations in position space; experiments average over a range of conditions under which particles are emitted, lattice simulations are for precisely fixed thermodynamic conditions; experiments cannot unambiguously separate initial state fluctuations from final-state (thermal) ones, lattice simulations only consider thermal fluctuations. (see [38] for a detailed discussion of this method and comparison with experimental data.) It is thus not entirely surpring that a recent analysis of net-kaon fluctuations using the R_{12} measure seem to indicate that the strangeness content of the final hadron yields may freeze out at somewhat ($\approx$10 MeV) higher temperature [39]. This could be attributed to their larger mass, which may let strange quarks lose their mobility slightly earlier when the QCD phase boundary is approached from above during the expansion of the fireball.

Overall, the experimental results are consistent with the principle that the fluctuations of conserved quantum numbers are frozen in at the moment when chemical reactions among hadrons freeze out. Since the quantum numbers probed by this method are carried by valence quarks, this observation is consistent with the idea that chemical freeze-out is related to valence-quark percolation. It is worth mentioning that the analyses for B and Q agree surprisingly well (see [37]), because electric charge can be transported by pions, which diffuse easily and could be able to modify net charge fluctuations until kinetic freeze-out at lower temperature.

4. Excluded Volume Bootstrap Model

As a schematic model of valence-quark percolation, let us investigate the statistical bootstrap model with excluded volume of Hagedorn and Rafelski [3]. The model is defined by the partition function,

$$Z(\beta,\mu,V) = \sum_{b=-\infty}^{\infty} e^{b\beta\mu} \int d^4p\, e^{-\beta\cdot p}\sigma(p,b,V), \tag{7}$$

where b is the baryon number of the hadron and $\sigma(\beta,b,V)$ denotes the level density for hadrons with a given baryon number b. $\beta = 1/T$ denotes the inverse temperature, μ are the chemical potentials for each quark flavor and b counts the baryon number for each quark flavor. Here, $\mu_u = \mu_d = \mu_B/3$ and $\mu_s = 0$ are assumed. The level density can be expressed as invariant phase space integral over the mass spectrum $\tau(m^2,b)$:

$$\sigma(p,b,V) = \delta^4(p)\delta_{b,0} +$$

$$\sum_{N=1}^{\infty}\frac{1}{N!}\delta^4\left(p-\sum_i p_i\right)\sum_{\{b_i\}}\delta_{b,\sum_i b_i}$$

$$\times\prod_{i=1}^{N}\int d^4 p_i\,\frac{2\Delta\cdot p}{(2\pi)^3}\tau(p_i^2,b_i), \tag{8}$$

where the first term corresponds to the vacuum state. Here, $\delta^4(\cdots)$ is the Dirac delta function, and $\delta_{x,y}$ is the Kronecker delta. The N-th term involves a sum over partitions of the baryon number b on N hadron clusters with masses $m_i^2 = p_i^2$. The four-vector $\Delta^\mu \equiv \Delta u^\mu$, where Δ is the unoccupied volume, denotes the volume remaining after excluding the proper volumes, $V_{\mathrm{cl},i}$, of all hadron clusters from the total fireball volume: $\Delta^\mu = V^\mu - \sum_i V_{\mathrm{cl},i}^\mu$.

Following Hagedorn and Rafelski [3], the requirement that any hadron cluster is composed of smaller hadron clusters is expressed by the bootstrap condition [3],

$$\sigma(p,b,V_{\mathrm{cl}}) = H\,\tau(p^2,b), \tag{9}$$

with the bootstrap constant $H = 0.724\,(\mathrm{GeV})^{-2}$. The bootstrap condition allows to generate the entire hadron cluster spectrum from a small number of "elementary" hadrons. In [3] only pions and nucleons are considered as input into the bootstrap equation; here, the entire set of quark model ground states is considered: the pseudoscalar and vector meson nonets and the baryon and antibaryon octets and decuplets. Counting spin and isospin degrees of freedom this comprises 148 "elementary" states, which are labeled by the index α.

Inserting these states into Equation (7) and applying the Boltzmann approximation leads to the input partition function:

$$\begin{aligned}\varphi(\beta,\mu) &= \sum_b e^{b\mu}\int d^4 p\, e^{-\beta\cdot p}\,H\sum_\alpha g_\alpha\delta_+(p^2 - m_\alpha^2)\\ &= 2\pi HT\sum_\alpha g_\alpha e^{b_\alpha\mu} m_\alpha K_1(m_\alpha/T),\end{aligned} \tag{10}$$

where g_α denotes the degeneracy of each state and b_α its baryon number counting only u and d valence quarks. For values of μ_B approaching the nucleon mass, Fermi–Dirac quantum corrections must be taken into account for nucleons, which was done in the plots shown below.

The complete single-cluster partition function, $\phi(\beta,\mu)$, generated by Equation (9), is related to φ by the implicit equation,

$$\varphi(\beta,\mu) = 2\phi(\beta,\mu) - e^{\phi(\beta,\mu)} + 1. \tag{11}$$

The full partition function is obtained from ϕ as

$$\ln Z(\beta,\mu,V) = -\frac{2\Delta}{(2\pi)^3 H}\frac{\partial\phi(\beta,\mu)}{\partial\beta}. \tag{12}$$

In order to determine the unoccupied volume, Δ, for a given hadron configuration within the fireball volume V, one needs to specify the proper volume of each cluster. Here, Ref. [3] is followed by using the MIT-Bag model relation $V_{\mathrm{cl}} = m_{\mathrm{cl}}/(4B)$, where m_{cl} is the cluster mass. This leads to the following equation for the unoccupied volume fraction [3]:

$$1 - \xi \equiv \frac{\Delta}{V} = \left(1 + \frac{\partial^2\phi/\partial\beta^2}{2(2\pi)^3 HB}\right)^{-1}. \tag{13}$$

Here the occupied (excluded) volume fraction, ζ, is introduced which is the parameter that controls percolation.

As shown in Ref. [3], $\partial^2 \phi / \partial \beta^2$ diverges when $\phi(\beta, \mu) = \ln 2$, and thus Δ vanishes: The entire fireball volume is occupied by a single hadronic cluster, which corresponds to the quark-gluon plasma. The line $\Delta = 0$ in the T–μ_B diagram delineates the boundary of the hadronic phase in the statistical bootstrap model. This line is denoted by $T_c(\mu_B)$ here, as shown in Figure 2. otherwise. The solid line in Figure 2 shows the phase boundary for the full set of ground state hadrons; for comparison, the dashed line shows the boundary when only pions and nucleons are considered as in Ref. [3]. The critical temperature, T_c, at $\mu_B = 0$ for the full set of ground state hadrons is $T_c(0) = 162$ MeV, in remarkably close coincidence with the pseudocritical temperature for the chiral phase transition found in lattice-QCD.

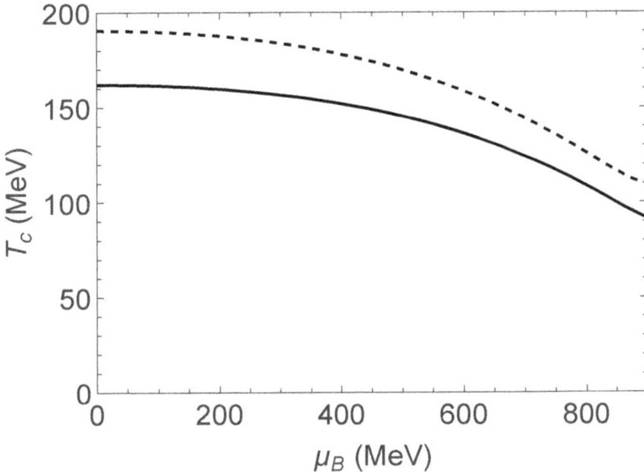

Figure 2. Critical temperature line, $T_c(\mu_B)$, delineating the boundary of hadronic matter in the excluded volume statistical bootstrap model. The solid line shows the boundary location when all quark model ground states are used at input into the statistical bootstrap; the dashed line shows the boundary when only pions and nucleons are considered as input.

The line $T_c(\mu_B)$ traces the location of singularities in the partition function of the statistical bootstrap model with excluded volume; it delineates the thermodynamic phase boundary. Percolating clusters of hadrons already exist at lower temperatures corresponding to an unoccupied volume fraction $\Delta/V > 0$. Since the exact value of the percolation threshold, $\zeta_c = 1 - \Delta_c/V$, is not known, several contour lines of equal occupied (excluded) volume fraction ζ are shown in Figure 3. We denote these lines by $T_\zeta(\mu_B, \zeta)$ where obviously $T_\zeta(\mu_B, 1) = T_c(\mu_B)$. According to the discussion in Section 2, the line $T_\zeta(\mu_B, \zeta_c)$ indicates the hadrochemical horizon, $T_h(\mu_B)$. Lacking precise knowledge of the dynamics of valence-quark percolation, the precise location of this horizon is uncertain, but it can be estimated that $T_h(\mu_B)$ lies between $T_\zeta(\mu_B, 0.6)$ and $T_\zeta(\mu_B, 0.8)$.

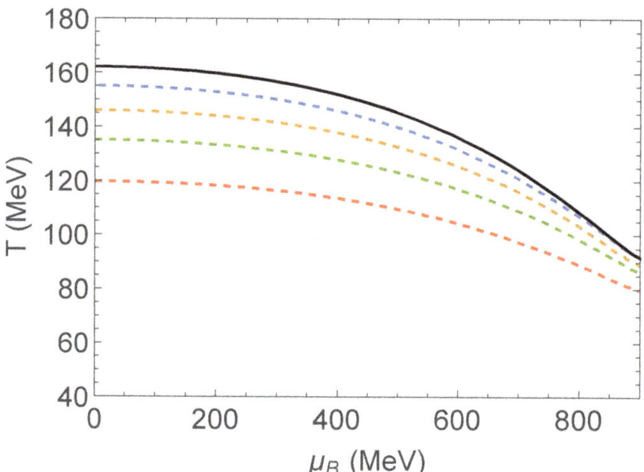

Figure 3. Iso-occupancy contour lines, $T_\zeta(\mu_B, \zeta)$, for hadronic matter in the statistical bootstrap model with excluded volume. The dashed lines show the contours for the occupation fractions $\zeta = 0.8, 0.6, 0.4$, and 0.2 (top to bottom). The solid black line corresponds to $T_\zeta(\mu_b, 1) = T_c(\mu_B)$, also shown in Figure 1. The hadrochemical horizon identified by the valence-quark percolation threshold is expected to lie between the orange ($\zeta = 0.6$) and blue ($\zeta = 0.8$) dashed lines.

5. Summary and Outlook

The chemical freeze-out line or chemical horizon, $T_{\text{chem}}(\mu_B)$, of temperature vs. baryochemical potential, serves as an experimentally measurable proxy for the phase boundary between matter composed of hadrons and quark matter. It is controlled by hadrochemical kinetics and, thus, somewhat sensitive to the overall size and expansion rate of the fireball. As chemical reactions among hadrons proceed primarily by quark (flavor) exchange, the chemical freeze-out line is expected to lie close to the valence-quark percolation line in the QCD phase diagram.

In the low net-baryon density region of the phase diagram the chemical horizon agrees within experimental uncertainties with the pseudocritical thermodynamic phase boundary, $T_c(\mu_B)$, as determined by lattice QCD. At high net-baryon density the data indicate that the chemical horizon lies increasingly below the phase boundary. This could be caused by either a slower expansion rate during the time when the matter is near the phase boundary, or by changing chemical reaction kinetics in the baryon-dominated regime, or both. The origin of the empirical rule—"Cleymans' Law"—that chemical freeze-out occurs along a line of constant energy per particle [28] over the entire measured range is still unknown. Some reminiscences of the late Jean Cleymans, who made fundamental contributions to the theory of relativistic heavy-ion collisions, are presented in Appendix A.

It would be interesting to extend the theoretical simulations of the chemical freeze-out line in hadronic transport models to hybrid models of the reaction that include a hydrodynamic quark-gluon plasma phase. It would also be interesting to construct a more detailed model of valence-quark percolation and identify the precise location of the critical percolation line. Finally, it would be interesting to construct and study observables for valence-quark percolation that can be studied on the lattice.

Note Added in Proof: After submission of the manuscript, I became aware of a publication by Fukushima, Kojo and Weise [40], in which a similar concept of valence-quark percolation through meson exchange is discussed for the transition from nuclear matter to quark matter at high baryon density.

Funding: This work was supported by the Office of Science of the U. S. Department of Energy (DOE) under Grant DE-FG02-05ER41367.

Data Availability Statement: Not applicable.

Acknowledgments: The author acknowledges support from Yale University for a sabbatical visit.

Conflicts of Interest: The author declares no conflict of interest.

Appendix A. Reminiscences of Jean Cleymans

I met Jean many times at conferences and always enjoyed the discussions with him. He was exemplary in being full of deep insights into the physics of relativistic heavy ion collisions and implications of experimental data and simultaneously gentle and modest. I also had multiple opportunities to interact with him during visits to Cape Town, which he had made his home in the mid-1980s. That was a time when it required personal courage to ignore world-wide calls for boycotts of South Africa, but Jean was convinced that he could work for the benefit of South African society by helping to educate the future leaders, who would have to assume influential positions in science and education once apartheid ended; and end it did in 1991, sooner than many expected.

Not long afterwards, I visited the University of Cape Town at Jean's invitation after a conference [41] that was organized by Horst Stöcker in honor of Walter Greiner's 60th birthday in Wilderness on the Indian Ocean coast of South Africa. During my visit, Jean arranged an excursion by boat for a few of us to Robben Island, the small island south of Cape Town where Nelson Mandela and other leaders of the African National Congress had been incarcerated from 1964 until 1982. At that time the site was not yet open to the public, and our small group had a personal tour of the island, the prison complex and the quarry in which Mandela and their fellow inmates had to labor under difficult conditions. It was an almost unreal experience to imagine the violence that was perpetrated at this site juxtaposed with its beautiful and tranquil nature, which included a small colony of penguins that nested at the shore. The visit gave us a sense of the incredible moral strength that allowed Mandela to maintain their conviction that reconciliation was possible after the end of apartheid through almost three decades of imprisonment. The *Truth and Reconciliation* process started when Mandela became President in 1994 and was in full swing during the time of our visit.

My last visit to Cape Town at Jean's invitation occurred in 2017 on the occasion of a trip to South Africa with a small delegation from Brookhaven National Laboratory (BNL). The purpose of our trip was to explore possible collaborations between BNL and various South African institutions, one of which was the University of Cape Town. Jean was as gracious host as ever, and he arranged for a dinner at the residence of the President of the University. It was clear that the University, which always had been open to students of color, had become an even more integrated institution of higher learning, which served its role of educating future leaders of the country that Jean had envisioned well. A younger generation of scientists working on relativistic heavy-ion physics is now carrying on his vision.

Openness and inclusivity often do not make life easier in the short term, but ultimately it is the only way by which societies can make lasting progress. Seeing this and working towards it with an eye to the long term was one of Jean's strengths, which is also reflected in the way he approached science. With this he served as a role model for many of his friends, younger colleagues, and students. It is his legacy.

References

1. Alford, M.G.; Schmitt, A.; Rajagopal, K.; Schäfer, T. Color superconductivity in dense quark matter. *Rev. Mod. Phys.* **2008**, *80*, 1455–1515. [CrossRef]
2. Andronic, A.; Blaschke, D.; Braun-Munzinger, P.; Cleymans, J.; Fukushima, K.; McLerran, L.D.; Oeschler, H.; Pisarski, R.D.; Redlich, K.; Stachel, J.; et al. Hadron production in ultra-relativistic nuclear collisions: Quarkyonic matter and a triple point in the phase diagram of QCD. *Nucl. Phys. A* **2010**, *837*, 65–86. [CrossRef]
3. Hagedorn, R.; Rafelski, J. Hot hadronic matter and nuclear collisions. *Phys. Lett. B* **1980**, *97*, 136. [CrossRef]

4. Bazavov, A.; Ding, H.T.; Hegde, P.; Kaczmarek, O.; Karsch, F.; Laermann, E.; Maezawa, Y.; Mukherjee, S.; Ohno, H.; Wagner, M.; et al. The QCD equation of state to $\mathcal{O}(\mu_B^6)$ from lattice QCD. *Phys. Rev. D* **2017**, *95*, 054504. [CrossRef]

5. Borsanyi, S.; Fodor, Z.; Guenther, J.N.; Kara, R.; Katz, S.D.; Parotto, P.; Pasztor, A.; Ratti, C.; Szabo, K.K. QCD Crossover at finite chemical potential from lattice simulations. *Phys. Rev. Lett.* **2020**, *125*, 052001. [CrossRef]

6. Hatsuda, T.; Tachibana, M.; Yamamoto, N.; Baym, G. New critical point induced by the axial anomaly in dense QCD. *Phys. Rev. Lett.* **2006**, *97*, 122001. [CrossRef]

7. Borsányi, S.; Fodor, Z.; Guenther, J.N.; Kara, R.; Katz, S.D.; Parotto, P.; Pásztor, A.; Ratti, C.; Szabó, K.K. Lattice QCD equation of state at finite chemical potential from an alternative expansion scheme. *Phys. Rev. Lett.* **2021**, *126*, 232001. [CrossRef]

8. Fu, W.j.; Pawlowski, J.M.; Rennecke, F. QCD phase structure at finite temperature and density. *Phys. Rev. D* **2020**, *101*, 054032. [CrossRef]

9. Critelli, R.; Noronha, J.; Noronha-Hostler, J.; Portillo, I.; Ratti, C.; Rougemont, R. Critical point in the phase diagram of primordial quark-gluon matter from black hole physics. *Phys. Rev. D* **2017**, *96*, 096026. [CrossRef]

10. Bazavov, A.; Ding, H.T.; Hegde, P.; Kaczmarek, O.; Karsch, F.; Karthik, N.; Laermann, E.; Lahiri, A.; Larsen, R.; Steinbrecher, P.; et al. Chiral crossover in QCD at zero and non-zero chemical potentials. *Phys. Lett. B* **2019**, *795*, 15–21. [CrossRef]

11. Magas, V.; Satz, H. Conditions for confinement and freezeout. *Eur. Phys. J. C* **2003**, *32*, 115–119. [CrossRef]

12. Armesto, N.; Braun, M.A.; Ferreiro, E.G.; Pajares, C. Percolation approach to quark-gluon plasma and J/psi suppression. *Phys. Rev. Lett.* **1996**, *77*, 3736–3738. [CrossRef] [PubMed]

13. Braun, M.A.; Pajares, C.; Ranft, J. Fusion of strings versus percolation and the transition to the quark gluon plasma. *Int. J. Mod. Phys. A* **1999**, *14*, 2689–2704. [CrossRef]

14. Kovner, A.; McLerran, L.D.; Weigert, H. Gluon production from nonAbelian Weizsacker-Williams fields in nucleus-nucleus collisions. *Phys. Rev. D* **1995**, *52*, 6231–6237. [CrossRef] [PubMed]

15. McLerran, L. Color glass condensate and glasma. *AIP Conf. Proc.* **2007**, *917*, 219–230. [CrossRef]

16. Chodos, A.; Jaffe, R.L.; Johnson, K.; Thorn, C.B. Baryon structure in the bag theory. *Phys. Rev. D* **1974**, *10*, 2599. [CrossRef]

17. Thomas, A.W.; Theberge, S.; Miller, G.A. Cloudy bag model of the nucleon. *Phys. Rev. D* **1981**, *24*, 216. [CrossRef]

18. Stauffer, D.; Aharony, A. *Introduction to Percolation Theory*; Taylor & Francis: London, UK, 1994. [CrossRef]

19. Grimaldi, C. Tree-ansatz percolation of hard spheres. *J. Chem. Phys.* **2017**, *147*, 074502. [CrossRef]

20. Desmond, K.W.; Weeks, E.R. Influence of particle size distribution on random close packing of spheres. *Phys. Rev. E* **2014**, *90*, 022204. [CrossRef]

21. Cleymans, J.; Satz, H. Thermal hadron production in high-energy heavy ion collisions. *Z. Phys. C* **1993**, *57*, 135–148. [CrossRef]

22. Suhonen, E.; Cleymans, J.; Redlich, K.; Satz, H. Hadron production ratios as probes of confinement and freezeout. *arXiv* **1993**, arXiv:hep-ph/9310345. [CrossRef]

23. Letessier, J.; Rafelski, J.; Tounsi, A. Strangeness and particle freezeout in nuclear collisions at 14.6 GeV A. *Phys. Lett. B* **1994**, *328*, 499–505. [CrossRef]

24. Koch, P.; Müller, B.; Rafelski, J. Strangeness production and evolution in quark gluon plasma. *Z. Phys. A* **1986**, *324*, 453–463. [CrossRef]

25. Koch, P.; Müller, B.; Rafelski, J. Strangeness in relativistic heavy ion collisions. *Phys. Rep.* **1986**, *142*, 167–262. [CrossRef]

26. Braun-Munzinger, P.; Stachel, J.; Wetterich, C. Chemical freezeout and the QCD phase transition temperature. *Phys. Lett. B* **2004**, *596*, 61–69. [CrossRef]

27. Harris, J.W.; Müller, B. The Search for the quark-gluon plasma. *Ann. Rev. Nucl. Part. Sci.* **1996**, *46*, 71–107. [CrossRef]

28. Cleymans, J.; Redlich, K. Unified description of freezeout parameters in relativistic heavy ion collisions. *Phys. Rev. Lett.* **1998**, *81*, 5284–5286. [CrossRef]

29. Borsanyi, S.; Endrodi, G.; Fodor, Z.; Jakovac, A.; Katz, S.D.; Krieg, S.; Ratti, C.; Szabo, K.K. The QCD equation of state with dynamical quarks. *J. High Energy Phys.* **2010**, *11*, 077. [CrossRef]

30. Adamczyk, L.; Adkins, J.K.; Agakishiev, G.; Aggarwal, M.M.; Ahammed, Z.; Ajitanand, N.N.; Alekseev, I.; Anderson, D.M.; Aoyama, R.; Aparin, A.; et al. [STAR Collaboration]. Bulk properties of the medium produced in relativistic heavy-ion collisions from the beam energy scan program. *Phys. Rev. C* **2017**, *96*, 044904. [CrossRef]

31. Becattini, F.; Manninen, J.; Gazdzicki, M. Energy and system size dependence of chemical freeze-out in relativistic nuclear collisions. *Phys. Rev. C* **2006**, *73*, 044905. [CrossRef]

32. Andronic, A.; Braun-Munzinger, P.; Stachel, J. Hadron production in central nucleus-nucleus collisions at chemical freeze-out. *Nucl. Phys. A* **2006**, *772*, 167–199. [CrossRef]

33. Flor, F.A.; Olinger, G.; Bellwied, R. Flavour and energy dependence of chemical freeze-out temperatures in relativistic heavy ion collisions from RHIC-BES to LHC energies. *Phys. Lett. B* **2021**, *814*, 136098. [CrossRef]

34. Reichert, T.; Inghirami, G.; Bleicher, M. Probing chemical freeze-out criteria in relativistic nuclear collisions with coarse grained transport simulations. *Eur. Phys. J. A* **2020**, *56*, 267. [CrossRef]

35. Becattini, F.; Cleymans, J.; Keranen, A.; Suhonen, E.; Redlich, K. Features of particle multiplicities and strangeness production in central heavy ion collisions between $1.7A$ and $158A$ GeV/c. *Phys. Rev. C* **2001**, *64*, 024901. [CrossRef]

36. Bazavov, A.; Ding, H.T.; Hegde, P.; Kaczmarek, O.; Karsch, F.; Laermann, E.; Mukherjee, S.; Petreczky, P.; Schmidt, C.; Smith, D.; et al. Freeze-out conditions in heavy ion collisions from QCD thermodynamics. *Phys. Rev. Lett.* **2012**, *109*, 192302. [CrossRef] [PubMed]

37. Borsanyi, S.; Fodor, Z.; Katz, S.D.; Krieg, S.; Ratti, C.; Szabo, K.K. Freeze-out parameters: Lattice meets experiment. *Phys. Rev. Lett.* **2013**, *111*, 062005. [CrossRef]
38. Ratti, C.; Bellwied, R. *The Deconfinement Transition of QCD. Theory Meets Experiment*; Springer Nature Switzerland AG: Cham, Switzerland, 2021. [CrossRef]
39. Bellwied, R.; Noronha-Hostler, J.; Parotto, P.; Portillo Vazquez, I.; Ratti, C.; Stafford, J.M. Freeze-out temperature from net-kaon fluctuations at energies available at the BNL Relativistic Heavy Ion Collider. *Phys. Rev. C* **2019**, *99*, 034912. [CrossRef]
40. Fukushima, K.; Kojo, T.; Weise, W. Hard-core deconfinement and soft-surface delocalization from nuclear to quark matter. *Phys. Rev. D* **2020**, *102*, 096017. [CrossRef]
41. Stoecker, H.; Gallmann, A.; Hamilton, J.H. (Eds.) *Structure of Vacuum and Elementary Matter, Proceedings of the International Conference on Nuclear Physics at the Turn of the Millennium*; World Scientific Co.: Singapore, 1997. [CrossRef]

Article

Enthusiasm and Skepticism: Two Pillars of Science—A Nonextensive Statistics Case

Constantino Tsallis [1,2,3]

1   Centro Brasileiro de Pesquisas Físicas and National Institute of Science and Technology of Complex Systems, Rua Xavier Sigaud 150, Rio de Janeiro 22290-180, Brazil; tsallis@cbpf.br
2   Santa Fe Institute, 1399 Hyde Park Road, Santa Fe, NM 87501, USA
3   Complexity Science Hub Vienna, Josefstädter Strasse 39, 1080 Vienna, Austria

Abstract: Science and its evolution are based on complex epistemological structures. Two of the pillars of such a construction definitively are enthusiasm and skepticism, both being ingredients without which solid knowledge is hardly achieved and certainly not guaranteed. Our friend and colleague Jean Willy André Cleymans (1944–2021), with his open personality, high and longstanding interest for innovation, and recognized leadership in high-energy physics, constitutes a beautiful example of the former. Recently, Joseph I. Kapusta has generously and laboriously offered an interesting illustration of the latter pillar, in the very same field of physics, concerning the very same theoretical frame, namely, nonextensive statistical mechanics and the nonadditive q-entropies on which it is based. I present here a detailed analysis, point by point, of Kapusta's 19 May 2021 talk and, placing the discussion in a sensibly wider and updated perspective, I refute his bold conclusion that indices q have no physical foundation.

Keywords: Jean Cleymans; Joseph Kapusta; thermodynamics; nonadditive entropies; nonextensive statistical mechanics; high-energy physics

Citation: Tsallis, C. Enthusiasm and Skepticism: Two Pillars of Science—A Nonextensive Statistics Case. *Physics* **2022**, *4*, 609–632. https://doi.org/10.3390/physics4020041

Received: 13 April 2022
Accepted: 6 May 2022
Published: 27 May 2022

1. Reminiscences Related to Jean Cleymans

I had the privilege of personally meeting Jean Cleymans in two occasions. The first one was at CERN (European Organization for Nuclear Research) for the Heavy Ion Forum held in 2013–2014 to discuss q-statistics in high-energy physics. His experience and scientific weight among his CERN colleagues were noticeable. That was my first visit to CERN, where I stayed a couple of days. The second occasion was during a visit, over several days, to the University of Cape Town, when I also had the opportunity to lecture and discuss, with him and his international research group, on the foundations and applications of nonadditive entropies and nonextensive statistical mechanics.

In the bibliography in [1], over 9000 articles, by nearly 16,000 authors, are registered on this research area, which started in 1988 [2]. Among them, more than one thousand concern high-energy physics (high-energy collisions, possible connections to quantum chromodynamics (QCD), thermofractals, solar physics, astronomy, astrophysics, black holes, cosmology, dark energy; theoretical, observational, experimental). I could find therein fourty two articles by Cleymans and co-authors [3–44], spanning over one decade. His longstanding *enthusiasm* about the subject, at least in what concerns high-energy physics, can be hardly doubted.

2. Analysis of Joseph Kapusta's Talk "A Primer on Tsallis Statistics for Nuclear and Particle Physics" (19 May 2021)

2.1. Preliminaries

On 19 May 2021, Joseph Kapusta delivered an online talk within the Theoretical Physics Colloquium series that Igor Shovkovy has been hosting at Arizona State University. The talk was entitled "A primer on Tsallis statistics for nuclear and particle physics", and it is accessible on the Internet [45]. Making it publicly accessible was, in some sense, a fortunate initiative. Indeed, in spite of the fact that my own name appears in the title of the talk, I never received notice of it, rather regretfully. Nevertheless, fortuitously, I could eventually hear, in 27 December 2021, Kapusta's talk when I became aware of its existence through an anonymous referee report on a project involving q-concepts.

However, why am I giving all these details? Basically, because I agree with the old dictate "silence gives consent" ("qui tacet consentit"), and by no means can I agree with nor remain silent regarding the final conclusions offered by the speaker of that specific talk. Indeed, in what follows, I intend to offer, essentially point by point, an analysis of and a reply to Kapusta's talk, not only in what concerns the logical implications of his statements and conclusions, but—quite appropriate under the present circumstances—also in what concerns the spirit of the "analyse du discours" [46–48]. This is why I focus here on the talk itself and not on the written version, which Kapusta published some months later [49]. Let me add for completeness that, as soon as I became aware of the existence of that talk on YouTube, I proposed an open academic debate between Kapusta and myself, possibly organized within the same or similar frame of the Arizona State University Colloquia series (or any other appropriate forum in fact), but I did not receive back a signal of operational interest. Unfortunately, an opportunity for a free and direct exchange of scientific ideas focusing on a currently active research area was lost.

Before starting, let me by all means make absolutely clear my genuine gratitude to Kapusta for having dedicated a sensible amount of his time and knowledge to analyze and better understand the theory that I proposed over three decades ago (in [2,50] and elsewhere). It is my understanding that, by so proceeding, Kapusta also attempted to serve science along its best lines, in this case through *skepticism*, as it becomes clear in what follows.

2.2. On the Title of Kapusta's Talk

The choice of the word "primer" in the title is suggestive. For this word one finds in the Merriam-Webster Dictionary:

- A small book for teaching children to read;
- A small introductory book on a subject;
- A short informative piece of writing.

  In the Cambridge Dictionary, one finds:

- A small book containing basic facts about a subject, used especially when you are beginning to learn about that subject;
- A basic text for teaching something.

It appears therefore that the word "primer" tends to induce the audience in the sense that basic undeniable facts will be presented, as opposed to opinions or beliefs. In classical philosophical Plato terms, "primer" induces us to think about "episteme" (from the Greek $\epsilon\pi\iota\sigma\tau\eta\mu\eta$, related to facts) and not about "doxa" (from the Greek $\delta\acute{o}\xi\alpha$, opinion). But, as the reader can verify, what one hears in various parts of the talk is a blurring between these two concepts, facts and opinion, a blurring between "text" and "context".

Along a similar vein, the expression "Tsallis and his collaborators" that Kapusta used several times is not exempted from ambiguity. Indeed, there are my collaborators—co-authors, quite frequently—and there are also thousands of scientists (see [1]) who became interested in the subject and have published, along the years, thousands of articles, some of which presenting aspects that I would surely endorse, and others whose statements are certainly not to be confused with my own.

2.3. Nonadditive Entropy S_q: Where It Comes from

In his talk, Kapusta declared a couple of times that he did not know where S_q came from. Although spread in the literature for decades, let us summarize in the present occasion where it comes from. Like the Boltzmann–Gibbs entropic functional, it is basically a *postulate*, as stated in [2]. There is, however, a simple rationale behind it. Let us describe it.

An entropic functional must be invariant under permutation of states; indeed, what matters are the probabilities $\{p_i\}$ and not the states themselves. The simplest manner which satisfies this property is to involve $\sum_i f(p_i)$, and the simplest form for $f(p_i)$ is just a power-law, i.e., $f(p_i) \propto p_i^q$ $(q \in \mathbb{R})$. Consistently, the entropic functional being constructed is assumed to be of the form $S_q(\{p_i\}) = \Phi(\sum_i p_i^q)$, with $\sum_i p_i = 1$. The simplest form for $\Phi(x)$ is to be linear, i.e., $S_q(\{p_i\}) = A(q) + B(q)\sum_i p_i^q$. Probabilistic certainty should make the entropy to vanish. Therefore, if it is possible for a state j to exist such that $p_j = 1$, then $\sum_i p_i^q = 1$, hence $A(q) + B(q) = 0$. Consequently, $S_q(\{p_i\}) = A(q)(1 - \sum_i p_i^q)$. Now, the whole intuition for the present postulate, inspired in multifractals as declared in [2], is to provide a *bias* for the state probability. More definitely, to compare p_i^q with p_i for any $p_i \in (0,1)$: $p_i^q > p_i$ if $q < 1$, $p_i^q < p_i$ if $q > 1$, and finally $p_i^q = p_i$ if $q = 1$. As one can see, $q = 1$ simply means absence of bias, and this is intuitively assumed to well correspond to the Boltzmann–Gibbs (GB) entropy variable, $S_{\mathrm{BG}} = k\sum_i p_i \ln(1/p_i)$. Therefore, $\lim_{q\to 1} A(q)$ must diverge in order to compensate for the fact that $\lim_{q\to 1}[1 - \sum_i p_i^q] = \lim_{q\to 1}[1 - \sum_i p_i p_i^{q-1}] = \lim_{q\to 1}[1 - \sum_i p_i e^{(q-1)\ln p_i}] = \lim_{q\to 1}\{1 - \sum_i p_i[1 + (q-1)\ln p_i]\} = \lim_{q\to 1}(q-1)[-\sum_i p_i \ln p_i] = 0$; moreover, $A(q)$ must have the dimensions of the constant k, hence be proportional to k. The simplest such possibility satisfying $\lim_{q\to 1} S_q\{p_i\} = S_{\mathrm{BG}}\{p_i\}$ is $A(q) = k/(q-1)$, and we are done. The entropy S_q is therefore defined by

$$
S_q(\{p_i\}) \equiv k\frac{1 - \sum_{i=1}^{W} p_i^q}{q - 1} = k\sum_{i=1}^{W} p_i \ln_q \frac{1}{p_i}
$$

$$
= -k\sum_{i=1}^{W} p_i^q \ln_q p_i = -k\sum_{i=1}^{W} p_i \ln_{2-q} p_i \quad (S_1 = S_{\mathrm{BG}}), \tag{1}
$$

where the (monotonically increasing) q-logarithmic function is defined through

$$
\ln_q z \equiv \frac{z^{1-q} - 1}{1 - q} \quad (\ln_1 z = \ln z), \tag{2}
$$

with $d\ln_q z/dz = 1/z^q$. The quantity $\sigma(p) \equiv \ln(1/p)$ is referred to as "surprise" [51] or "unexpectedness" [52]. It vanishes when the probability p equals unity and diverges when the probability $p \to 0$. One can consistently define q-surprise or q-unexpectedness as $\sigma_q(p) \equiv \ln_q(1/p)$, hence $\sigma_q(1) = 0$ and $\sigma_q(0)$ attains its maximum (infinity if $q \leq 1$). With this definition, S_q can be rewritten as follows:

$$
S_q = k\sum_{i=1}^{W} p_i \ln_q(1/p_i) \equiv k\langle\ln_q(1/p_i)\rangle = k\langle\sigma_q\rangle, \tag{3}
$$

where $\langle\dots\rangle$ denotes the mean value. Equation (3) satisfies in fact the most general form for a "trace-form" entropic functional, $S_\Psi(\{p_i\}) = k\sum_i p_i \ln_\Psi(1/p_i)$, $\ln_\Psi(z)$ being a generic monotonically increasing function which satisfies $\ln_\Psi(1) = 0$; for details, see [50,53].

In addition, an entropic functional $S(\{p_i\}; \eta)$ is said "composable" if it satisfies, for two *probabilistically independent* systems A and B, the property $S(A + B)/k = F(S(A)/k, S(B)/k; \eta)$, where η is a set of fixed indices characterizing the functional (e.g., for S_q, it is $\eta \equiv q$); the notation $\{0\}$ is used here to indicate absence of any such index. $F(x, y; \{\eta\})$ is assumed to be a smooth function of (x, y), which depends on a (typically small) set of universal indices $\{\eta\}$, defined in such a way that $F(x, y; \{0\}) = x + y$ (additivity), which corresponds to the Boltzmann–Gibbs entropy. Furthermore, $F(x, y; \{\eta\})$ is assumed to satisfy $F(x, 0; \{\eta\}) = x$ (null-composability), $F(x, y; \{\eta\}) = F(y, x; \{\eta\})$ (symmetry), $F(x, F(y, z; \{\eta\}); \{\eta\}) = F(F(x, y; \{\eta\}), z; \{\eta\})$ (associativity); for details and thermodynamical motivation, see [50,53].

The Enciso–Tempesta theorem [54] proves that S_q is the *unique* entropic functional which simultaneously is a trace form, composable, and includes the Boltzmann–Gibbs form as a particular instance. These properties play a crucial role in the validity, for *all values* of q, of the Einstein requirement for likelihood factorization [55]. This important property is consistent with the fact that the Boltzmann–Gibbs statistical mechanics is *sufficient but not necessary* for the validity of thermodynamics and its Legendre structure.

The word "postulate" used above deserves a comment. For the Boltzmann–Gibbs entropy, it is possible to replace its postulated form by a set of four axioms. That was first shown by Shannon in his celebrated uniqueness theorem [56,57], and later on, in a more elegant form, by Khinchin in his own uniqueness theorem [58,59]. Both theorems have been q-generalized into Santo's uniqueness theorem [60] and Abe's uniqueness theorem [61], respectively. Let us finally mention that, in addition to the above uniqueness theorems (Santos in 1997, Abe in 2000, and Enciso and Tempesta in 2017), S_q is unique in other relevant senses as well, namely, the Topsøe factorizability in game theory [62], the Amari–Ohara–Matsuzoe conformally invariant geometry [63] and the Biro–Barnafoldi–Van thermostat universal independence [64].

2.4. Additivity versus Extensivity

All the way along his talk, Kapusta did not distinguish—clearly enough, and even at all—the concepts of "extensivity" and "additivity", applicable to both entropy and energy. This is quite unfortunate since this distinction ought to be made in any introductory talk on the subject. Indeed, it plays a foundational role in nonextensive statistical mechanics. Let us address now these two important notions, focusing specifically on entropic additivity and entropic extensivity.

Following O. Penrose [65], an entropic functional $S(\{p_i\})$ is said "additive" if, for two *probabilistically independent* systems A and B (i.e., $p_{ij}^{A+B} = p_i^A p_j^B$), one verifies $S(A+B) = S(A) + S(B)$, in other words, if

$$S(\{p_i^A p_j^B\}) = S(\{p_i^A\}) + S(\{p_j^B\}) \tag{4}$$

is verified.

Otherwise, $S(\{p_i\})$ is said "nonadditive". It immediately follows that S_{BG} is additive. In contrast, S_q satisfies

$$\frac{S_q(A+B)}{k} = \frac{S_q(A)}{k} + \frac{S_q(B)}{k} + (1-q)\frac{S_q(A)}{k}\frac{S_q(B)}{k}, \tag{5}$$

hence

$$S_q(A+B) = S_q(A) + S_q(B) + \frac{1-q}{k}S_q(A)S_q(B). \tag{6}$$

Therefore, unless $(1-q)/k \to 0$, S_q is nonadditive.

Let us now address the other important entropic concept, namely, extensivity. An entropy $S(N)$ is said "extensive" if a specific entropic functional is applied to a specific class of many-body systems with $N = L^d$ particles, where L is its dimensionless linear size and d its spatial dimension, and satisfies the thermodynamical expectation

$$0 < \lim_{N\to\infty} \frac{S(N)}{N} < \infty, \tag{7}$$

hence, $S(N) \propto N$ for $N \gg 1$. Therefore, entropic additivity only depends on the entropic functional, whereas entropic extensivity depends on *both* the chosen entropic functional *and* the system itself (i.e., its constituents and the correlations among them).

Let us illustrate this fundamental distinction through four, among infinitely many, equal-probability typical examples of $W(N)$ $(N \to \infty)$, where W is the total number of possibilities whose probability does not vanish.

- Exponential class $W(N) \sim A\mu^N$ $(A > 0; \mu > 1)$:
  This is the typical case within the Boltzmann–Gibbs theory. One gets: $S_{\mathrm{BG}}(N) = k \ln W(N) \sim N \ln \mu + \ln A \propto N$, therefore, S_{BG} is extensive, as thermodynamically required.

- Power-law class $W(N) \sim BN^\rho$ $(B > 0; \rho > 0)$:
  One should not use S_{BG} since it implies $S_{\mathrm{BG}}(N) = k \ln W(N) \sim \rho \ln N + \ln B \propto \ln N$, thus violating thermodynamics. One verifies instead that $S_{q=1-1/\rho}(N) = k \ln_{q=1-1/\rho} W(N) \propto N$, as thermodynamically required.

- Stretched exponential class $W(N) \sim C\nu^{N^\gamma}$ $(C > 0; \nu > 1; 0 < \gamma < 1)$:
  In this case, no value of q exists which would yield an extensive entropy S_q. One can instead use $S_\delta \equiv k \sum_{i=1}^{W} p_i [\ln 1/p_i]^\delta$ [50] with $\delta = 1/\gamma$. Indeed, $S_{\delta=1/\gamma}(N) = k[\ln W(N)]^{1/\gamma} \propto N$, as thermodynamically required.

- Logarithmic class $W(N) \sim D \ln N$ $(D > 0)$:
  In this case, no values of (q, δ) exist which are able to yield an extensive entropy $S_{q,\delta} \equiv k \sum_i p_i [\ln_q(1/p_i)]^\delta$ [66]. Instead, one can use the Curado entropy [53], $S_\lambda^C(N) = k\left[e^{\lambda W(N)} - e^\lambda\right]$ with $\lambda = 1/D$. Indeed, one can verify that $S_{\lambda=1/D}^C(N) \propto N$, as thermodynamically required.

These four paradigmatic classes are described in Figure 1.

SYSTEMS $W(N)$ (equiprobable)	ENTROPY S_{BG} (ADDITIVE)	ENTROPY S_q $(q \neq 1)$ (NONADDITIVE)	ENTROPY S_δ $(\delta \neq 1)$ (NONADDITIVE)	ENTROPY S_λ^C $(\lambda > 0)$ (NONADDITIVE)
$\sim A\mu^N$ $(A>0, \mu>1)$	EXTENSIVE	NONEXTENSIVE	NONEXTENSIVE	NONEXTENSIVE
$\sim B N^\rho$ $(B>0, \rho>0)$	NONEXTENSIVE	EXTENSIVE $(q=1-1/\rho)$	NONEXTENSIVE	NONEXTENSIVE
$\sim C\nu^{N^\gamma}$ $(C>0, \nu>1, 0<\gamma<1)$	NONEXTENSIVE	NONEXTENSIVE	EXTENSIVE $(\delta=1/\gamma)$	NONEXTENSIVE
$\sim D \ln N$ $(D>0)$	NONEXTENSIVE	NONEXTENSIVE	NONEXTENSIVE	EXTENSIVE $(\lambda=1/D)$

Figure 1. Typical behaviors of $W(N)$ (number of nonzero-probability states of a system with N random variables) in the $N \to \infty$ limit and entropic functionals which, under the assumption of *equal probabilities* for all states with nonzero probability, yield extensive entropies for specific values of the corresponding (nonadditive) entropic indices. Concerning the exponential class $W(N) \sim A\mu^N$, S_{BG} is not the unique entropy that yields entropic extensivity; the (additive) Renyi entropic functional, $S_q^R \equiv k(\ln \sum_i p_i^q)/(1-q)$, also is extensive for all values of q. Analogously, concerning the stretched-exponential class $W(N) \sim C\nu^{N^\gamma}$, the (nonadditive) entropic functional S_δ is not unique. All the entropic families illustrated contain S_{BG} as a particular case, excepting the Curado entropy, S_λ^C, which is appropriate for the logarithmic class $W(N) \sim D \ln N$. In the limit $N \to \infty$, the inequalities $\mu^N \gg \nu^{N^\gamma} \gg N^\rho \gg \ln N \gg 1$ are satisfied, hence $\lim_{N\to\infty} \nu^{N^\gamma}/\mu^N = \lim_{N\to\infty} N^\rho/\mu^N = \lim_{N\to\infty} \ln N/\mu^N = 0$. This exhibits that, in all these nonadditive cases, the occupancy of the full phase space corresponds essentially to a *zero* Lebesgue measure, similarly to a whole class of (multi)fractals. If the equal-probabilities hypothesis is not satisfied, a specific analysis becomes necessary and the results might be different. Taken from [50].

At this point, it is pertinent to remind Einstein's 1949 comment [67]: "A theory is the more impressive the greater the simplicity of its premises is, the more different kinds of things it relates, and the more extended is its area of applicability. Therefore, the deep impression that classical thermodynamics made upon me. It is the only physical theory of universal content concerning which I am convinced that, within the framework of applicability of its basic concepts, it will never be overthrown".

To better understand the strength of these words, a metaphor can be used. Within Newtonian mechanics, there is the known Galilean composition of velocities, $v_{13} = v_{12} + v_{23}$. In special relativity, this law is generalized into $v_{13} = [v_{12} + v_{23}]/[1 + v_{12}v_{23}/c^2]$, where c is the speed of light, thus violating additivity. Why did Einstein abandon the simple *linear* composition of Galileo? Because he had a higher goal, namely, to unify mechanics and Maxwell's electromagnetism, and, for this, he had to impose the invariance with regard to the Lorentz transformation. One can, thus, see the violation of the linear Galilean composition as a small price to pay for a major endeavor. Similarly, what is expressed in Figure 1, is that q-generalizing the *linear* composition law of S_{BG}, with regard to independent systems, into the nonlinear composition (5) may be seen as a small price to pay for a major endeavor, namely, to always satisfy the Legendre structure of thermodynamics. However, it is mandatory to register here that such viewpoint is nevertheless not free from controversy, in spite of its simplicity. For example, the known expression of Bekenstein and Hawking for the entropy for a black hole is proportional to its surface instead of to its volume, therefore violating the above requirement *if* the black-hole is assumed to be a three-dimensional object; see [66].

Before leaving this point, it is fair to mention that Kapusta's lack of distinction between entropic extensivity and entropic additivity would not be particularly surprising, were it not for his own perception of his talk as a "primer". In any case, such confusion constitutes, still today, no exception in the literature even if my co-author Murray Gell-Mann and myself started around 2003 to emphasize the distinction in my book [50] and in our paper [68].

2.5. On the Nature of the Constraints Used for Entropic Optimization

The relevance of the invariance, under a zero-energy shift at *fixed* inverse effective temperature β_q, of the distribution of probabilities extremizing S_q took, from [2] through [69] to [70], close to ten years to become clarified. As pedagogically explained by Kapusta, this process led, through types I and II, to the type III or third path, which was specially focused in his talk. The equivalence, under appropriate transformations of q and, consistently, of the Lagrange parameters (α, β), associated with these three paths was later on established in [71,72]. Indeed, the possibility of linking S_{2-q} (instead of S_q) with the extremizing q-exponential distribution and *linear* constraints such as $\sum_i p_i E_i = \langle E \rangle$ (instead of the *nonlinear* constraint $\sum_i p_i^q E_i / \sum_i p_i^q = \langle E \rangle_q$) was later on used in [73–82] on the basis of identities (1), and has nowadays become quite popular due to its operational simplicity. This link is based on a nonlinear Fokker–Planck Equation (with nonlinear exponent $(2-q)$) in the presence of a confining potential [83,84]. This equation has, as its basic stationary state, a q-Gaussian and not a $(2-q)$-Gaussian. Further examples exhibiting the same type of duality are available at [85,86].

Let us mention at this stage that several central related issues were regretfully ignored in Kapusta's talk, namely:

- The peculiar way the nonlinear constraint of type III was phrased by the speaker left floating in the audience (which even repeated this later on, with the tacit agreement of the speaker) that this assumption was violating the theory of probabilities. There is *no* such thing.

    Indeed, the above *nonlinear* constraint in p_i is completely equivalent to a *linear* constraint expressed in terms of $\sum_i P_i^{(q)} E_i$ where $P_i^{(q)} \equiv p_i^q / \sum_j p_j^q$ (hence $p_i = (P_i^{(q)})^{1/q} / \sum_j (P_j^{(q)})^{1/q}$, with $\sum_i p_i = \sum_i P_i^{(q)} = 1$) is the so-called "escort distribution," defined in the theory of probabilities; see [87] and references therein.

- The requirement for system-independence for the adopted extremization procedure consists of having for both the norm constraint, expressed in terms of $\sum_i p_i$, and the energy constraint, expressed in terms of $\sum_i P_i^{(q)} E_i$, *one and the same* upper admissible value for q. Let us illustrate this through an example. If the set of eigenvalues $\{E_i\}$ are nondegenerate, the extremizing q-exponential distribution with $q > 1$ asymptotically behaves as $1/E_i^{1/(q-1)}$. Therefore, its norm is well defined up to $q = 2$. The same happens with the constraint $\sum_i P_i^{(q)} E_i$. Indeed, it asymptotically behaves as $E_i / E_i^{q/(q-1)} = 1/E_i^{1/(q-1)}$, hence it has the same upper bound admissibility, i.e., once again $q = 2$. In strong variance, if one were to use here the usual $\sum_i p_i E_i$ constraint, the asymptotic behavior would be given by $E_i / E_i^{1/(q-1)}$, hence its upper admissible value would be $q = 3/2$, which *differs* from the norm admissibility value, $q = 2$. The general importance of this point is lengthily discussed in [88].

- In the q-generalized central limit theorem and Lévy–Gnedenko's limit theorem [89,90], the escort mean values emerge naturally from the mathematical operations themselves.
- All mean values must be mathematically defined for the fat-tailed $q > 1$ distributions. This does happen with the appropriate escort averages but fails with the normal linear averages. A simple numerical example with q-Gaussians and its astonishing practical consequences are illustrated in [91].

2.6. On Ad Hoc Constraints for Optimizing the Entropy

Kapusta referred, in a kind of appreciative style, to a 2004 argument by Zanette and Montemurro [92], which would "disqualify" S_q as a "physical" entropy. It refers to the trivial mathematical fact that by adjusting in an ad hoc manner the constraints under which the entropy optimization is to be done, one can obtain virtually *any* desired distribution. This mathematical feature applies to *all* admissible entropic functionals (including S_{BG}) and has been an overcome issue for a long time. Indeed, already in 1983, Montroll and Shlesinger [93] exhibited how the nontrivial constraint can be adapted in order to obtain Lévy distributions from the *classical Boltzmann–Gibbs* entropy. The author's epistemological dissatisfaction is evident in the paper. Indeed, one is forced to *first* know the distribution that is aimed, and *then*, through a sort of reverse engineering, write down the nontrivial constraint in order to get it. This is actually possible for any entropic functional, as mentioned above, but provides no useful information at all. Constraints within information theory are not to be freely manipulated: they must reflect constants of motion and similar quantities. For a simple real random variable x, they must restrict to robust constraints such as the values of $\langle x \rangle$ and/or of $\langle x^2 \rangle$. It is up to the entropic functional, and *not* to the ad hoc constraints, to do *the most meaningful and nontrivial job*.

We argue here that, since it is known to be so for *any* entropic form and, in particular, for the (additive) Boltzmann–Gibbs entropy, S_{BG} (see [94]), the critique brings absolutely no novelty to the area. In other words, it has *nothing* special to do with the entropy S_q. In favor of the constraints based on the usual simple variables (typically, the averaging of the random variable x_i or of x_i^2, where x_i is to be identified according to the nature of the system), we argue, and this for *all* entropic forms, that they can hardly be considered as arbitrary, as the authors of [92] seemingly consider. Indeed, once the natural variables of the system have been identified (e.g., constants of motion of the system, such as the energy for Hamiltonian systems), the variable itself and, in some occasions, its square are the most basic quantities to be constrained. Such constraints are used in hundreds (perhaps thousands) of useful applications outside (and also inside) thermodynamical systems, along the information theory lines of Jaynes and Shannon, and more recently of A. Plastino and others. Furthermore, this is so for S_{BG}, S_q, and any other entropic form. Rebuttals of the Zanette and Montemurro outdated criticism can be found in [50,95,96].

As a final comment, let us mention that statistical mechanics is much more that just a stationary-state (e.g., thermal equilibrium) distribution. Indeed, under exactly the same constraints, the optimization of S_{BG} and $(S_{\mathrm{BG}})^3$ yields the same distribution. This is not a sufficient reason for using $(S_{\mathrm{BG}})^3$ instead of S_{BG} in a thermostatistical theory which must also satisfy various thermodynamical requirements, including its Legendre-transformation structure.

2.7. Thermodynamics and Legendre Transformations

Before focusing on the Legendre structure itself, let us review some long-known facts concerning long-range interactions. Let us consider a d-dimensional classical many-body system with, e.g., attractive two-body isotropic interactions decaying with a dimensionless distance $r \geq 1$ as $-A/r^\alpha$ ($A > 0$, $\alpha \geq 0$), and with an infinitely repulsive potential for $0 \leq r \leq 1$. At zero temperature T, the total kinetic energy vanishes, and the potential energy per particle is proportional to $\int_1^\infty dr\, r^d r^{-\alpha}$. This quantity converges if $\alpha/d > 1$ and diverges otherwise. These two regimes are from now on referred to as "short-range" and "long-range" interactions, respectively; see Figure 2.

Furthermore, they can be shown to, respectively, correspond, within the Boltzmann–Gibbs theory, to *finite* and *divergent* partition functions. This is indeed the point that was addressed by Gibbs himself [97]: "In treating of the canonical distribution, we shall always suppose the multiple integral in equation (92) [the partition function, as it is called nowadays] to have a finite value, as otherwise the coefficient of probability vanishes, and the law of distribution becomes illusory. This will exclude certain cases, but not such apparently, as will affect the value of our results with respect to their bearing on thermodynamics. It will exclude, for instance, cases in which the system or parts of it can be distributed in unlimited space [...]. It also excludes many cases in which the energy can decrease without limit, as when the system contains material points which attract one another inversely as the squares of their distances. [...]. For the purposes of a general discussion, it is sufficient to call attention to the assumption implicitly involved in the formula (92)".

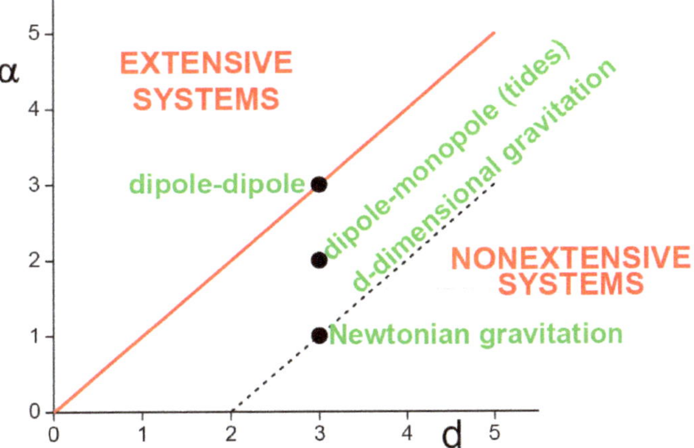

Figure 2. The so-called "extensive systems" ($\alpha/d > 1$ for the classical ones with the exponent α and dimension d) correspond to an extensive total energy and typically involve *absolutely convergent* series, whereas the so-called "nonextensive systems" ($0 \leq \alpha/d < 1$ for the classical ones) correspond to a superextensive total energy and typically involve *divergent* series. The marginal systems ($\alpha/d = 1$ here) typically involve *conditionally convergent* series, which therefore depend on the boundary conditions, i.e., typically on the external shape of the system. Capacitors constitute a notorious example of the $\alpha/d = 1$ case. The standard Lennard–Jones gas is located at $(d, \alpha) = (3, 6)$, thus belonging to the extensive class of systems. Taken from [53].

In a vein slightly differing from the standard Boltzmann–Gibbs recipe, which would demand integration up to infinity in $\int_1^\infty dr\, r^d r^{-\alpha}$, let us assume that the N-particle system is roughly homogeneously distributed within a limited sphere. Then, the potential energy per particle scales as follows:

$$\frac{U_{pot}(N)}{N} \propto -A \int_1^{N^{1/d}} dr\, r^{d-1}\, r^{-\alpha} = -\frac{A}{d} N^* \,, \tag{8}$$

with

$$N^* \equiv \frac{N^{1-\alpha/d}-1}{1-\alpha/d} = \ln_{\alpha/d} N \sim \begin{cases} \dfrac{1}{\alpha/d-1} & \text{if}\quad \alpha/d > 1; \\[2ex] \ln N & \text{if}\quad \alpha/d = 1; \\[2ex] \dfrac{N^{1-\alpha/d}}{1-\alpha/d} & \text{if}\quad 0 \le \alpha/d < 1. \end{cases} \tag{9}$$

Therefore, in the $N \to \infty$ limit, $U_{pot}(N)/N$ approaches a constant ($\propto -A/(\alpha - d)$) if $\alpha/d > 1$, and diverges like $N^{1-\alpha/d}/(1-\alpha/d)$ if $0 \le \alpha/d < 1$ (it diverges logarithmically if $\alpha/d = 1$). In other words, the total potential energy is extensive for short-range interactions ($\alpha/d > 1$), and nonextensive for long-range interactions ($0 \le \alpha/d \le 1$). Satisfactorily enough, Equation (9) recovers the characterization with $\int_1^\infty dr\, r^d r^{-\alpha}$ in the limit $N \to \infty$, but it has the great advantage of providing, for *finite N*, a *finite* value. This fact is now shown to enable us to properly scale the macroscopic quantities in the thermodynamic limit ($N \to \infty$), for *all* values of $\alpha/d \ge 0$ and not only for $\alpha/d > 1$ (as in textbooks of thermodynamics, even if the point is too frequently not made explicitly).

The mathematical structure of classical thermodynamics is based on the Legendre transforms. It is not sufficiently realized that thermodynamics does not depend on microscopic details *only for short-range* interactions. As is illustrated here below, it does depend on quantities such as (α, d) for long-range interactions. Quoting Landsberg [98]: "The presence of long-range forces causes important amendments to thermodynamics, some of which are not fully investigated as yet".

Let us consider a d-dimensional homogeneous and isotropic classical fluid constituted by magnetic particles in thermodynamical equilibrium. Its Gibbs free energy is then:

$$\begin{aligned} G(N,T,p,\mu,H) &= U(N,T,p,\mu,H) - TS(N,T,p,\mu,H) \\ &+ pV(N,T,p,\mu,H) - \mu N - HM(N,T,p,\mu,H), \end{aligned} \tag{10}$$

where (T,p,μ,H) correspond, respectively, to the temperature, pressure, chemical potential and external magnetic field, U is the internal energy, S is the entropy, V is the volume, N is the number of particles, and M the magnetization.

If the interactions are short-ranged, i.e., if $\alpha/d > 1$, one can divide this equation by N and then take the $N \to \infty$ limit. One straightforwardly obtains:

$$g(T,p,\mu,H) = u(T,p,\mu,H) - Ts(T,p,\mu,H) + pv(T,p,\mu,H) - \mu - Hm(T,p,H) \,, \tag{11}$$

where $g(T,p,\mu,H) \equiv \lim_{N\to\infty} G(N,T,p,\mu,H)/N$, and analogously for the other variables of the equation.

If the interactions are long-ranged instead, i.e., if $0 \le \alpha/d \le 1$, all the terms of expression (11) diverge, hence, thermodynamically speaking, they are nonsense. Consequently, the generically correct procedure for all values of $\alpha/d \ge 0$, must conform to the following lines:

$$\begin{aligned} \lim_{N\to\infty} \frac{G(N,T,p,\mu,H)}{NN^*} &= \lim_{N\to\infty} \frac{U(N,T,p,\mu,H)}{NN^*} - \lim_{N\to\infty}\left[\frac{T}{N^*}\frac{S(N,T,p,\mu,H)}{N}\right] \\ &+ \lim_{N\to\infty}\left[\frac{p}{N^*}\frac{V(N,T,p,\mu,H)}{N}\right] - \lim_{N\to\infty}\frac{\mu}{N^*} - \lim_{N\to\infty}\left[\frac{H}{N^*}\frac{M(N,T,p,\mu,H)}{N}\right] \,, \end{aligned} \tag{12}$$

hence,

$$g(T^\star, p^\star, \mu^\star, H^\star) = u(T^\star, p^\star, \mu^\star, H^\star) - T^\star s(T^\star, p^\star, \mu^\star, H^\star)$$
$$+ p^\star v(T^\star, p^\star, \mu^\star, H^\star) - \mu^\star - H^\star m(T^\star, p^\star, \mu^\star, H^\star), \qquad (13)$$

where the definitions of $T^\star$ and of all the other variables are self-explanatory (e.g., $T^\star \equiv \lim_{N \to \infty}[T/N^\star]$ and $s(T^\star, p^\star, \mu^\star, H^\star) \equiv \lim_{N \to \infty}[S(N, T, p, \mu, H)/N]$). In other words, in order to have *finite* thermodynamic equations of states, one must, in general, express them in the $(T^\star, p^\star, \mu^\star, H^\star)$ variables. If $\alpha/d > 1$, this procedure recovers the usual equations of states, and the usual *extensive* (G, U, S, V, N, M) and *intensive* (T, p, μ, H) thermodynamic variables. However, if $0 \le \alpha/d \le 1$, the situation is more complex, and *three* instead of the traditional *two* classes of thermodynamic variables emerge. One may call them *extensive* (S, V, N, M), *pseudoextensive* (G, U) (superextensive in the present case) and *pseudointensive* (T, p, μ, H) (superintensive in the present case) variables. All the energy-type thermodynamical variables (G, F, U), F being the Helmholtz free energy, give rise to pseudoextensive ones, whereas those which appear in the usual Legendre thermodynamical pairs give rise to pseudointensive ones (T, p, μ, H) and extensive ones (S, V, N, M). Let us emphasize that (S, V, N, M) are extensive in *all* cases; see Figure 3. The exactness of all these scalings has been repeatedly verified in the literature for diverse concrete physical systems, such as fluids, magnets, polymers and percolation; see [50] and references therein.

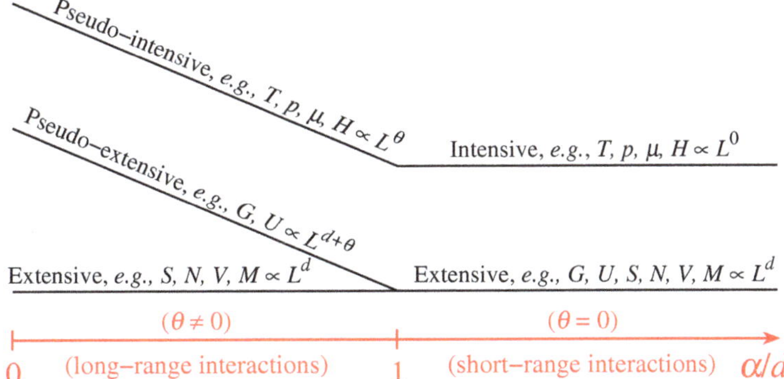

Figure 3. Representation of the different scaling regimes for classical d-dimensional systems. For attractive long-range interactions (i.e., $0 \le \alpha/d \le 1$, α characterizing the interaction range in a potential with the form $1/r^\alpha$; for example, Newtonian gravitation corresponds to $(d, \alpha) = (3, 1)$) one can distinguish *three* classes of thermodynamic variables, namely: those, scaling with L^θ, named "pseudointensive" (L is a characteristic linear length, θ depends on the class of systems); those, scaling with $L^{d+\theta}$ with $\theta = d - \alpha$, the pseudoextensive ones (the energies); and those, scaling with L^d (which are always extensive). For short-range interactions (i.e., $\alpha > d$) one has $\theta = 0$ and the energies recover their standard L^d extensive scaling, falling within the same class of S, N, V, M, etc. (see text for notations of the thermodynamic variables), whereas the previous pseudointensive variables become truly intensive ones (independent of L); this is the region, with only *two* classes of variables, that is covered by the traditional textbooks on thermodynamics. For more details, see [50,53,66,99–101].

Let us also emphasize that, consistently:

- The ratio of any two pseudointensive variables $(T, p, \mu, H, \ldots)$, e.g., p/T, is intensive in all cases;
- The ratio of any pseudoextensive variable (G, F, U) with any pseudointensive variable, e.g., U/T, is extensive in all cases;
- A most important implication is that, in expressions such as $e_q^{-\beta_q \mathcal{H}_N}$ where $\mathcal{H}_N$ is an N-body Hamiltonian, the argument $\beta_q \mathcal{H}_N$ is extensive in all cases. This plays a crucial role in the possible q-generalization of what is currently referred to as the large

deviation theory. Indeed, the extensivity of $\beta_q \, \mathcal{H}_N$ appears to mirror, in all cases, the extensivity of the total entropy involved in $r_q N$, r_q being the ratio function (defined within the Large Deviation probability $P(N) = P(0)e_q^{-r_q N}$), seemingly always related to some relative nonadditive entropy *per particle* [102–106].

2.8. Boltzmann Equation

Kapusta also focused on the current Boltzmann equation associated with the $a + b \leftrightarrow c + d$ reaction. He specifically presented and discussed

$$
\begin{aligned}
\frac{df_a}{dt} = & \int \frac{d^3 p_b}{(2\pi)^3} \frac{d^3 p_c}{(2\pi)^3} \frac{d^3 p_d}{(2\pi)^3} \Big\{ \frac{1}{1+\delta_{cd}} W(c+d \to a+b) f_c f_d \times [1 + (-1)^{2s_a} f_a][1 + (-1)^{2s_b} f_b] \\
& - \frac{1}{1+\delta_{ab}} W(a+b \to c+d) f_a f_b \times [1 + (-1)^{2s_c} f_c][1 + (-1)^{2s_d} f_d] \Big\},
\end{aligned}
\tag{14}
$$

where s_i and p_i are the spin and 3-momentum of the ith particle, t is the time, and δ_{ij} is the Kronecker delta, where $i = a, b, c,$ and d. In thermal equilibrium, $df_a/dt = 0$, f_a has the usual Boltzmann–Gibbs expression, and the energy conservation, $\omega_a + \omega_b = \omega_c + \omega_d$ is guaranteed, both in the quantum and the classical cases. In the classical limit, one finds the usual expression, $f = e^{-\beta\omega}$.

Kapusta conveniently pointed out that replacing, in the Boltzmann equation, the traditional molecular chaos hypothesis (*stosszahlansatz*) factorization, $f_c f_d$ by $[f_c^{1-q} + f_d^{1-q} - 1]^{1/(1-q)}$ (and analogously for $f_a f_b$) [107,108], leads to the q-exponential distribution as the stationary state, also satisfying simultaneously the energy conservation, $\omega_a + \omega_b = \omega_c + \omega_d$. However, he shared that, essentially, he did not know where that specific replacement came from.

Ulrich Heinz, in the audience, shared with friendly complicity that "After so many papers that I have rejected on Tsallis statistics, I am really happy to hear your talk". About that, not having been communicated in any way that this talk was going to happen online so that I could participate myself, I have no other comment than to say that it is allowed to think that, maybe, not all those rejections were justified. On the other hand, interestingly enough, Heinz also shared that he could, however, "live" with a properly generalized Boltzmann equation.

Let us clarify, at this point, that $[f_c^{1-q} + f_d^{1-q} - 1]^{1/(1-q)}$ is known to correspond, for $q \neq 1$, to a specific (and apparently not rare at all in nature) class of strong correlations between the relevant random variables.

The product xy of two real numbers has been conveniently generalized as the following q-product [109,110]:

$$
x \otimes_q y \equiv e_q^{\ln_q x + \ln_q y} = \left[x^{1-q} + y^{1-q} - 1 \right]_+^{\frac{1}{1-q}} \quad (x \geq 0, \ y \geq 0),
\tag{15}
$$

where $[\ldots]_+ = [\ldots]$ if $[\ldots] > 0$ and vanishes otherwise. Let us list some of its main properties:

- It recovers the standard product as the $q = 1$ particular instance, i.e.,

$$
x \otimes_1 y = xy;
\tag{16}
$$

- It is commutative for all values of q, i.e.,

$$
x \otimes_q y = y \otimes_q x;
\tag{17}
$$

- It is additive under q-logarithm for all values of q, i.e.,

$$
\ln_q(x \otimes_q y) = \ln_q x + \ln_q y,
\tag{18}
$$

(referred to as "extensivity"), whereas let us remind that

$$\ln_q(x\,y) = \ln_q x + \ln_q y + (1-q)(\ln_q x)(\ln_q y) \tag{19}$$

(referred to as "nonadditivity").
Consistently,

$$e_q^x \otimes_q e_q^y = e_q^{x+y}, \tag{20}$$

whereas

$$e_q^x \, e_q^y = e_q^{x+y+(1-q)xy}; \tag{21}$$

- It is associative for all values of q, i.e.,

$$x \otimes_q (y \otimes_q z) = (x \otimes_q y) \otimes_q z = x \otimes_q y \otimes_q z = (x^{1-q} + y^{1-q} + z^{1-q} - 2)^{1/(1-q)}; \tag{22}$$

- It admits unity for all values of q, i.e.,

$$x \otimes_q 1 = x; \tag{23}$$

- It admits zero under certain conditions, namely

$$x \otimes_q 0 = \begin{cases} 0 & \text{if } (q \geq 1 \text{ and } x \geq 0) \text{ or if } (q < 1 \text{ and } 0 \leq x \leq 1), \\[2ex] \left(x^{1-q} - 1\right)^{\frac{1}{1-q}} & \text{if } q < 1 \text{ and } x > 1; \end{cases} \tag{24}$$

- It is distributive with regard to the generalized associative sum analyzed in [111], namely,

$$x \oplus^{(q)} y \equiv e_q^{\ln\left[e^{\ln_q x} + e^{\ln_q y}\right]} = \left\{1 + (1-q)\ln\left[e^{\frac{x^{1-q}-1}{1-q}} + e^{\frac{y^{1-q}-1}{1-q}}\right]\right\}^{1/(1-q)}. \tag{25}$$

In other words, the q-product (15) satisfies all the standard requirements of a full algebraic structure.

Consistently, on the basis of this product, it is possible to generalize, for $q \geq 1$, the Fourier transform of a non-negative function $f(x)$ as follows [89,90]:

$$F_q[f](\xi) \equiv \int dx\, e_q^{i\xi x} \otimes_q f(x) F_q[f](\xi) = \int_{-\infty}^{\infty} dx\, e_q^{i\xi x[f(x)]^{q-1}} f(x). \tag{26}$$

It is transparent that this transformation is, for $q \neq 1$, *nonlinear*. Indeed, if $f(x) \to \lambda f(x)$, λ being any constant, one verifies that $F_q[\lambda f](\xi) \neq \lambda F_q[f](\xi)$. This generalization of the standard Fourier transform $F_1[f](\xi)$ was introduced in order to have a remarkable *property*: it transforms q-Gaussians into q-Gaussians. Indeed, one verifies:

$$F_q\left[B_q\,\sqrt{\beta}\,e_q^{-\beta x^2}\right](\xi) = e_{q_1}^{-\beta_1 \xi^2}, \tag{27}$$

where

$$(q_1, \beta_1) = \left(\frac{1+q}{3-q}, \frac{3-q}{8\,\beta^{2-q}B_q^{2(q-1)}}\right) \quad (1 \leq q < 2), \tag{28}$$

B_q being an appropriate normalizing quantity. Within this frame, and others as well, the central limit theorem has been generalized [89,90], showing that, while averaging a large number of random variables within a wide class of *nonlocal correlations* (yet only partially explored), q-Gaussians emerge as attractors in the space of distributions. This provides an epistemological basis for understanding why there are so many q-Gaussians (and, consistently, so many q-exponentials) in nature; for more details and proofs, see [112]. Moreover, in what concerns the physical significance of the q-product (15) in relation with the Boltzmann transport equation, detailed arguments along a different line are presented as well in [113].

Another relevant line of mathematical research concerns the relations within various sets of q-indices (e.g., q-triplets) that very frequently emerge in complex systems [68,114].

Let us also mention that, in the context of relaxation times related to the Boltzmann equation, Heinz inquired about an interesting question, namely, whether there is any argument yielding q-statistics as an outcome of corresponding dynamics. On general grounds, this not an easy question. However, it is surely useful to have in mind that, in the same way the solution of $dx/dt = -x/\tau$ (with $x(0) = 1$) is $x = e^{-t/\tau}$, the solution of $dx/dt = -x^q/\tau_q$ (with $x(0) = 1$) is $x = e_q^{-t/\tau_q}$ [115].

At this point, it might be useful to mention that possible q-generalizations of the connections with the Kadanoff–Baym approach have been advanced in [116,117]. Finally, a first step focusing on the Bogoliubov–Born–Green–Kirkwood–Yvon (BBGKY) hierarchy has been introduced in [118].

2.9. On the Second Principle of Thermodynamics

Masoud Shokri, in the audience, inquired about the validity of the second principle for $q \neq 1$. By all means, it appears to be valid in the same way as for the Boltzmann–Gibbs case ($q = 1$). The following (nonexhaustive) list of arguments that are available in the literature can be mentioned:

- It has been proved that a detailed balance implies the time irreversibility of S_q [119].
- The celebrated Clausius inequality, $\delta Q/T \leq dS$, is consistent with the second principle. It has been shown [120] that it remains valid as it stands for $q > 1$ as well.
- The validity of the H-theorem has been proved for a wide class of classical d-dimensional many-body overdamped systems including repulsive short-range two-body interactions [73–82].
- The H-theorem has been proved for S_q by imposing Galilean-invariance on a Boltzmann lattice model which discretizes the Navier–Stokes equation. The unique value of q is determined by the structure of the Bravais lattice that is being focused on; for more details, see [121,122].
- The time evolution of the entropy for low-dimensional maps, or, actually, the Pesin identities, do certainly *not* prove the validity of the second principle. However, the fact that the behavior is quite similar (see [123–125] and references therein) for, e.g., the logistic map at the most chaotic value of its external parameter (with $q = 1$) and at the Feigenbaum point (with $q = 0.2445\ldots$) does provide a suggestive indication.

2.10. On the Zeroth Principle of Thermodynamics

Boris Tomasik, in the audience, inquired about the concept of thermal contact among systems in equilibrium, thus tacitly about its transitivity. This highly interesting issue points essentially towards the content of the zeroth principle.

In various overdamped classical d-dimensional many-body systems including short-range repulsive interactions, it has been possible to analytically calculate and numerically verify the validity of q-statistics, focusing on space and velocity distributions, equations of states, Carnot cycle, H-theorem, effective temperature, *and* the zeroth principle of thermodynamics [73–82]. For example, for repulsive interactions proportional to $1/r^\alpha$, it can be proved [82] that the distribution of positions is a q-Gaussian with $q = 1 - \alpha/d < 1$, which, in the limit $\alpha/d \to 1$, recovers the type-II superconductor result $q = 0$ [73–75]. This class of systems constitute a rare case where nearly all q-thermostatistical quantities can be analytically calculated.

However, this may be considered as the infancy of a general proof of the validity of the zeroth principle. Its general discussion remains to be done, and it is related, within q-statistics, to the concept of "effective temperature" itself. Indeed, in the above overdamped systems, two different temperatures coexist, namely, the *kinetic* and *effective* temperatures. The former is based on the distributions of velocities, whereas the latter is based on the spatial distribution. This is in fact quite analogous to what happens in granular packing, sandpiles and the like, where one is forced to distinguish the kinetic temperature from the structural one [126]. This is in variance with Kapusta's repeated expression referring to "the true temperature", without really explaining in physically neat and indisputable terms what "temperature" (or "inverse β") he is referring to. It is allowed to think that he only has in his mind the usual kinetic concept of temperature, thus ignoring that, in many complex systems, more than one temperature typically emerges; see, e.g., [73–82].

2.11. Indices q: First-Principle Characterization of Universality Classes, or Merely Efficient Fitting Parameters?

Kapusta's ultimate question consists of establishing whether or not the indices q have a first-principle physical basis. He eventually answers by the negative. Let us point out here that a definitively opposite perspective emerges through the (nonexhaustive) list of counter-examples, either from first principles or at a clear-cut mesoscopic level, in diverse complex systems, including high-energy collisions. Let us emphasize that it is well understood that the analytical determination of the q values is, in most of the cases—but fortunately not all!—mathematically intractable.

- First-principle quantum field theory calculation of q for a one-dimensional quantum many-body Hamiltonian system: One such example is the quantum phase transition at $T = 0$ criticality for the $d = 1$ first-neighbor Ising ferromagnet in the presence of a transverse magnetic field, and similar quantum systems characterized by a central charge, $c \geq 0$. A subsystem of large linear size L of an infinitely long such chain satisfies $S_{\mathrm{BG}}(L) \propto \ln L$, which is nonextensive and therefore violates the Legendre structure of thermodynamics. It turns out, however, that a unique value of q exists such that $S_q(L) \propto L$, thus satisfying thermodynamics. Its value is given by [127]

$$q = \frac{\sqrt{9 + c^2} - 3}{c} , \tag{29}$$

which is depicted in Figure 4. As one can see, q characterizes here the thermodynamically admissible entropies associated with the criticality universality classes of many-body systems, defined by the central charge c.

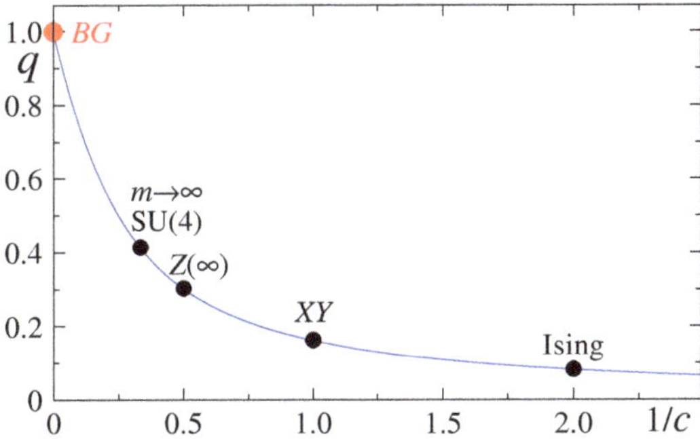

Figure 4. The index q as a function of the inverse central charge, $1/c$. The universality classes of some specific models are indicated; for example, $c = 1/2$ corresponds to the universality class of the short-range finite-spin Ising ferromagnet in the presence of a (critical) transverse field, at temperature $T = 0$. The Boltzmann–Gibbs value $q = 1$ is recovered in the $c \to \infty$ limit. Taken from [55].

- Zero-Lebesgue occupancy of the probability space in the $N \to \infty$ limit of a system with strongly correlated random binary variables [68]: it is verified that S_q is extensive for a unique value of q, namely,

$$q = 1 - 1/d \,, \tag{30}$$

where d is the width, $\forall N$, of the strip with nonvanishing probabilities.

- Overdamped d-dimensional many-body system with short-range repulsive interactions decaying as $1/r^\alpha$ ($\alpha \geq d$): The space attractor is a q-Gaussian with [82]

$$q = 1 - \alpha/d \,. \tag{31}$$

The limit $\alpha/d = 1$ recovers the superconductor type-II space distribution, which has been proved to correspond to $q = 0$ [73].

- Nonlinear dynamical systems exhibit various direct connections with the time evolution of the entropy and with its consequences. There are, in this respect, two important classes of chaotic behavior, namely: *strong chaos*, characterized by *exponential* sensitivity to the initial conditions (referred to, for classical systems, as having a *positive* maximal Lyapunov exponent); and *weak chaos*, characterized by *subexponential* (frequently, power-law) sensitivity to the initial conditions (referred to, for classical systems, as having a *vanishing* maximal Lyapunov exponent). Let us focus here on an important issue, namely, the time evolution of the entropy while exploring the system's phase space. Let us illustrate this issue with a paradigmatic *dissipative* system, namely, the logistic map, $x_{t+1} = 1 - a x_t^2$ ($x_t \in [-1, 1]$; $a \in [0, 2]$; $t = 0, 1, 2, \ldots$). For $a = 2$, the system is strongly chaotic since the sensitivity to the initial conditions satisfies $\xi(t) \equiv \lim_{\Delta x(0) \to 0} \Delta x(t)/\Delta x(0) = e^{\lambda t}$ with a Lyapunov exponent $\lambda = \ln 2 > 0$. Consistently, if one starts at $t = 0$ from a set of M initial conditions in an arbitrarily chosen single window with W of them equally partitioning the interval $[-1, 1]$, one obtains the entropy production per unit time,

$$K_{\mathrm{BG}} \equiv \lim_{t \to \infty} \lim_{W \to \infty} \lim_{M \to \infty} \frac{S_{\mathrm{BG}}(t)/k}{t} = \lambda > 0 \,, \tag{32}$$

thus verifying the Pesin identity. One consistently verifies that, for $q > 1$, $K_q \equiv \lim_{t \to \infty} \lim_{W \to \infty} \lim_{M \to \infty} \frac{S_q(t)/k}{t}$ vanishes, whereas, for $q < 1$, it diverges. In other words, $q = 1$ is the unique value of the index for which $S_q(t)$ asymptotically increases *linearly* with time.

If the same operations are applied at the edge of chaos corresponding to the double-bifurcation accumulation point (weak chaos) of the z-logistic map ($x_{t+1} = 1 - a|x_t|^z$ with $z > 0$) and its topologically equivalent one-dimensional dissipative maps, one verifies that $\xi = e_q^{\lambda_q t}$ with $\lambda_q = 1/(1-q)$ [125]. It follows that, for $t \to \infty$, ξ diverges subexponentially, namely as the power-law $\xi \propto t^{1/(1-q)}$. Moreover, it can be shown, through scaling arguments [128], that

$$\frac{1}{1-q} = \frac{1}{\alpha_{\min}} - \frac{1}{\alpha_{\max}} \,, \tag{33}$$

where $\alpha_{\min}$ and $\alpha_{\max}$ refer to the (concave) multifractal function $f(\alpha)$, defined in the interval $[\alpha_{\min}, \alpha_{\max}]$ with $f(\alpha_{\min}) = f(\alpha_{\max}) = 0$ [87].

If, for the standard logistic map ($z = 2$), one actually focuses on the Feigenbaum point (also referred to as the Feigenbaum–Coullet–Tresser point) $a = a_c \equiv 1.40115518909\ldots,$

which corresponds to a vanishing Lyapunov exponent λ (i.e., weak chaos), one does verify that $\xi = e_q^{\lambda_q t}$ with $\lambda_q = 1/(1-q)$ [125], as well as

$$\frac{1}{1-q} = \frac{1}{\alpha_{\min}} - \frac{1}{\alpha_{\max}} = \frac{\ln \alpha_F}{\ln 2}, \tag{34}$$

where α_F is the so called Feigenbaum universal constant. From this relation, it follows that

$$q = 0.244487701341282066198\ldots \tag{35}$$

(*1018* exact digits are known [50]). Analogously to the Boltzmann–Gibbs case, a q-generalized Pesin identity $K_q = \lambda_q$ is verified. Consistently, K_q numerically appears to vanish for q above that special value and to diverge for q below that value. In other words, this is the unique value of the index q, for which $S_q(t)$ asymptotically increases *linearly* with time, thus yielding a *finite* entropy production per unit time.

- Still in the area of nonlinear dynamical systems, now with $N \geq 1$ degrees of freedom $\{x_1, x_2, \ldots, x_N\}$, scaling arguments [129] are consistent with

$$\frac{1}{1-q_{\text{entropy}}} = \sum_{k=1}^{N} \frac{1}{1-q_k}, \tag{36}$$

where q_{entropy} is the value of q for which the entropy S_q production per unit time is *finite*, and $\{q_k\}$ characterize the q_k-exponential sensitivity to the initial conditions, respectively, associated with the variable x_k ($1 \geq q_1 \geq q_2 \geq \cdots \geq q_N > -\infty$). This quite general relation is verified in the following illustrative cases:
(i) It suffices that one (q_1) of these q-indices equals unity and $q_{\text{entropy}} = 1$ hence $S_{q_{\text{entropy}}} = S_{\text{BG}}$;
(ii) If all $\{q_k\}$ are equal and smaller than unity, then:

$$q_{\text{entropy}} = 1 - \frac{1-q_1}{N} \quad (q_1 < 1), \tag{37}$$

hence $N \to \infty$ implies $q_{\text{entropy}} = 1$; Equations (30) and (31) as well as the model (with N exchangeable random variables) introduced and discussed in [106,130] correspond to this case;
(iii) If $N = 1$, then $q_{\text{entropy}} = q_1$, which corresponds to the cases (32)–(35) here above.

- Another important issue is the central limit theorem attractor in the space of distributions when time-averaging a single coordinate of the system. Within this theorem, $y \equiv \sum_{t=1}^{T} x_t$ is defined. For example, for the logistic map with $a = 2$ one has that, after proper scaling and centering, the distribution $P(y)$ is given by a Gaussian, according to the classical central limit theorem, whereas, for $a = a_c$, it is seemingly given by a q-Gaussian with $q > 1$.

Let us further illustrate this issue with a paradigmatic *conservative* system, namely, the standard map, introduced by Chirikov in 1979. This area-preserving map turns out to be relevant within a variety of physical situations such as particle confinement in magnetic traps, particle dynamics in accelerators, comet dynamics, the ionization of Rydberg atoms, electron magnetotransport. It is defined as follows:

$$\begin{aligned} p_{i+1} &= p_i - K \sin x_i \quad (\mathrm{mod}\, 2\pi) \quad (K \geq 0) \\ x_{i+1} &= x_i + p_{i+1} \quad (\mathrm{mod}\, 2\pi) \end{aligned} \tag{38}$$

Each (x, p) point yields a Lyapunov exponent $\lambda^{(1)} = -\lambda^{(2)} \geq 0$. Next, along the central limit theorem lines, let us define the following quantity:

$$\bar{y} \equiv \sum_{i=1}^{T} (x_i^{(j)} - \langle x \rangle), \tag{39}$$

with

$$\langle x \rangle \equiv \frac{1}{M}\frac{1}{T}\sum_{j=1}^{M}\sum_{i=1}^{T} x_i^{(j)}, \tag{40}$$

where $M \to \infty$ is the number of initial conditions and $T \to \infty$ is the number of iterations for each of those M initial conditions. The limiting $K = 0$ case (hence a linear map, though with a highly nontrivial set of stable orbits) exhibits zero Lyapunov exponent in the entire phase space. It has been studied [131] and the attractor is a q-Gaussian with

$$q = 2. \tag{41}$$

This is in notorious variance with the attractor corresponding to $K \gg 1$, where the Lyapunov exponents corresponding to almost all the points of the phase space are positive, Boltzmann's molecular chaos hypothesis certainly is valid, and the corresponding attractor is a Gaussian, i.e., $q = 1$, in conformity with the standard central limit theorem.

- An interesting short-range-interacting d-dimensional ferromagnetic system is that whose symmetry is dictated by rotations in n dimensions, i.e., the so called $O(n)$ symmetry ($n = 2$ corresponds to the XY model, $n = 3$ corresponds to the Heisenberg model, and so on; $n \to \infty$ corresponds to the spherical model). As soon as one focuses on the kinetics of point defects during a quenching from a high temperature to a zero temperature for the $d = n$ model, its theoretical description can be done in terms of a time-dependent Ginzburg–Landau equation [132,133]. The distribution of the vortex velocity turns out to be a q-Gaussian with

$$q = \frac{d+4}{d+2}. \tag{42}$$

The index q decreases from two to one when d increases from zero to infinity. It is certainly intriguing, although yet unexplained, the fact that the value of q given in Equation (42) separates finite from infinite variance for d-dimensional q-Gaussians.

- For the coherent noise model for earthquakes and biological extinction, it has been possible to prove [134,135] the emergence of a q-Gaussian with

$$q = \frac{\tau+2}{\tau}, \tag{43}$$

where τ is the index characterizing the (asymptotically) power-law distribution of the avalanches.

- Cold atoms. Lutz predicted in 2003 [136] that the velocity distributions of cold atoms in dissipative optical lattices should be q-Gaussians with

$$q = 1 + \frac{44\,E_R}{U_0}, \tag{44}$$

where E_R and U_0 are, respectively, the recoil energy and the potential depth. The prediction was verified three years later [137,138].

- Granular matter. The following scaling law was predicted in 1996 [84] (see also [83]):

$$\mu = \frac{2}{3-q}, \tag{45}$$

where q is the index of the q-Gaussian distribution of $d = 1$ fluctuations and μ is the exponent associated with anomalous diffusion (i.e., the quadratic position x^2 scales like t^μ; $q = 1$ recovers the known scaling for Brownian motion normal diffusion, i.e., $\mu = 1$). This scaling relation was experimentally verified in 2015 [139], within a $\pm 2\%$ precision along a wide experimental range, for vertically confined grains under horizontal shear.

- In some simple models, such as the so-called "inelastic Maxwell model," analytic calculations can be performed; see, e.g., [140]). The velocity distribution that is obtained, from the microscopic dynamics of the system of cooling experiments for a spatially uniform gas whose temperature is monotonically decreasing with time, is given by (see [141] and references therein) a q-Gaussian with

$$q = 3/2. \tag{46}$$

- Lattice Boltzmann models for fluids. The incompressible Navier–Stokes equation has been considered [121] on a discretized D-dimensional Bravais lattice of coordination number b. It is further assumed that there is a single value for the particle mass, and also for speed. The basic requirement for the lattice Boltzmann model is to be Galilean invariant (i.e., invariant under a change of inertial reference frame) like the Navier–Stokes equation itself. It has been proved [121] that an H-theorem is satisfied for a trace-form entropy only if it is S_q with

$$q = 1 - \frac{2}{D}. \tag{47}$$

Therefore, $q < 1$ in all cases ($q > 0$ if $D > 2$, and $q < 0$ if $D < 2$), and approaches unity from below in the $D \to \infty$ limit. This study has been generalized by allowing multiple masses and multiple speeds. Galilean invariance once again mandates [122] an entropy of the form S_q, with a unique value of q determined by a transcendental equation involving the dimension and symmetry properties of the Bravais lattice as well as the multiple values of the masses and of the speeds. Of course, Equation (47) is recovered for the particular case of a single mass and a single speed. Finally, Equation (47) may be seen as one more verification of the structure reflected in Equation (37).

- In the area of high-energy collisions, a highly interesting development based on first-principle Yang–Mills/QCD grounds became available a couple of years ago [142]. It yielded

$$\frac{1}{q-1} = \frac{11}{3}N_c - \frac{2}{3}N_f, \tag{48}$$

where N_c is the number of colors and N_f the number of flavors. If $N_c = N_f/2 = 3$, one finds $1/(q-1) = 7$, hence $q = 8/7 \simeq 1.14$, which is amazingly close to the LHC (Large Hadron Collider) values indicated in the figure of [143] exhibiting the fittings (along very many experimental decades; notice also that the fitting temperature $T = 0.13$ GeV practically corresponds to the mass of pions, ubiquitous in proton–proton collisions) with which Kapusta begun his talk. Furthermore, if $N_c = N_f = 3$ is used, $q = 10/9 \simeq 1.11$ is obtained which, interestingly enough, coincides with the value phenomenologically advanced in 2000 by Walton and Rafelski [144] in their approach of a heavy quark diffusing in a quark–gluon plasma.

By the way, let me mention at this point that some years ago I briefly met in Brazil with the CERN researcher J. R. Ellis. I asked him whether he was aware that the q-exponential distribution ubiquitously emerged in high-energy collision data. He answered that he was aware. I then asked him whether he thought that this fact could be deduced from QCD. He answered that the value of q possibly yes, but the whole distribution probably not. The Deppman–Megias–Menezes relation (48) is by all means relevant along that line of thought; so are also the Wilk–Wlodarczyk, Beck and Beck–Cohen connections between q, fluctuations and degrees of freedom [145–147].

3. Final Remarks

Kapusta started his presentation by declaring "I am going to get to the bottom of this". He said that "q is not a fundamental or an intrinsic quantity to be interpreted in terms of new statistics" and "should not be considered a fundamental constant", and, while conceding that it constitutes an efficient fitting parameter, he concluded by stating that the index q "is not a fundamental quantity *in any sense*". Along the same line, Giorgio Torrieri enthusiastically congratulated a "really great talk" and declared, in a very happy tone, "I will forward the video to many people".

Before ending, it might be useful to make explicit the following analogy. Within the Newtonian theory of the planetary system, there is *only one* fundamental fitting parameter, namely, the gravitational constant, G. All the rest about the orbits and motion of planets, asteroids, comets, etc., would be uniquely determined *if* (and what a big *if* is this one) we knew all the involved masses and their initial conditions at some time of the system's history. It is our human impossibility of having not only this information but also access to an unthinkably powerful computer to allow the corresponding Newtonian equations to be blamed if the task cannot be completed, certainly not the Newtonian theory. Still, this theory showed to be capable of reproducing the Kepler's empirical laws, in particular the elliptic analytic form of the orbits of the planets. Using these general mathematical forms, astronomers have been able to quite precisely determine the orbit of say Mars by introducing as many fitting parameters as necessary to close the description.

Similarly, nonextensive statistical mechanics has *only one* fundamental fitting parameter, namely, the Boltzmann constant, k, shared in fact with Boltzmann–Gibbs statistical mechanics. All the rest would be uniquely determined *if* (and what a big *if* is this one) we were able to analytically scrutinize the usually intractable mathematics associated with the microscopic (or at least the mesoscopic) dynamics of the nonlinear system so that the necessary indices q are determined from mechanical/electromagnetic first principles. It is our human impossibility of handling this formidable mathematics to be blamed if the task cannot be completed, certainly not the nonextensive thermostatistical theory! Still, this theory showed to be capable of providing useful analytical forms such as q-exponentials and q-Gaussians, as well as H-theorems, generalized central limit theorems, among others, which make possible the desired study of a specific complex system by introducing as many fitting parameters as necessary to close the description.

The present paper replies to Joseph Kapusta's talk and hopefully puts things in a further updated and more reliable perspective, proving in particular that the q indices surely are definitively more than possible convenient fitting parameters. As the reader has had the opportunity to directly verify here by themselves, the "bottom of this" is quite further deeper. Allow me to finish by reminding that, in fact, theoretical physics, like anything else, always appears to admit some new deeper steps down, below what was previously thought to be the bottom line (e.g., the nowadays celebrated emergence of relativity from Newtonian mechanics and, almost simultaneously, of quantum mechanics as well). Quoting Henry David Thoreau (*Walden; or, Life in the Woods*): "It is never too late to give up our prejudices. No way of thinking or doing, however ancient, can be trusted without proof. What everybody echoes or in silence passes by as true today may turn out to be falsehood tomorrow, mere smoke of opinion, which some had trusted for a cloud that would sprinkle fertilizing rain on their fields."

An oral reply [148] to the content of Kapusta's talk was delivered on April 12, 2022 at a seminar, hosted by the Santa Fe Institute, New Mexico, and chaired by Sidney Redner.

Funding: Partially supported by CNPq and Faperj (Brazilian agencies).

Acknowledgments: I am grateful for useful remarks from A. Deppman, E. Megias, A.R. Plastino, D. Rocha and R.S. Wedemann.

References

1. Group of Statistical Physics. Nonextensive Statistical Mechanics and Thermodynamics. Bibliography. Available online: http://tsallis.cat.cbpf.br/biblio.htm (accessed on 3 May 2022).
2. Tsallis, C. Possible generalization of Boltzmann-Gibbs statistics. *J. Stat. Phys.* **1988**, *52*, 479–487. [CrossRef]
3. Cleymans, J.; Hamar, G.; Levai, P.; Wheaton, S. Near-thermal equilibrium with Tsallis distributions in heavy ion collisions. *J. Phys. G Nucl. Part. Phys.* **2009**, *36*, 064018. [CrossRef]

4.   Cleymans, J. Is strangeness chemically equilibrated? *arXiv* **2010**, arXiv:1001.3002. [CrossRef]
5.   Cleymans, J. Recent developments around chemical equilibrium. *J. Phys. G Nucl. Part. Phys.* **2010**, *37*, 094015. [CrossRef]
6.   Cleymans, J.; Worku, D. The Tsallis distribution and transverse momentum distributions in high-energy physics. *arXiv* **2011**, arXiv:1106.3405. [CrossRef]
7.   Cleymans, J. The thermal model at the Large Hadron Collider. *Acta Phys. Pol. B* **2012**, *43*, 563–570. [CrossRef]
8.   Cleymans, J.; Worku, D. The Tsallis distribution in proton-proton collisions at $\sqrt{s} = 0.9$ TeV at the LHC. *J. Phys. G Nucl. Part. Phys.* **2012**, *39*, 025006. [CrossRef]
9.   Cleymans, J.; Worku, D. Relativistic thermodynamics: Transverse momentum distributions in high-energy physics. *Eur. Phys. J. A* **2012**, *48*, 160. [CrossRef]
10.   Azmi, M.D.; Cleymans, J. Transverse momentum distributions at the LHC and Tsallis thermodynamics. *arXiv* **2013**, arXiv:1310.0217. [CrossRef]
11.   Azmi, M.D.; Cleymans, J. Transverse momentum distributions in p-Pb collisions and Tsallis thermodynamics. *arXiv* **2013**, arXiv:1311.2909. [CrossRef]
12.   Cleymans, J. The Tsallis distribution at the LHC. *arXiv* **2012**, arXiv:1210.7464. [CrossRef]
13.   Cleymans, J. The Tsallis distribution for p-p collisions at the LHC. *J. Phys. Conf. Ser.* **2013**, *455*, 012049. [CrossRef]
14.   Cleymans, J.; Lykasov, G.I.; Parvan, A.S.; Sorin, A.S.; Teryaev, O.V.; Worku, D. Systematic properties of the Tsallis distribution: Energy dependence of parameters in high energy p-p collisions. *Phys. Lett. B* **2013**, *723*, 351–354. [CrossRef]
15.   Cleymans, J. The Tsallis Distribution at the LHC. Talk at CERN Heavy Ion Forum, 1 February 2013, Geneva, Switzerland. Available online: https://indico.cern.ch/event/232225/ (accessed on 3 May 2022).
16.   Cleymans, J. The Tsallis Distribution. ALICE Matters 15 April 2013. Available online: https://alicematters.web.cern.ch/JeanCleyman (accessed on 3 May 2022).
17.   Cleymans, J. The Tsallis Distribution at the LHC. Talk at CERN Heavy Ion Forum, 13 January 2014, Geneva, Switzerland. Available online: https://indico.cern.ch/event/285968/ (accessed on 3 May 2022).
18.   Cleymans, J. The Tsallis distribution at the LHC: Phenomenology. *AIP Conf. Proc.* **2014**, *1625*, 31–37. [CrossRef]
19.   Cleymans, J. Systematic properties of the Tsallis distribution: Energy dependence of parameters. *J. Phys. Conf. Ser.* **2014**, *509*, 012099. [CrossRef]
20.   Cleymans, J. The Tsallis distribution at the LHC. *EPJ Web Conf.* **2014**, *70*, 00009. [CrossRef]
21.   Azmi, M.D.; Cleymans, J. Transverse momentum distributions in proton-proton collisions at LHC energies and Tsallis thermodynamics. *J. Phys. G Nucl. Part. Phys.* **2014**, *41*, 065001. [CrossRef]
22.   Cleymans, J.; Azmi, M.D. Large transverse momenta and Tsallis thermodynamics. *J. Phys. Conf. Ser.* **2016**, *668*, 012050. [CrossRef]
23.   Azmi, M.D.; Cleymans, J. The Tsallis distribution at large transverse momenta. *Eur. Phys. J. C* **2015**, *75*, 430. [CrossRef]
24.   Marques, L.; Cleymans, J.; Deppman, A. Description of high-energy pp collisions using Tsallis thermodynamics: Transverse momentum and rapidity distributions. *Phys. Rev. D* **2015**, *91*, 054025. [CrossRef]
25.   Deppman, A.; Marques, L.; Cleymans, J. Longitudinal properties of high energy collisions. *J. Phys. Conf. Ser.* **2015**, *623*, 012009. [CrossRef]
26.   Khuntia, A.; Sahoo, P.; Garg, P.; Sahoo, R.; Cleymans, J. Speed of sound in hadronic matter using non-extensive statistics. *Proc. DAE–BRNS Symp. Nucl. Phys.* **2015**, *60*, 744–745. Available online: http://sympnp.org/proceedings/60/ (accessed on 3 May 2022). [CrossRef]
27.   Thakur, D.; Tripathy, S.; Garg, P.; Sahoo, R.; Cleymans, J. Indication of a differential freeze-out in proton-proton and heavy-ion collisions at RHIC and LHC energies. *Adv. High Energy Phys.* **2016**, *2016*, 4149352. [CrossRef]
28.   Bhattacharyya, T.; Cleymans, J.; Khuntia, A.; Pareek, P.; Sahoo, R. Radial flow in non-extensive thermodynamics and study of particle spectra at LHC in the limit of small $(q - 1)$. *Eur. Phys. J. A* **2016**, *52*, 30. [CrossRef]
29.   Khuntia, A.; Sahoo, P.; Garg, P.; Sahoo, R.; Cleymans, J. Speed of sound in hadronic matter using non-extensive Tsallis statistics. *Eur. Phys. J. A* **2016**, *52*, 292. [CrossRef]
30.   Bhattacharyya, T.; Cleymans, J.; Mogliacci, S. Analytic results for the Tsallis thermodynamic variables. *Phys. Rev. D* **2016**, *94*, 094026. [CrossRef]
31.   Bhattacharyya, T.; Cleymans, J.; Khuntia, A.; Pareek, P.; Sahoo, R. Small $(q - 1)$ expansion of the Tsallis distribution and study of particle spectra at LHC. In Proceedings of the 61st Annual Conference of the South African Institute of Physics (SAIP206), Johannesburg, South Africa, 4–8 July 2016; pp. 463–468. Available onine: https://inspirehep.net/literature/1720205 (accessed on 3 May 2022).
32.   Cleymans, J.; Azmi, M.D.; Parvan, A.S.; Teryaev, O.V. The parameters of the Tsallis distribution at the LHC. *EPJ Web Conf.* **2017**, *137*, 11004. [CrossRef]
33.   Bhattacharyya, T.; Cleymans, J.; Garg, P.; Kumar, P.; Mogliacci, S.; Sahoo, R.; Tripathy, S. Applications of the Tsallis statistics in high energy collisions. *J. Phys. Conf. Ser.* **2017**, *878*, 012016. [CrossRef]
34.   Bhattacharyya, T.; Cleymans, J. Non-extensive Fokker–Planck transport coefficients of heavy quarks. *arXiv* **2017**, arXiv:1707.08425. [CrossRef]
35.   Khuntia, A.; Tripathy, S.; Sahoo, R.; Cleymans, J. Multiplicity dependence of non-extensive parameters for strange and multi-strange particles in proton-proton collisions at $\sqrt{s} = 7$ TeV at the LHC. *Eur. Phys. J. A* **2017**, *53*, 103. [CrossRef]

36. Parvan, A.S.; Teryaev, O.V.; Cleymans, J. Systematic comparison of Tsallis statistics for charged pions produced in pp collisions. *Eur. Phys. J. A* **2017**, *53*, 102. [CrossRef]
37. Cleymans, J. On the use of the Tsallis distribution at LHC energies. *J. Phys. Conf. Ser.* **2017**, *779*, 012079. [CrossRef]
38. Bhattacharyya, T.; Cleymans, J.; Marques, L.; Mogliacci, S.; Paradza, M.W. On the precise determination of the Tsallis parameters in proton-proton collisions at LHC energies. *J. Phys. G Nucl. Part. Phys.* **2018**, *45*, 055001. [CrossRef]
39. Bhattacharyya, T.; Cleymans, J.; Mogliacci, S.; Parvan, A.S.; Sorin, A.S.; Teryaev, O.V. Non extensivity of the QCD p_T-spectra. *Eur. Phys. J. A* **2018**, *54*, 222. [CrossRef]
40. Khuntia, A.; Sharma, H.; Tiwari, S.K.; Sahoo, R.; Cleymans, J. Radial flow and differential freeze-out in proton-proton collisions at $\sqrt{s}$ = 7 TeV at the LHC. *Eur. Phys. J. A* **2019**, *55*, 3. [CrossRef]
41. Cleymans, J.; Paradza, M.W. Determination of the chemical potential in the Tsallis distribution at LHC energies. *arXiv* **2020**, arXiv:2010.05565. [CrossRef]
42. Cleymans, J.; Paradza, M.W. Tsallis statistics in high energy physics: Chemical and thermal freeze-outs. *Physics* **2020**, *2*, 654–664. [CrossRef]
43. Azmi, M.D.; Bhattacharyya, T.; Cleymans, J.; Paradza, M. Energy density at kinetic freeze-out in Pb-Pb collisions at the LHC using the Tsallis distribution. *J. Phys. G Nucl. Part. Phys.* **2020**, *47*, 045001. [CrossRef]
44. Rath, R.; Khuntia, A.; Sahoo, R.; Cleymans, J. Event multiplicity, transverse momentum and energy dependence of charged particle production, and system thermodynamics in pp collisions at the Large Hadron Collider. *J. Phys. G Nucl. Part. Phys.* **2020**, *47*, 055111. [CrossRef]
45. Kapusta, J.I. A Primer on Tsallis Statistics for Nuclear and Particle Physics. 19 May 2021. Available online: https://www.youtube.com/watch?v=CmyXk1Xkvcg (accessed on 3 May 2022).
46. Maingueneau, D. *Genèses du Discours*; Mardaga: Bruxelles, Brussels, 1984.
47. Pêcheux, M. *Les Vérités de La Palice*; Maspero: Paris, France, 1975.
48. Foucault, M. *L' Ordre du Discours*; Gallimard: Paris, France, 1971.
49. Kapusta, J.I. Perspective on Tsallis statistics for nuclear and particle physics. *Int. J. Mod. Phys. E* **2021**, *30*, 2130006. [CrossRef]
50. Tsallis, C. *Nonextensive Statistical Mechanics. Approaching a Complex World*, 1st ed.; Springer: New York, NY, USA, 2009. [CrossRef]
51. Watanabe, S. *Knowing and Guessing*; Wiley: New York, NY, USA, 1969. [CrossRef]
52. Barlow, H. Conditions for versatile learning, Helmholtz's unconscious inference, and the task of perception. *Vision. Res.* **1990**, *30*, 1561–1571. [CrossRef]
53. Tsallis, C. Entropy. *Encyclopedia* **2022**, *2*, 264–300. [CrossRef]
54. Enciso, A.; Tempesta, P. Uniqueness and characterization theorems for generalized entropies. *J. Stat. Mech.* **2017**, *2017*, 123101. [CrossRef]
55. Tsallis, C.; Haubold, H.J. Boltzmann-Gibbs entropy is sufficient but not necessary for the likelihood factorization required by Einstein. *EPL (Europhys. Lett.)* **2015**, *110*, 30005. [CrossRef]
56. Shannon, C.E. A Mathematical theory of communication. *Bell Syst. Tech. J.* **1948**, *27*, 379–423. [CrossRef]
57. Shannon, C.E. A Mathematical theory of communication. Part III. *Bell Syst. Tech. J.* **1948**, *27*, 623–656. [CrossRef]
58. Khinchin, A.I. The entropy concept of probability theory. *Uspekhi Matem. Nauk* **1953**, *8*, 3–20. (In Russian) [CrossRef]
59. Khinchin, A.I. *Mathematical Foundations of Information Theory*; Silverman, R.A., Friedman, M.D., Translators; Dover: New York, NY, USA, 1957; pp. 1–28.
60. dos Santos, R.J.V. Generalization of Shannon' s theorem for Tsallis entropy. *J. Math. Phys.* **1997**, *38*, 4104–4107. [CrossRef]
61. Abe, S. Axioms and uniqueness theorem for Tsallis entropy. *Phys. Lett. A* **2000**, *271*, 74–79. [CrossRef]
62. Topsøe, F. Factorization and escorting in the game-theoretical approach to non-extensive entropy measures. *Phys. A Stat. Mech. Appl.* **2006**, *365*, 91–95. [CrossRef]
63. Amari, S.; Ohara, A.; Matsuzoe, H. Geometry of deformed exponential families: Invariant, dually-flat and conformal geometries. *Phys. A Stat. Mech. Appl.* **2012**, *391*, 4308–4319. [CrossRef]
64. Biro, T.S.; Barnafoldi, G.G.; Van, P. New entropy formula with fluctuating reservoir. *Phys. A Stat. Mech. Appl.* **2015**, *417*, 215–220. [CrossRef]
65. Penrose, O. *Foundations of Statistical Mechanics: A Deductive Treatment*; Pergamon: Oxford, UK, 1970; p. 167. [CrossRef]
66. Tsallis, C.; Cirto, L.J.L. Black hole thermodynamical entropy. *Eur. Phys. J. C* **2013**, *73*, 2487. [CrossRef]
67. Einstein, A. Autobiographical Notes. In *Albert Einstein: Historical and Cultural Perspectives*; Holton, G., Elkana, Y., Eds.; Dover Publications: Mineola, NY, USA, 1997; p. 227.
68. Tsallis, C.; Gell-Mann, M.; Sato, Y. Asymptotically scale-invariant occupancy of phase space makes the entropy S_q extensive. *Proc. Natl. Acad. Sci. USA* **2005**, *102*, 15377. [CrossRef]
69. Curado, E.M.F.; Tsallis, C. Generalized statistical mechanics: Connection with thermodynamics. *J. Phys. A Math. Gen* **1991**, *24*, L69–L72; Erratum in *J. Phys. A Math. Gen.* **1992**, *25*, 1019. [CrossRef] [PubMed]
70. Tsallis, C.; Mendes, R.S.; Plastino, A.R. The role of constraints within generalized nonextensive statistics. *Phys. A Stat. Mech. Appl.* **1998**, *261*, 534–554. [CrossRef]
71. Ferri, G.L.; Martínez, S.; Plastino, A. Equivalence of the four versions of Tsallis's statistics. *J. Stat. Mech. Theory Exp.* **2005**, *2005*, P04009. [CrossRef]

72. Ferri, G.L.; Martínez, S.; Plastino, A. The role of constraints in Tsallis' nonextensive treatment revisited. *Phys. A Stat. Mech. Appl.* **2005**, *347*, 205–220. [CrossRef]
73. Andrade, J.S., Jr.; da Silva, G.F.T.; Moreira, A.A.; Nobre, F.D.; Curado, E.M.F. Thermostatistics of overdamped motion of interacting particles. *Phys. Rev. Lett.* **2010**, *105*, 260601. [CrossRef]
74. Ribeiro, M.S.; Nobre, F.D.; Curado, E.M.F. Classes of N-Dimensional nonlinear Fokker–Planck equations associated to Tsallis entropy. *Entropy* **2011**, *13*, 1928–1944. [CrossRef]
75. Ribeiro, M.S.; Nobre, F.D.; Curado, E.M.F. Time evolution of interacting vortices under overdamped motion. *Phys. Rev. E* **2012**, *85*, 021146. [CrossRef]
76. Curado, E.M.F.; Souza, A.M.C.; Nobre, F.D.; Andrade, R.F.S. Carnot cycle for interacting particles in the absence of thermal noise. *Phys. Rev. E* **2014**, *89*, 022117. [CrossRef]
77. Andrade, R.F.S.; Souza, A.M.C.; Curado, E.M.F.; Nobre, F.D. A thermodynamical formalism describing mechanical interactions. *EPL (Europhys. Lett.)* **2014**, *108*, 20001. [CrossRef] [PubMed]
78. Nobre, F.D.; Curado, E.M.F.; Souza, A.M.C.; Andrade, R.F.S. Consistent thermodynamic framework for interacting particles by neglecting thermal noise. *Phys. Rev. E* **2015**, *91*, 022135. [CrossRef] [PubMed]
79. Ribeiro, M.S.; Casas, G.A.; Nobre, F.D. Second law and entropy production in a nonextensive system. *Phys. Rev. E* **2015**, *91*, 012140. [CrossRef]
80. Vieira, C.M.; Carmona, H.A.; Andrade, J.S., Jr.; Moreira, A.A. General continuum approach for dissipative systems of repulsive particles. *Phys. Rev. E* **2016**, *93*, 060103(R). [CrossRef] [PubMed]
81. Ribeiro, M.S.; Nobre, F.D. Repulsive particles under a general external potential: Thermodynamics by neglecting thermal noise. *Phys. Rev. E* **2016**, *94*, 022120. [CrossRef] [PubMed]
82. Moreira, A.A.; Vieira, C.M.; Carmona, H.A.; Andrade, J.S., Jr.; Tsallis, C. Overdamped dynamics of particles with repulsive power-law interactions. *Phys. Rev. E* **2018**, *98*, 032138. [CrossRef]
83. Plastino, A.R.; Plastino, A. Non-extensive statistical mechanics and generalized Fokker–Planck equation. *Phys. A Stat. Mech. Appl.* **1995**, *222*, 347–354. [CrossRef]
84. Tsallis, C.; Bukman, D.J. Anomalous diffusion in the presence of external forces: Exact time-dependent solutions and their thermostatistical basis. *Phys. Rev. E* **1996**, *54*, R2197. [CrossRef]
85. Ribeiro, M.S.; Tsallis, C.; Nobre, F.D. Probability distributions extremizing the nonadditive entropy S_δ and stationary states of the corresponding nonlinear Fokker–Planck equation. *Phys. Rev. E* **2013**, *88*, 052107. [CrossRef]
86. Ribeiro, M.S.; Nobre, F.D.; Tsallis, C. Probability distributions and associated nonlinear Fokker–Planck equation for the two-index entropic form $S_{q,\delta}$. *Phys. Rev. E* **2014**, *89*, 052135. [CrossRef]
87. Beck, C.; Schögl, F. *Thermodynamics of Chaotic Systems*; Cambridge University Press: New York, NY, USA, 1993. [CrossRef] [PubMed]
88. Tsallis, C.; Plastino, A.R.; Alvarez-Estrada, R.F. Escort mean values and the characterization of power-law-decaying probability densities. *J. Math. Phys.* **2009**, *50*, 043303. [CrossRef] [PubMed]
89. Umarov, S.; Tsallis, C.; Steinberg, S. On a q-central limit theorem consistent with nonextensive statistical mechanics. *Milan J. Math.* **2008**, *76*, 307–328. [CrossRef]
90. Umarov, S.; Tsallis, C.; Gell-Mann, M.; Steinberg, S. Generalization of symmetric α-stable Lévy distributions for $q > 1$. *J. Math. Phys.* **2010**, *51*, 033502. [CrossRef]
91. Thistleton, W.; Marsh, J.A.; Nelson, K.; Tsallis, C. Generalized Box-Muller method for generating q-Gaussian random deviates. *IEEE Trans. Inf. Theory* **2007**, *53*, 4805–4810. [CrossRef]
92. Zanette, D.H.; Montemurro, M.A. A note on non-thermodynamical applications of non-extensive statistics. *Phys. Lett. A* **2004**, *324*, 383–387. [CrossRef] [PubMed]
93. Montroll, E.W.; Shlesinger, M.F. Maximum entropy formalism, fractals, scaling phenomena, and $1/f$ noise: A tale of tails. *J. Stat. Phys.* **1983**, *32*, 209–230. [CrossRef]
94. Englman, R. Maximum entropy principles in fragmentation data analysis. In *High-Pressure Shock Compression of Solids II. Dynamic Fracture and Fragmentation*; Davison, L., Grady, D.E., Shahinpoor, M., Eds.; Springer: New York, NY, USA, 1996; pp. 264–281. [CrossRef]
95. Plastino, A.R.; Plastino, A.; Soffer, B.H. Ambiguities in the forms of the entropic functional and constraints in the maximum entropy formalism. *Phys. Lett. A* **2007**, *363*, 48–52. [CrossRef]
96. Tsallis, C. What should a statistical mechanics satisfy to reflect nature? *Phys. D Nonlinear Phenom.* **2004**, *193*, 3–34. [CrossRef]
97. Gibbs, J.W. *Elementary Principles in Statistical Mechanics*; Charles Scribner's Sons: New York, NY, USA, 1902. Available online: https://www.gutenberg.org/files/50992/50992-pdf.pdf (accessed on 3 May 2022). [CrossRef]
98. Landsberg, P.T. *Thermodynamics and Statistical Mechanics*; Dover Publications: New York, NY, USA, 1990. [CrossRef]
99. Tsallis, C.; Cirto, L.J.L. Thermodynamics is more powerful than the role to it reserved by Boltzmann-Gibbs statistical mechanics. *Eur. Phys. J. Spec. Top.* **2014**, *223*, 2161–2175. [CrossRef]
100. Tsallis, C. Approach of complexity in nature: Entropic nonuniqueness. *Axioms* **2016**, *5*, 20. [CrossRef]
101. Tsallis, C. Beyond Boltzmann-Gibbs-Shannon in physics and elsewhere. *Entropy* **2019**, *21*, 696. [CrossRef]
102. Ruiz, G.; Tsallis, C. Towards a large deviation theory for strongly correlated systems. *Phys. Lett. A* **2012**, *376*, 2451–2454. [CrossRef]

103. Touchette, H. Comment on "Towards a large deviation theory for strongly correlated systems". *Phys. Lett. A* **2013**, *377*, 436–438. [CrossRef] [PubMed]
104. Ruiz, G.; Tsallis, C. Reply to Comment on "Towards a large deviation theory for strongly correlated systems". *Phys. Lett. A* **2013**, *377*, 491–495. [CrossRef]
105. Tirnakli, U.; Tsallis, C.; Ay, N. Approaching a large deviation theory for complex systems. *Nonlinear Dyn.* **2021**, *106*, 2537–2546. [CrossRef]
106. Tirnakli, U.; Marques, M.; Tsallis, C. Entropic extensivity and large deviations in the presence of strong correlations. *Phys. D Nonlinear Phenom.* **2022**, *431*, 133132. [CrossRef]
107. Lima, J.A.S.; Silva, R.; Plastino, A.R. Nonextensive thermostatistics and the H-theorem. *Phys. Rev. Lett.* **2001**, *86*, 2938–2941. [CrossRef]
108. Lavagno, A. Relativistic nonextensive thermodynamics. *Phys. Lett. A* **2002**, *301*, 13–18. [CrossRef]
109. Nivanen, L.; Le Méhauté, A.; Wang, Q.A. Generalized algebra within a nonextensive statistics. *Rep. Math. Phys.* **2003**, *52*, 437–444. [CrossRef]
110. Borges, E.P. A possible deformed algebra and calculus inspired in nonextensive thermostatistics. *Phys. A Stat. Mech. Appl.* **2004**, *340*, 95–101; Corrigendum in *Phys. A Stat. Mech. Appl.* **2021**, *581*, 126206. [CrossRef]
111. Borges, E.P.; da Costa, B.G. Deformed mathematical objects stemming from the q-logarithm function. *Axioms* **2022**, *11*, 138. [CrossRef]
112. Umarov, S.; Tsallis, C. *Mathematical Foundations of Nonextensive Statistical Mechanics*; World Scientific: Singapore, 2022. [CrossRef]
113. Megias, E.; Lima, J.A.S.; Deppman, A. Transport equation for small systems and the nonadditive entropy. *Mathematics* **2022**, *10*, 1625. [CrossRef]
114. Gazeau, J.-P.; Tsallis, C. Möbius transforms, cycles and q-triplets in statistical mechanics. *Entropy* **2019**, *21*, 1155. [CrossRef]
115. Tsallis, C.; Bemski, G.; Mendes, R.S. Is re-association in folded proteins a case of nonextensivity? *Phys. Lett. A* **1999**, *257*, 93–98. [CrossRef]
116. Rajagopal, A.K.; Mendes, R.S.; Lenzi, E.K. Quantum statistical mechanics for nonextensive systems: Prediction for possible experimental tests. *Phys. Rev. Lett.* **1998**, *80*, 3907–3910. [CrossRef]
117. Lenzi, E.K.; Mendes, R.S.; Rajagopal, A.K. Quantum statistical mechanics for nonextensive systems. *Phys. Rev. Lett.* **1999**, *59*, 1398. [CrossRef]
118. Chavanis, P.H. Hamiltonian and Brownian systems with long-range interactions: III. The BBGKY hierarchy for spatially inhomogeneous systems. *Phys. A Stat. mech. Appl.* **2008**, *387*, 787–805. [CrossRef]
119. Mariz, A.M. On the irreversible nature of the Tsallis and Renyi entropies. *Phys. Lett. A* **1992**, *165*, 409–411. [CrossRef]
120. Abe, S.; Rajagopal, A.K. Validity of the second law in nonextensive quantum thermodynamics. *Phys. Rev. Lett.* **2003**, *91*, 120601. [CrossRef]
121. Boghosian, B.M.; Love, P.J.; Coveney, P.V.; Karlin, I.V.; Succi, S.; Yepez, J. Galilean-invariant lattice-Boltzmann models with H-theorem. *Phys. Rev. E* **2003**, *68*, 025103(R). [CrossRef]
122. Boghosian, B.M.; Love, P.; Yepez, J.; Coveney, P.V. Galilean-invariant multi-speed entropic lattice Boltzmann models. *Phys. D Nonlinear Phenom.* **2004**, *193*, 169–181. [CrossRef]
123. Latora, V.; Baranger, M. Kolmogorov-Sinai entropy rate versus physical entropy. *Phys. Rev. Lett.* **1999**, *82*, 520–523. [CrossRef] [PubMed]
124. Latora, V.; Baranger, M.; Rapisarda, A.; Tsallis, C. The rate of entropy increase at the edge of chaos. *Phys. Lett. A* **2000**, *273*, 97–103. [CrossRef] [PubMed]
125. Baldovin, F.; Robledo, A. Nonextensive Pesin identity-Exact renormalization group analytical results for the dynamics at the edge of chaos of the logistic map. *Phys. Rev. E* **2004**, *69*, 045202(R). [CrossRef]
126. Coniglio, A.; Fierro, A.; Herrmann, H.J.; Nicodemi, M. (Eds.) *Unifying Concepts in Granular Media and Glasses*; Elsevier Science: Amsterdam, The Netherlands, 2004. [CrossRef]
127. Caruso, F.; Tsallis, C. Nonadditive entropy reconciles the area law in quantum systems with classical thermodynamics. *Phys. Rev. E* **2008**, *78*, 021102. [CrossRef]
128. Lyra, M.L.; Tsallis, C. Nonextensivity and multifractality in low-dimensional dissipative systems. *Phys. Rev. Lett.* **1998**, *80*, 53. [CrossRef]
129. Ananos, G.F.J.; Baldovin, F.; Tsallis, C. Anomalous sensitivity to initial conditions and entropy production in standard maps: Nonextensive approach. *Eur. Phys. J. B* **2005**, *46*, 409. [CrossRef]
130. Hanel, R.; Thurner, S.; Tsallis, C. Limit distributions of scale-invariant probabilistic models of correlated random variables with the q-Gaussian as an explicit example. *Eur. Phys. J. B* **2009**, *72*, 263–268. [CrossRef]
131. Bountis, A.; Veerman, J.J.P.; Vivaldi, F. Cauchy distributions for the integrable standard map. *Phys. Lett. A* **2020**, *384*, 126659. [CrossRef]
132. Mazenko, G.F. Vortex velocities in the $O(n)$ symmetric time-dependent Ginzburg-Landau model. *Phys. Rev. Lett.* **1997**, *78*, 401–404. [CrossRef]
133. Qian, H.; Mazenko, G.F. Vortex dynamics in a coarsening two-dimensional XY model. *Phys. Rev. E* **2003**, *68*, 021109. [CrossRef]
134. Celikoglu, A.; Tirnakli, U.; Queiros, S.M.D. Analysis of return distributions in the coherent noise model. *Phys. Rev. E* **2010**, *82*, 021124. [CrossRef]

135. Celikoglu, A.; Tirnakli, U. Earthquakes, model systems and connections to q-statistics. *Acta Geophys.* **2012**, *60*, 535–546. [CrossRef]
136. Lutz, E. Anomalous diffusion and Tsallis statistics in an optical lattice. *Phys. Rev. A* **2003**, *67*, 051402(R). [CrossRef] [PubMed]
137. Douglas, P.; Bergamini, S.; Renzoni, F. Tunable Tsallis distributions in dissipative optical lattices. *Phys. Rev. Lett.* **2006**, *96*, 110601. [CrossRef] [PubMed]
138. Lutz, E.; Renzoni, F. Beyond Boltzmann-Gibbs statistical mechanics in optical lattices. *Nat. Phys.* **2013**, *9*, 615–619. [CrossRef]
139. Combe, G.; Richefeu, V.; Stasiak, M.; Atman, A.P.F. Experimental validation of nonextensive scaling law in confined granular media. *Phys. Rev. Lett.* **2015**, *115*, 238301. [CrossRef]
140. Baldassari, A.; Marconi, U.M.B.; Puglisi, A. Influence of correlations on the velocity statistics of scalar granular gases. *Europhys. Lett. (EPL)* **2002**, *58*, 14–20. [CrossRef]
141. Sattin, F. Derivation of Tsallis' statistics from dynamical equations for a granular gas. *J. Phys. A Math. Gen.* **2003**, *36*, 1583–1591. [CrossRef]
142. Deppman, A.; Megias, E.; Menezes, D.P. Fractals, nonextensive statistics, and QCD. *Phys. Rev. D* **2020**, *101*, 034019. [CrossRef]
143. Wong, C.Y.; Wilk, G.; Cirto, L.J.L.; Tsallis, C. From QCD-based hard-scattering to nonextensive statistical mechanical descriptions of transverse momentum spectra in high-energy pp and $p\bar{p}$ collisions. *Phys. Rev. D* **2015**, *91*, 114027. [CrossRef]
144. Walton, D.B.; Rafelski, J. Equilibrium distribution of heavy quarks in Fokker–Planck dynamics. *Phys. Rev. Lett.* **2000**, *84*, 31. [CrossRef]
145. Wilk, G.; Wlodarczyk, Z. Interpretation of the nonextensivity parameter q in some applications of Tsallis statistics and Lévy distributions. *Phys. Rev. Lett.* **2000**, *84*, 2770–2773. [CrossRef]
146. Beck, C. Dynamical foundations of nonextensive statistical mechanics. *Phys. Rev. Lett.* **2001**, *87*, 180601. [CrossRef]
147. Beck, C.; Cohen, E.G.D. Superstatistics. *Phys. A Stat. Mech. Appl.* **2003**, *322*, 267–275. [CrossRef]
148. Tsallis, C. Nonadditive Entropies and Statistical Mechanics at the Edge of Chaos: Cornerstones. The Santa Fe Institute YouTube. Available online: https://www.youtube.com/watch?v=uQGN2PThukk (accessed on 3 May 2022). [CrossRef]

Article

Nonextensive Statistics in High Energy Collisions

Lucas Q. Rocha [1], Eugenio Megías [2], Luis A. Trevisan [3], Khusniddin K. Olimov [4], Fuhu Liu [5] and Airton Deppman [1,*]

[1] Instituto de Física, Rua do Matão 1371-Butantã, São Paulo 05580-090, Brazil; quinsan29@usp.br
[2] Departamento de Física Atómica, Molecular y Nuclear and Instituto Carlos I de Física Teórica y Computacional, Universidad de Granada, Avenida de Fuente Nueva s/n, 18071 Granada, Spain; emegias@ugr.es
[3] Department of Mathemathics and Statistics, Universidade Estadual de Ponta Grossa, Av. Carlos Cavalcante 4748, Ponta Grossa 84031-990, Brazil; luisaugustotrevisan@uepg.br
[4] Physical-Technical Institute of Uzbekistan Academy of Sciences, Chingiz Aytmatov Street, 2B, Tashkent 100084, Uzbekistan; khkolimov@gmail.com
[5] State Key Laboratory of Quantum Optics and Quantum Optics Devices & Collaborative Innovation Center of Extreme Optics, Institute of Theoretical Physics, Shanxi University, Wucheng Road, Taiyuan 030006, China; fuhuliu@sxu.edu.cn
* Correspondence: deppman@usp.br

Abstract: The present paper reports on the methods of the systematic analysis of the high-energy collision distributions—in particular, those adopted by Jean Cleymans. The analysis of data on high-energy collisions, using non-extensive statistics, represents an important part of Jean Cleymans scientific activity in the last decade. The methods of analysis, developed and employed by Cleymans, are discussed and compared with other similar methods. As an example, analyses of a set of the data of proton-proton collisions at the center-of-mass energies, $\sqrt{s} = 0.9$ and 7 TeV, are provided applying different methods and the results obtained are discussed. This line of research has the potential to enlarge our understanding of strongly interacting systems and to be continued in the future.

Keywords: non-extensive statistics; high-energy collisions; multiparticle production; transversal momentum distribution

Citation: Rocha, L.Q.; Megias, E.; Trevisan, L.A.; Olimov, K.K., Liu, F.; Deppman, A. Nonextensive Statistics in High Energy Collisions. *Physics* **2022**, *4*, 659–671. https://doi.org/10.3390/physics4020044

Received: 12 April 2022
Accepted: 27 May 2022
Published: 9 June 2022

Publisher's Note: MDPI stays neutral with regard to jurisdictional claims in published maps and institutional affiliations.

1. Introduction

One of the most evident features of high-energy collision (HEC) experimental data is the observation of the q-exponential distributions of the energy and momentum of the created particles. The experimental evidence motivated many studies on the use of Tsallis statistics in the multiparticle production process [1]. The role played by Jean Cleymans in the efforts to clarify the non-extensive aspects of the HEC can hardly be overestimated. He was a leading researcher on the Hadron Resonance Gas model [2,3] approach to the high-energy phenomena, and soon assumed a fundamental role in the investigations of the non-extensive generalization of the model using the Tsallis statistics [4].

One of Cleymans' best-known contributions to the study of the non-extensive distributions presents the formula for the distribution, derived from the non-additive entropy by using the thermodynamical relations [5]. With this formula, Cleymans performed a series of studies with several collaborators, developing systematic analyses of the experimental data and finding important patterns in the behavior of the parameters of the formula [6–9]. The investigations on the physical meaning of the non-extensive statistics involved systematic research for different energies and different particle multiplicities [7,10,11]. Cleymans was also involved in predictions of the outcomes for future experiments; the latter were largely confirmed by the experimental data.

One fundamental contribution to the investigation of the hadronization of the quark–gluon plasma was the theoretical determination of the critical line for the confined–deconfined regimes of the hadronic matter [12]. This study was subsequently generalized to include the non-extensive statistics [13]. With this work, Cleymans opened the opportunity to apply the knowledge, gathered in HEC, to other fields, such as neutron–star modeling. This illustrates the importance and the reach of Jean Cleymans' work. In one of his last papers [6], Cleymans and his collaborators present an accurate analysis of the non-extensive distributions for a wide range of collision energies and for several particle species. The study includes a detailed analysis of the covariances of the fitting parameters, largely extending previous investigations. Such an accurate analysis is of relevance in view of the predictions, by a thermofractal model of quantum chromodynamics (QCD), that the value for the entropic parameter, q, is calculated in terms of the fundamental parameters of the theory, namely, the number of colors, N_c, and the number of flavors, N_f, by a famous relation, $(q-1)^{-1} = (11/3)N_c - (4/3)N_f/2$, resulting to $q = 8/7$ [14]. Cleymans' study represents a rigorous test for the theoretical prediction [9]. The line of research, to which Cleymans contributed so extensively, is believed to continue to develope for many years to come, contributing to our knowledge about the strongly interacting systems.

In this paper, a systematic analysis of the HEC data is conducted in the style that Cleymans used to perform in his late years. The formula by Cleymans for the transverse momentum (p_T) distribution is used to fit the experimental data at different energies and for different particle species. The results are compared with the fits, obtained by other formulas. The analysis, presented here, is restricted to proton-proton (pp) collisions, since, for larger systems, other phenomena can interfere, such as the collective flow. The inclusion of this effect in the statistical description of the multiparticle production is possible, but it demands additional parameters, which we want to avoid at this stage of the investigation.

2. Momentum Distributions

The main formula for p_T-distribution is

$$\frac{d^2N}{dp_T\,dy} = gV\frac{p_T m_T \cosh y}{(2\pi)^2}\left(1 + (q-1)\frac{m_T \cosh y - \mu}{T}\right)^{\frac{-q}{q-1}}, \tag{1}$$

where N is the number of particles, V is the volume g, p, y, and μ denote the the particle degeneracy, momentum, rapidity, and chemical potential, respectively, and m_T is the particle transverse mass, $m_T = \sqrt{p_T^2 + m_0^2}$, with m_0 the remaining mass of the particle. Equation (1) was shown by Cleymans and Worku [5] to be a direct consequence of the Tsallis non-additive entropy when the correct thermodynamical relations are applied. In what follows, Equation (1) is addressed as the Cleymans formula.

All parameters in the Cleymans formula present well-understood physical meanings, and the insistence of Cleymans in the use of the correct formula shows that he always considered the Tsallis statistics to play an underlying physical role in the high-energy processes. Many studies, presenting analyses of HEC data using Cleymans' approach, have appeared during the last few years [15–23].

It is typical, however, to find different forms of the same expression being used to fit high-energy collision distributions [24]. Setting $y = 0$ and $\mu = 0$ in Equation (1):

$$\frac{d^2N}{dp_T dy} = \frac{dN}{dy}\frac{(n-1)(n-2)p_T}{nT[nT + m_0(n-2)]}\left(1 + \frac{m_T - m_0}{nT}\right)^{-n}, \tag{2}$$

where the chemical potential is fixed to the particle mass, i.e., $\mu = m_0$, and the power exponent, $n = q/(q-1)$. However, the multiplication factor, m_T, is absent in Equation (2), giving space to a dependence on the temperature, T. The term, dN/dy, represents the rapidity density, and it is generally calculated in a small window around $y = 0$, being assumed to be an arbitrary constant in the fits. In what follows, Equation (2) is addressed as the Levy–Tsallis formula.

Another form of the distribution is the so called Tsallis–Pareto formula [25],

$$\frac{d^2N}{dp_Tdy} = C \times p_T^{a_o} \times \left(1 + \frac{m_T - m_0}{nT}\right)^{-n},\tag{3}$$

where C and a_o are adjustable constants, which are obtained at $y = 0$ and $\mu = 0$.

The three formulas are similar in many aspects, and they all fit the data well enough. The only important difference is the factor m_T, which is changed into a temperature dependence normalization in Equation (2) and completely disappears in Equation (3). In short, $p_T m_T$ in Equation (1) is changed to p_T in Equation (2) and $p_T^{a_o}$ in Equation (3). To better understand the formulas and the approximations involved, it is constructive to calculate Equation (1) from the first principles, and then relate the result to other formulas; for a discussion about the behavior of the frequently used formulas, see Ref. [26].

The fundamental quantity is the number of one-particle configurations in the phase space according to Tsallis statistics, which is given by

$$dN = NgV\frac{d^4p}{(2\pi)^4}\left(1 + (q-1)\frac{E-\mu}{T}\right)^{\frac{-q}{q-1}}\delta(p^2 - m_0^2)\,\Theta(E)\,,\tag{4}$$

where N is the number of particles per event, $E = p_0$ and p are the particle energy and four-momentum, respectively, with $p^2 = p_0^2 - \vec{p}^2$. Integrating over p_0, and dividing by the total number of the particles produced in the event, one gets the probability density:

$$\frac{1}{N}d^3N = gV\frac{d^3p}{(2\pi)^3}\frac{1}{2E}\left(1 + (q-1)\frac{E-\mu}{T}\right)^{\frac{-q}{q-1}}.\tag{5}$$

The rapidity y is defined by the relation,

$$e^y = \frac{E + p_3}{m_T}\,,\tag{6}$$

where p_3 is the longitudinal momentum. From definition (6), $E = m_T \cosh y$. The independent coordinates of the momentum can be used to replace p_3 by y. The definition of the rapidity is equivalent to defining

$$\begin{cases} E = m_T \cosh y\,, \\ p_3 = m_T \sinh y\,, \end{cases}\tag{7}$$

therefore, $p_3/E = \tanh y \equiv \beta$ and $dp_3 = m_T \cosh y\, dy = E\, dy$, where $\beta = v/c$ with v being the velocity of a beam in the rest frame and c the speed of light.

The volume element is $d^3p = 2\pi p_T dp_T dp_3$; then, using Equation (5) and the relations (7), one finds:

$$\frac{1}{N}\frac{d^2N}{dp_Tdy} = \frac{E}{N}\frac{d^2N}{dp_Tdp_3} = \frac{gVp_T}{2(2\pi)^2}\left(1 + (q-1)\frac{m_T \cosh y - \mu}{T}\right)^{-\frac{q}{q-1}}.\tag{8}$$

One can see that Equation (8) is similar to Equation (3) with $a_o = 1$ and differs from Equation (1) by an absence of the multiplicative term $m_T \cosh y$. Equation (8) also differs from Equation (2), since the first equation has a dependence on the chemical potential that is not present in the second one. A detailed discussion about the Lorentz transformation of the transverse momentum distribution can be found in Ref. [18]. In what follows, Equation (8) is addressed as the Lorentz invariant cross-section formula.

In the analysis, a care should be taken the correct double differential distribution is used in terms of momentum components or in terms of rapidity, since this leads to different formulas. Moreover, the invariant differential distribution should be used. In the analysis below, the quality of the fits, obtained with the Formulas (1)–(3) and (8) is investigated, and the parameters obtained are compared in order to identify the correspondences. The non-extensive formulas, used here, are able to fit data of transverse momentum distributions over many orders of magnitude, and can be related to physical phenomena that need further investigation [1]. In the following, a brief discussion about the behavior of the parameters obtained with the different formulas is given, while a detailed consideration of the physical aspects of the results is left for the future.

3. Analysis

In the present analysis, the free parameters in Equations (1)–(3) are obtained by fits to the distributions from high-energy (pp) collision data, reported in Refs. [27,28]. For Equation (3), two different methods are used for fitting the data distributions, namely, one using the parameter a_0 free to be adjusted to the measurements, and another method fixing a_0: $a_0 = 1$. The results of the fits for different particle species and energies, as fitted by each of the Formulas (1)–(3), analyzed here, are displayed in Figure 1. The best-fit parameters for each formula are displayed in Tables 1–5. Note that the errors of the multiplicative constant that result from the fits of the curves for the last three sets of data are much larger than the errors for the other cases.

Table 1. Best-fit parameters of the Cleymans Formula (1). The data represent proton-proton collisions at the center-of-mass energies, $\sqrt{s} = 0.9$ and 7 TeV from Refs. [27,28]. The particle masses are taken from [29]. In the last column, the χ^2-statistics values are given per the number of degrees of freedom (ndf).

$\sqrt{s}$ (TeV)	Particle	q	T (GeV)	m_0 (GeV/c^2)	gV (fm^3)	μ (GeV)	χ^2/ndf
0.9	π^+	1.148 ± 0.005	0.091 ± 0.001	0.139570	$(4.07 \pm 0.04) \times 10^3$	0.141 ± 0.002	3.67/29
0.9	π^-	1.145 ± 0.005	0.076 ± 0.002	0.139570	$(1.80 \pm 0.02) \times 10^4$	0.031 ± 0.004	2.19/29
0.9	K^+	1.176 ± 0.015	0.092 ± 0.005	0.49368	$(1.02 \pm 0.02) \times 10^3$	0.20 ± 0.02	5.34/23
0.9	K^-	1.16 ± 0.01	0.084 ± 0.006	0.49368	$(2.33 \pm 0.04) \times 10^3$	0.129 ± 0.026	3.50/23
0.9	p	1.16 ± 0.02	0.09 ± 0.01	0.938272	774 ± 16	0.44 ± 0.05	7.43/21
0.9	$\bar{p}$	1.13 ± 0.02	0.10 ± 0.01	0.938272	730 ± 20	0.36 ± 0.06	7.78/20
0.9	π^0	1.14 ± 0.03	0.08 ± 0.04	0.134977	$(1 \pm 4) \times 10^8$	-0.05 ± 0.32	0.51/9
7	π^0	1.148 ± 0.005	0.13 ± 0.10	0.134977	$(0.5 \pm 2.7) \times 10^7$	0.2 ± 0.6	0.94/29
7	η	1.15 ± 0.03	0.1 ± 0.2	0.54751	$(0.2 \pm 1.8) \times 10^7$	0.1 ± 1.2	0.09/9

Table 2. Best-fit parameters of the Levy–Tsallis Formula (2). For details, see Table 1.

$\sqrt{s}$ (TeV)	Particle	q	T (GeV)	m_0 (GeV/c^2)	dN/dy	χ^2/ndf
0.9	π^+	1.148 ± 0.008	0.126 ± 0.003	0.139570	1.49 ± 0.02	3.07/30
0.9	π^-	1.142 ± 0.008	0.128 ± 0.003	0.139570	1.48 ± 0.02	1.84/30
0.9	K^+	1.21 ± 0.02	0.159 ± 0.009	0.49368	0.184 ± 0.004	5.41/24
0.9	K^-	1.19 ± 0.02	0.162 ± 0.009	0.49368	0.182 ± 0.004	3.59/24
0.9	p	1.19 ± 0.03	0.17 ± 0.01	0.938272	0.083 ± 0.002	7.43/21
0.9	$\bar{p}$	1.14 ± 0.03	0.19 ± 0.01	0.938272	0.079 ± 0.002	7.75/21
0.9	π^0	1.15 ± 0.04	0.13 ± 0.05	0.134977	$(9 \pm 5) \times 10^4$	0.47/10
7	π^0	1.171 ± 0.007	0.14 ± 0.01	0.134977	$(17 \pm 3) \times 10^4$	1.17/30
7	η	1.17 ± 0.04	0.23 ± 0.05	0.54751	$(15 \pm 5) \times 10^3$	0.09/10

Table 3. Best-fit parameters of the Tsallis–Pareto Formula (3), where the parameter a_0 is adjustable. For details, see Table 1.

$\sqrt{s}$ (TeV)	Particle	q	T (GeV)	m_0 (GeV/c^2)	C [(GeV/c)$^{-a_0-1}$]	a_0	χ^2/ndf
0.9	π^+	1.15 ± 0.01	0.12 ± 0.02	0.139570	43 ± 18	1.1 ± 0.2	2.69/29
0.9	π^-	1.148 ± 0.009	0.12 ± 0.01	0.139570	43 ± 17	1.1 ± 0.2	1.18 /29
0.9	K^+	1.22 ± 0.03	0.13 ± 0.04	0.49368	2 ± 1	1.2 ± 0.4	5.02/23
0.9	K^-	1.20 ± 0.03	0.14 ± 0.04	0.49368	1.94 ± 1.20	1.3 ± 0.4	3.12/23
0.9	p	1.13 ± 0.07	0.23 ± 0.07	0.938272	0.24 ± 0.08	0.6 ± 0.4	6.48/20
0.9	$\bar{p}$	1.08 ± 0.07	0.27 ± 0.07	0.938272	0.19 ± 0.06	0.6 ± 0.3	6.27/20
0.9	π^0	1.1 ± 0.3	0.5 ± 1.1	0.134977	$(0.8 \pm 2.7) \times 10^5$	-1 ± 3	0.34/9
7	π^0	1.14 ± 0.02	0.09 ± 0.03	0.134977	$(3 \pm 4) \times 10^7$	2 ± 1	0.90/29
7	η	1.17 ± 0.06	0.2 ± 0.3	0.54751	$(0.7 \pm 2.0) \times 10^5$	1 ± 3	0.09/9

Table 4. Best-fit parameters of the Tsallis–Pareto Formula 3, where the parameter $a_0 = 1$ is fixed. For details, see Table 1.

G (TeV)	Particle	q	T (GeV)	m_0 (GeV/c^2)	C [(GeV/c)$^{-a_0-1}$]	χ^2/ndf
0.9	π^+	1.148 ± 0.008	0.126 ± 0.003	0.139570	33.4 ± 0.8	3.07/30
0.9	π^-	1.142 ± 0.008	0.128 ± 0.003	0.139570	32.7 ± 0.7	1.84/30
0.9	K^+	1.21 ± 0.02	0.159 ± 0.009	0.49368	1.30 ± 0.07	5.41/24
0.9	K^-	1.19 ± 0.02	0.162 ± 0.009	0.49368	1.30 ± 0.06	3.59/24
0.9	p	1.19 ± 0.03	0.17 ± 0.01	0.938272	0.34 ± 0.02	7.43/21
0.9	$\bar{p}$	1.14 ± 0.03	0.19 ± 0.01	0.938272	0.31 ± 0.02	7.75/21
0.9	π^0	1.15 ± 0.04	0.13 ± 0.05	0.134977	$(2 \pm 2) \times 10^6$	0.47/10
7	π^0	1.171 ± 0.007	0.14 ± 0.01	0.134977	$(31 \pm 9) \times 10^5$	1.17/30
7	η	1.17 ± 0.04	0.23 ± 0.05	0.54751	$(7 \pm 4) \times 10^4$	0.09/10

Table 5. Best-fit parameters of the Lorentz invariant cross-section Formula (8). For details, see Table 1.

$\sqrt{s}$ (TeV)	Particle	q	T (GeV)	m_0 (GeV/c^2)	gV (fm^3)	μ (GeV)	χ^2/ndf
0.9	π^+	1.148 ± 0.008	0.115 ± 0.003	0.139570	$(3.97 \pm 0.06) \times 10^3$	-0.062 ± 0.009	3.07/29
0.9	π^-	1.142 ± 0.008	0.124 ± 0.003	0.139570	$(2.43 \pm 0.04) \times 10^3$	-0.017 ± 0.007	1.84/29
0.9	K^+	1.21 ± 0.02	0.107 ± 0.009	0.49368	789 ± 20	0.08 ± 0.04	5.41/23
0.9	K^-	1.19 ± 0.02	0.111 ± 0.009	0.49368	825 ± 20	0.06 ± 0.04	3.60/23
0.9	p	1.19 ± 0.03	0.10 ± 0.01	0.938272	442 ± 8	0.41 ± 0.06	7.43/20
0.9	$\bar{p}$	1.14 ± 0.03	0.11 ± 0.02	0.938272	$(1.03 \pm 0.08) \times 10^3$	0.23 ± 0.08	7.75/20
0.9	π^0	1.15 ± 0.04	0.1 ± 0.2	0.134977	$(0.1 \pm 1.4) \times 10^8$	-0.08 ± 1.12	0.47/9
7	π^0	1.171 ± 0.007	0.12 ± 0.02	0.134977	$(2 \pm 3) \times 10^7$	-0.10 ± 0.15	1.17/29
7	η	1.17 ± 0.04	0.2 ± 0.4	0.54751	$(0.6 \pm 8.4) \times 10^6$	0.1 ± 2.2	0.09/9

The large errors are due to the fact that, in these sets of data, the experimental data do not cover the region of the peak of the corresponding curves, as can be observed in the right panels of Figure 1. In these cases, the determination of the multiplicative constant, C, leads to a large number of combinations of T and C that can fit the data. The fitting procedure was performed with the use of the ROOT package, and the covariance matrix was obtained by using the MIGRAD routine in this package.

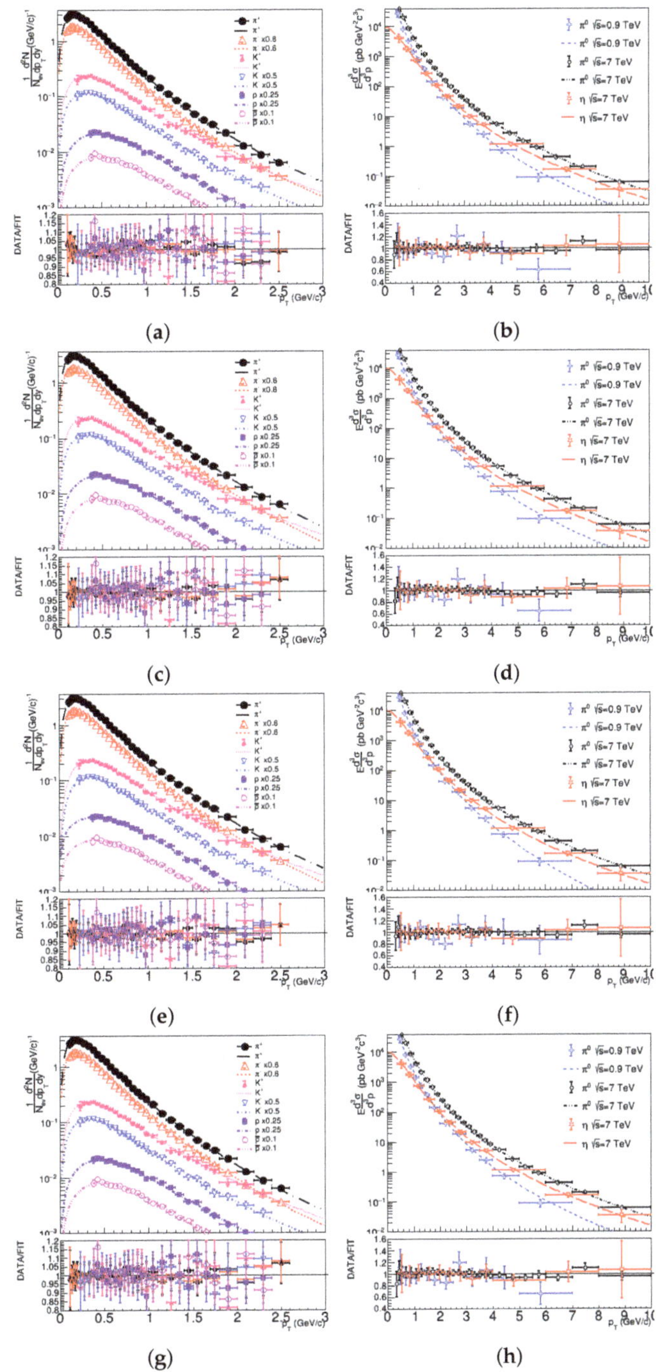

Figure 1. The double-differential transverse momentum distributions with the fits by (**a**,**b**) Equation (1), (**c**,**d**) Equation (2), and (**e**–**h**) Equation (3), with the parameter a_o left free (**e**,**f**) and $a_o = 1$ (**g**,**h**). The data are from proton-proton collisons at the center-of-mass energies, $\sqrt{s} = 0.9$ and 7 TeV; taken from [27,28].

As one can observe from the χ^2 values of the best fits, shown in Tables 1–5, all formulas fit the data well enough, except the small differences in the parameterization. One could expect that Equations (2) and (3) with $a_0 = 1$ would give similar results because the only difference is the incorporation of the temperature, T, in an overall normalization constant, C. This expectation is indeed observed in the results of the fits in Tables 2 and 4. Since T does not depend on p_T, this difference would be of no significance for the fits using Formulas (2) and (3). On the contrary, the Cleymans–Worku Formula (1) does have an additional dependence in the multiplication factor through the term m_T. The effects of the factors p_T and m_T on the distributions can be understood notifying that

$$m_T = \sqrt{p_T^2 + m_0^2} = m_0\sqrt{1 + (p_T/m_0)^2}, \tag{9}$$

which is approximately constant and equal to m_0 for $p_T \ll m_0$. Therefore, only in the low $p_T \ll m_0$ range of transverse momentum, the additional factor, m_T, is insignificant for the fits. The temperature, T, is well determined in the sets of data that cover the peak in the p_T-distribution.

The systematic analysis of the parameters is performed by comparing the behavior of the parameters q and T, extracted from the fit. The relation between n and q is used where necessary. The results of the analysis of the mass dependencies of the parameters are displayed in Figure 2. One has to remember, that the different exponents in the Cleymans formula and in the other formulas will lead to different numerical values for q and T.

Some authors, as in Refs. [1,18,25], adopt $n = 1/(q' - 1)$, and in such cases, it is possible to relate the values q_C and T_C from the Cleymans formula to the values q' and T' from the other formulas by the relations,

$$\begin{cases} q_C = 1/(2 - q') \\ T_C = q'T' \end{cases}. \tag{10}$$

One observes that the results for the best-fit value of the parameter q are similar for all formulas used, with results fluctuating around the expected theoretical value of $q = 1.143$. Using the relation above, one may expect $q_C = 1.16$ for the Cleymans formula. As can be observed in Figure 2a,b, there is no clear dependence observed on the particle mass or collision energy. The behavior of the parameter T, however, is different and shows a dependence on the particle mass for all formulas, except that for the Cleymans formula. This result might be related to the fact that, in the Cleymans formula, the chemical potential is a free parameter, while in the other formulas, it is the constant particle mass.

A theoretical study, generalizing Hagedorn's self-consistent thermodynamics by the inclusion of the Tsallis statistics, leads to the non-extensive self-consistent thermodynamics [30], which predicts that the ratio $T/(q - 1)$ should be independent of the the collision energy or particle species. In Figure 2c, this result is demonstrated by all formulas used, except the Tsallis–Pareto formula with the free parameter a_0.

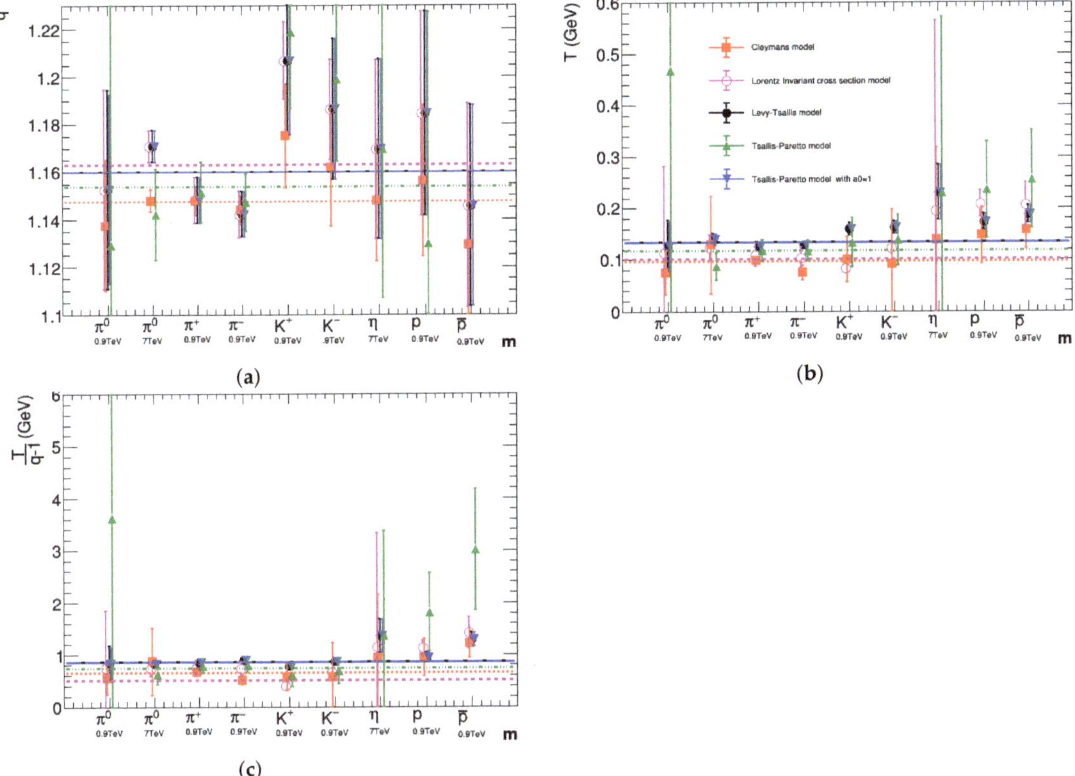

Figure 2. Behavior of the Tsallis exponent, q, temperature, T, and the ratio, $T/(q-1)$, as a function of the particle species. extracted from the fits with different formulas. The relation, $n = (q-1)^{-1}$, used where necessary (see the text for details). (**a**) Tsallis exponent, q (**b**) temperature, T (**c**) the ratio, $T/(q-1)$.

Contrary to what would be expected, the results of the analysis of the parameters show that the behavior of the parameters q and T is similar for Equations (1) and (2), but the values of the parameters differ from the values, obtained by Equation (3), mainly, for the heavier particles. Significantly smaller differences are observed from the comparison of the fits for other particles.

One can observe that, despite the plain curves, obtained with the formulas, used in the present analysis, the fitting procedure is not as direct. There are strong correlations among the fitting parameters, which can be observed by an analysis of the covariances. The introduction of the chemical potential as a free parameter makes the problem worse. Actually, it is always possible to obtain an equivalent fitting by fixing $\mu = 0$ if the multiplicative constant and the temperature are both adjustable. This is an indication of the limits of an analysis, based exclusively on the p_T-distribution data. Additional hypotheses or complementary analyses of, e.g., the ratio of the particle species yields, are necessary to determine unambiguously all the parameters.

Whereas the differences in the formulas, used in this analysis, are subtle when observed through the obtained best-fit parameters, the analysis of the covariances among the parameters shows clear differences depending on the formula used. Figure 3 shows the correlation among the parameters q and T obtained for each particle species and collision energy and for all formulas, used in this analysis. The ellipses show the 5% confidence level of the fitting, demonstrating a strong correlation between these two parameters. The theoretical values are indicated in Figure 3 and one can see that the expected values are attained within errors for all formulas. One can also observe that the covariances change significantly from one formula to another, except of the Levy–Tsallis formula and the Tsallis–Pareto formula with $a_o = 1$. This was expected since these two formulas differ only by the inclusion of a multiplicative term that depends on q and T in the Levy–Tsallis formula.

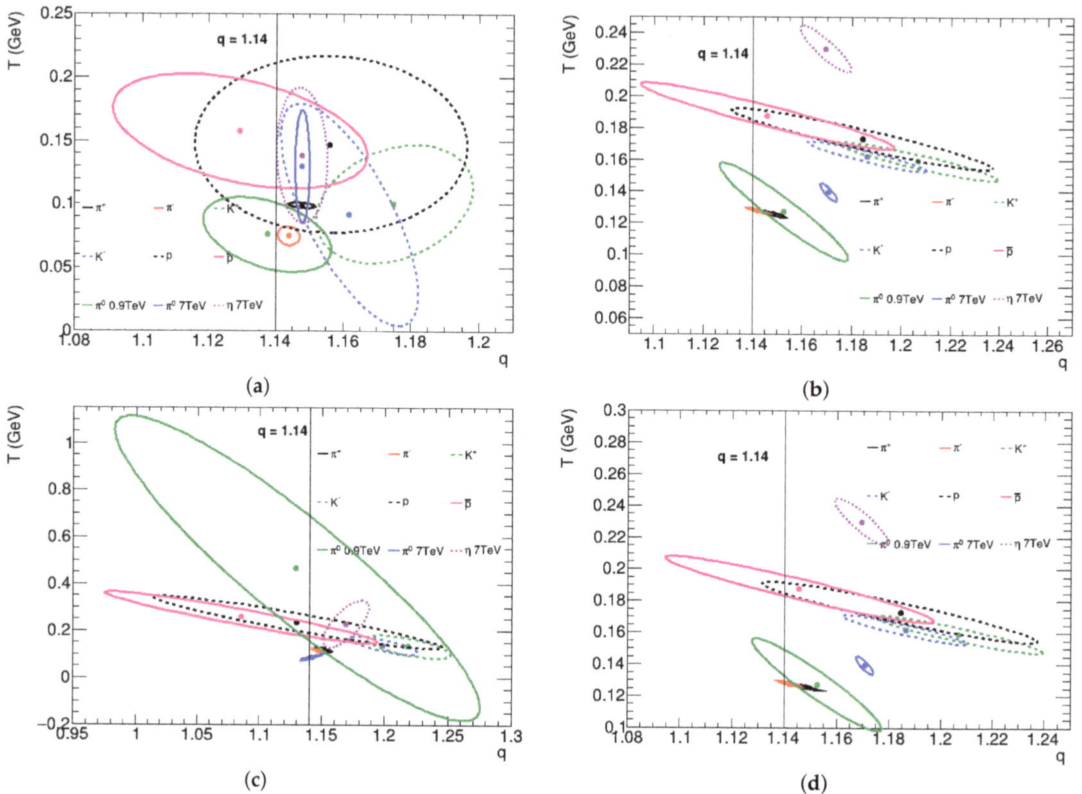

Figure 3. Joint confidence region on the T–q plane in the parameter space, with significance, $\alpha = 5\%$. The regions are obtained by using **(a)** Equation (1), **(b)** Equation (2), and **(c,d)** Equation (3), with **(c)** the parameter a_o left free to adjust and **(d)** fixed to the value $a_o = 1$. The results for $\sqrt{s} = 7$ TeV are indicated by the asterisks and the results for $\sqrt{s} = 0.9$ TeV are indicated by the solid circles for the centers of the ellipses.

A final analysis is performed by using the Lorentz invariant cross-section Formula (8). Observe that the multiplication factor, p_T, exponent, $a_o = 1$, and the chemical potential remains as an adjustable parameter. This formula allows us to test the hypothesis made above, i.e., that the variations in the temperature with the particle species, observed in some of the fits, are due to the fact that, in those cases, the chemical potential is fixed to the particle mass, while in the Cleymans formula, this parameter is an adjustable.

The results of the fittings with the Lorentz invariant cross-section formula are presented in Figure 4, and one observes that the data are well fitted also in this case. The best-fit parameters are listed in Table 5, and one can observe that the reduced χ^2, corresponding to the fitting, is small and comparable to those, obtained with other formulas.

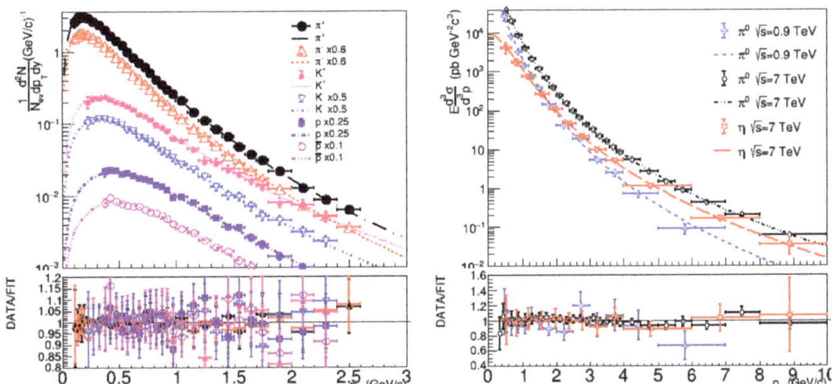

Figure 4. The double-differential transverse momentum distributions, fitted by the Lorentz invariant cross-section (8). The data are from [27,28].

The analysis of the parameters, obtained with the Lorentz invariant cross-section formula, are shown in Figure 2 by open circles. Whereas the results for q do not change significantly with respect to other formulas, one observes that the temperature, T, with the Lorentz invariant cross-section formula represents a behavior similar to that, observed with the Cleymans formula. Therefore, this result confirms the hypothesis that the dependence on T is due to the fact that the chemical potential is fixed in some of the formulas. The ratio, $T/(q-1)$, remains almost constant, as observed in all other formulas with the multiplicative term proportional to p_T.

In Figure 5, the chemical potential is shown for different particle species and collision energies along with the covariances between the parameters T and q for the fit with the Lorentz invariant cross-section formula. One can see that the chemical potential does not present any evident dependence on the particle mass or on the collision energy, but may depend on the baryonic number, since it is slightly higher for the cases of the proton and the anti-proton. This is in contrast with the assumptions in the Pareto–Tsallis and Levy–Tsallis formulas, where the chemical potential is assumed to be equal to the particle mass.

Except for proton and anti-proton, the chemical potentials are small or compatible with the value $\mu = 0$ for all other particles, independently of the collision energy. This result is in agreement with the findings of Cleymans and collaborators [20], where it is shown that the correlations between the parameters T, μ, and the multiplicative constant, V, allow one to set $\mu = 0$ and still obtain the same result with different values for T and V. Using the Lorentz invariant cross-section formula, it was verified that fixing $\mu = 0$ leads to the same fitted curve, the best-fit value for q does not change, and the values for T and V change according to the relations, found in Ref. [20] for all particles species, analyzed in the present paper. Let us note that the ellipses showing the covariance of the parameters T and q, shown in Figure 5, are similar to those obtained for the Cleymans formula. This is an indication that the adjustable chemical potential parameter has an influence on the result for the parameter T, as discussed above, and, thus, interferes with the correlation between the fit results for T and q. To note also is that there is a shift in the centroid of the ellipses with respect to those of the Cleymans' formula. This shift results from the different behavior of the multiplicative factor of the formulas with the transverse mass, m_T.

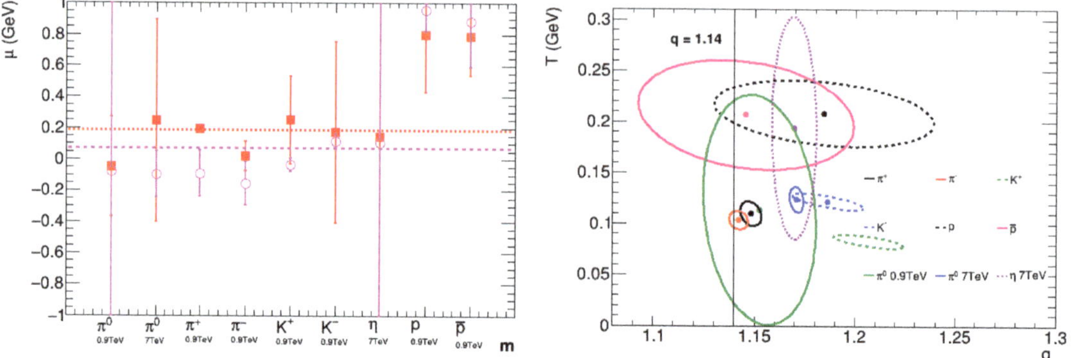

Figure 5. Left: the dependence of the chemical potential, μ, on the particle species. The results are for the Cleymans Formula (1) (solid squares) and for the Lorentz invariant cross-section Formula (8) (open circles). The dotted and dashed lines represent the average value for the chemical potential, obtained by the Cleymans formula and the Lorentz invariant cross-section formula, respectively. **Right**: the covariances in the T–q plane in the parameter space, corresponding to Lorentz invariant cross-section Formula. The results for $\sqrt{s} = 7$ TeV are indicated by the asterisks and the results for $\sqrt{s} = 0.9$ TeV are indicated by the solid circles for the centers of the ellipses.

4. Conclusions

In this paper, the analysis of the high-energy proton-proton collison data is performed using the methods of the non-extensive distributions developed by Jean Cleymans. The results are compared through different transverse momentum distribution formulas. It is shown that all formulas fit well the data, while giving different best-fit parameters.

The best-fit values of the parameters are discussed and, for the parameter q, the results are compared to the expected values according to the a fractal approach to quantum chromodynamics (QCD) [14]. In most cases, the predicted fractal approach value is attained by the different distributions; however, it is important to consider the covariances of the parameters, as shown in Figures 3 and 5 (Right).

It is not the objective of this study to make definitive conclusions, based on the results obtained. A more comprehensive analysis to make firmer conclusions is needed and is planned for the future. However, the present paper illustrates the importance of this type of analysis, to great extent developed by Jean Cleymans, for the understanding of the complex phenomena taking place during the multiparticle production in high-energy collisions.

A new phenomenon has emerged [31] in the analysis of the particle distributions using the non-extensive statistics. Therefore, a correct and accurate description of the distributions, finding the hidden patterns in the obtained parameters, will allow us to advance further in the understanding of the QCD at the non-perturbative regime.

We would like to remark here that the present analysis needs to be continued by the inclusion of additional data sets in order to arrive at firmer conclusions. This will allow us to perform statistical tests for theoretical predictions. However, the procedure adopted here, analyzing exclusively the transverse momentum, p_T, distributions, is not able to provide a complete determination of all the parameters. The point that any fit can be equally reproduced by fixing the chemical potential and allowing the temperature and the multiplicative factor to adjust freely is an indication that one needs additional hypotheses or complementary analyses of, e.g., the ratio of the yields of different particle species, in order to completely determine all parameters in the formulas. Nevertheless, one conclusion that can be drawn from this partial analysis is that a thermodynamical approach describes the multiparticle production process, and, therefore, the hypothesis of thermal equilibrium of the system at the freeze-out stage is robust.

Let us recall that in the non-extensive statistics, the temperature fluctuations play a major role, as has been demonstrated by many studies [24,32,33]. This aspect has implications in the equilibrium described by the Tsallis statistics, which must accommodate these fluctuations. This issue is considered in Ref. [34].

Author Contributions: All authors equally contributed to the present work. All authors have read and agreed to the published version of the manuscript.

Funding: The work of L.R. was supported by the *Coordenação de Aperfeiçoamento de Pessoal de Nível Superior—Brasil (CAPES)*—Finance Code 001, grant 88887.641069/2021-00, by Project PROEX. The work of A.D. was partially supported by the CNPq-Brazil, grant 304244/2018-0, by Project INCT-FNA Proc. No. 464 898/2014-5, and by FAPESP grant 2016/17612-7. The work of E.M. is supported by the project PID2020-114767GB-I00 financed by MCIN/AEI/10.13039/50110001103, by the FEDER/Junta de Andalucía-Consejería de Economía y Conocimiento 2014–2020 Operational Program under Grant A-FQM-178-UGR18, by Junta de Andalucía under Grant FQM-225, by the Consejería de Conocimiento, Investigación y Universidad of the Junta de Andalucía and European Regional Development Fund (ERDF) under Grant SOMM17/6105/UGR, and by the Ramón y Cajal Program of the Spanish MCIN under Grant RYC-2016-20678. The work of F.-H.L. was supported by the National Natural Science Foundation of China under Grant Nos. 12147215, 12047571, and 11575103, the Shanxi Provincial Natural Science Foundation under Grant Nos. 20210302124036 and 201901D111043, the Scientific and Technological Innovation Programs of Higher Education Institutions in Shanxi (STIP) under Grant No. 201802017, and the Fund for Shanxi "1331 Project" Key Subjects Construction. The work of K.K.O. was supported by the Ministry of Innovative Development of Uzbekistan within the framework of fundamental project No. F3-20200929146 on analysis of open data on proton-proton and heavy-ion collisions at the LHC.

Data Availability Statement: Not applicable.

Conflicts of Interest: The authors declare no conflict of interest.

References

1. Wilk, G.; Włodarczyk, W. Quasi-power laws in multiparticle production processes. *Chaos Solit. Fract.* **2015**, *81*, 487–496. [CrossRef]
2. Cleymans, J.; Hippolyte, B.; Paradza, M.W.; Sharma, N. Hadron resonance gas model and high multiplicities in p-p, p-Pb and Pb-Pb collisions at the LHC. *Int. J. Mod. Phys.* **2019**, *28*, 1940002. [CrossRef]
3. Oeschler, H.; Cleymans, J.; Hippolyte, B.; Redlich, K.; Sharma, N. Ratios of strange hadrons to pions in collisions of large and small nuclei. *Eur. Phys. J.* **2017**, *77*, 584. [CrossRef]
4. Tsallis, C. Possible Generalization of the Boltzmann-Gibbs Statistics. *J. Stat. Phys.* **1988**, *52*, 479–487. [CrossRef]
5. Cleymans, J.; Worku, D. The Tsallis distribution in proton-proton collisions at root s = 0.9 TeV at the LHC. *J. Phys.-Nucl. Part. Phys.* **2012**, *39*, 025006. [CrossRef]
6. Cleymans, J.; Paradza, M.W. Tsallis statistics in high energy physics: Chemical and thermal freeze-outs. *Physics* **2020**, *2*, 654–664. [CrossRef]
7. Rath, R.; Khuntia, A.; Sahoo, R.; Cleymans, J. Event multiplicity, transverse momentum and energy dependence of charged particle production, and system thermodynamics in pp collisions at the large hadron collider. *J. Phys. G Nucl. Part. Phys.* **2020**, *47*, 055111. [CrossRef]
8. Khuntia, A.; Sharma, H.; Tiwari, S.K.; Sahoo, R.; Cleymans, J. Radial flow and differential freeze-out in proton-proton collisions at $\sqrt{s} = 7$ TeV at the LHC. *Eur. Phys. J. A* **2019**, *55*, 3. [CrossRef]
9. Bhattacharyya, T.; Cleymans, J.; Marques, L.; Mogliacci, S.; Paradza, M.W. On the precise determination of the Tsallis parameters in proton–proton collisions at LHC energies. *J. Phys. G* **2018**, *45*, 055001. [CrossRef]
10. Cleymans, J.; Man Lo, P.; Redlich, K.; Sharma, N. Multiplicity dependence of (multi)strange baryons in the canonical ensemble with phase shift corrections. *Phys. Rev. C* **2021**, *103*, 014904. [CrossRef]
11. Sharma, N.; Cleymans, J.; Hippolyte, B. Thermodynamic limit in high-multiplicity proton-proton collisions at $\sqrt{s} = 7$ TeV. *Adv. High Energy Phys.* **2019**, *2019*, 5367349. [CrossRef]
12. Bugaev, K.; Sagun, V.; Ivanytskyi, A.; Nikonov, E.; Cleymans, J.; Mishustin, I.; Zinovjev, G.; Bravina, L.V.; Zabrodin, E.E. Separate freeze-out of strange particles and the quark-hadron phase transition. *EPJ Web Conf.* **2018**, *182*, 02057. [CrossRef]
13. Azmi, M.D.; Bhattacharyya, T.; Cleymans, J.; Paradza, M. Energy density at kinetic freeze-out in Pb-Pb collisions at the LHC using the Tsallis distribution. *J. Phys. G Nucl. Part. Phys.* **2020**, *47*, 045001. [CrossRef]
14. Deppman, A.; Megias, E.; Menezes, D.P. Fractals, nonextensive statistics, and QCD. *Phys. Rev.* **2020**, *101*, 034019. [CrossRef]
15. Sena, I.; Deppman, A. Systematic analysis of p(T)-distributions in p plus p collisions. *Eur. Phys. J. A* **2013**, *49*, 17. [CrossRef]
16. Liu, F.H.; Gao, Y.Q.; Li, B.C. Comparing two-Boltzmann distribution and Tsallis statistics of particle transverse momentums in collisions at LHC energies. *Eur. Phys. J. A* **2014**, *50*, 123. [CrossRef]

17. Khandai, P.K.; Sett, P.; Shukla, P.; Singh, V. System size dependence of hadron p_T spectra in p+p and Au+Au collisions at $\sqrt{s_{NN}}$ = 200 GeV. *J. Phys. G Nucl. Part Phys.* **2014**, *41*, 025105. [CrossRef]
18. Wong, C.Y.; Wilk, G. Tsallis fits to p_T spectra and multiple hard scattering in pp collisions at the LHC. *Phys. Rev. D* **2013**, *87*, 114007. [CrossRef]
19. Marques, L.; Andrade, E.; Deppman, A. Nonextensivity of hadronic systems. *Phys. Rev. D* **2013**, *87*, 114022. [CrossRef]
20. Cleymans, J.; Lykasov, G.I.; Parvan, A.S.; Sorin, A.S.; Teryaev, O.V.; Worku, D. Systematic properties of the Tsallis Distribution: Energy Dependence of Parameters in High-Energy p-p Collisions. *Phys. Lett. B* **2013**, *723*, 351–354. [CrossRef]
21. Azmi, M.D.; Cleymans, J. Transverse momentum distributions in proton–proton collisions at LHC energies and tsallis thermodynamics. *J. Phys. G Nucl. Part. Phys.* **2014**, *41*, 065001. [CrossRef]
22. Rybczynski, M.; Wlodarczyk, Z. Tsallis statistics approach to the transverse momentum distributions in p-p collisions. *Eur. Phys. J. C* **2014**, *74*, 2785. [CrossRef]
23. Marques, L.; Cleymans, J.; Deppman, A. Description of high-energy pp collisions using Tsallis thermodynamics: Transverse momentum and rapidity distributions. *Phys. Rev. D* **2015**, *91*, 054025. [CrossRef]
24. Bíró, G.; Barnaföldi, G.G.; Biró, T.S. Tsallis-thermometer: A QGP indicator for large and small collisional systems. *J. Phys. G Nucl. Part. Phys.* **2020**, *47*, 105002. [CrossRef]
25. Shen, K.; Barnafoldi, G.G.; Biro, T.S. Hadronization within the non-extensive approach and the evolution of the parameters. *Eur. Phys. J. A* **2019**, *55*, 126.
26. Zheng, H.; Zhu, L.; Bonasera, A. Systematic analysis of hadron spectra in $p + p$ collisions using tsallis distributions. *Phys. Rev. D* **2015**, *92*, 074009. [CrossRef]
27. ALICE Collaboration. Neutral pion and η meson production in proton–proton collisions at $\sqrt{s}$ = 0.9 TeV and $\sqrt{s}$ = 7 TeV. *Phys. Lett. B* **2012**, *717*, 162–172. [CrossRef]
28. ALICE Collaboration. Production of pions, kaons and protons in pp collisions at $\sqrt{s}$ = 900 GeV with ALICE at the LHC. *Eur. Phys. J. C* **2011**, *71*, 1655. [CrossRef]
29. Griffiths, D.J. *Introduction to Elementary Particles*; John Wiley & Sons, Inc.: New York, NY, USA, 1987. Available online: http://nuclphys.sinp.msu.ru/books/b/Griffiths.pdf (accessed on 25 May 2022).
30. Deppman, A. Self-consistency in non-extensive thermodynamics of highly excited hadronic states. *Physica A* **2012**, *391*, 6380–6385. [CrossRef]
31. Wilk, G.; Włodarczyk, W. Some intriguing aspects of multiparticle production processes. *Int. J. Mod. Phys. A* **2018**, *33*, 1830008. [CrossRef]
32. Wilk, G.; Wlodarczyk, Z. The imprints of nonextensive statistical mechanics in high-energy collisions. *Chaos Solit. Fract.* **2002**, *13*, 581–594. [CrossRef]
33. Hanel, R.; Thurner, S.; Gell-Mann, M. Generalized entropies and the transformation group of superstatistics. *Proc. Nat. Acad. Sci. USA* **2011**, *108*, 6390–6394. [CrossRef]
34. Megias, E.; Lima, J.A.S.; Deppman, A. Transport equation for small systems and the nonadditive entropy. *Mathematics* **2022**, *20*, 1625. [CrossRef]

Article

Nuclear Modification Factor in Small System Collisions within Perturbative QCD Including Thermal Effects [†]

Lucas Moriggi [1] and Magno Machado [2,*]

[1] Universidade Estadual do Centro-Oeste (UNICENTRO), Campus Cedeteg, Guarapuava 85015-430, Brazil; lucasmoriggi@unicentro.br

[2] Instituto de Física, Universidade Federal do Rio Grande do Sul, Porto Alegre 90010-150, Brazil

[*] Correspondence: magnus@if.ufrgs.br

[†] This paper is dedicated to the memory of the late Jean Cleymans.

Abstract: In this paper, the nuclear modification factors, R_{xA}, are investigated for pion production in small system collisions, measured by PHENIX experiment at RHIC (Relativistic Heavy Ion Collider). The theoretical framework is the parton transverse momentum k_T-factorization formalism for hard processes at small momentum fraction, x. Evidence for collective expansion and thermal effects for pions, produced at equilibrium, is studied based on phenomenological parametrization of blast-wave type in the relaxation time approximation. The dependencies on the centrality and on the projectile species are discussed in terms of the behavior of Cronin peak and the suppression of R_{xA} at large transverse momentum, p_T. The multiplicity of produced particles, which is sensitive to the soft sector of the spectra, is also included in the present analysis.

Keywords: k_T-factorization approach; parton saturation phenomenon; geometric scaling in hadron–hadron collisions; high-energy collisions; multiparticle production; transverse momentum distribution

Citation: Moriggi, L.; Machado, M. Nuclear Modification Factor in Small System Collisions within Perturbative QCD Including Thermal Effects. *Physics* **2022**, *4*, 787–799. https://doi.org/10.3390/physics4030050

Received: 23 May 2022
Accepted: 21 June 2022
Published: 18 July 2022

Publisher's Note: MDPI stays neutral with regard to jurisdictional claims in published maps and institutional affiliations.

1. Introduction

The transverse momentum (p_T) spectra of charged and neutral particles have been experimentally studied in proton–proton (pp), proton–nucleus (pA), and nucleus–nucleus (AA) collisions from different perspectives. First of all, the role played by this observable is highlighted as a measure of the partonic interactions in the perturbative quantum chromodynamics (pQCD) regime. One can extract from the particle spectra relevant information about the initial state dynamics of the colliding system. On the other hand, the p_T spectrum has been used as a probe of the collective behavior developed by the hydrodynamic expansion of the hot environment created in heavy ion collisions. A key aspect of the problem is the emergence of thermal behavior observed even in small systems such as pp and pA collisions [1–6]. The nuclear effects are quantified by the nuclear modification factor, R_{xA}, which is obtained by the ratio between the multiplicity of produced particles in the collision of a projectile X off a nucleus A and the scaling on the number N_{coll} of binary collisions, $N_{\text{coll}}(d^3 N_{pp}/d^3 p)$, with N_{pp} being the yield in pp collisions. The latter is expected in the case of an absence of final state nuclear medium effects. From the experimental point of view, the nuclear modification factor for small systems presents a suppression in the small transverse momentum region ($p_T \sim 1$ GeV), followed by an enhancement in the intermediate momentum region ($p_T \sim 2$–5 GeV), and it finally goes to unity at large p_T. This behavior is known as the Cronin effect [7], which has been explained by different theoretical models [8–18].

The scenario of hadron production from the decay of minijets described by k_T-factorization formalism, where k_T denotes the parton transverse momentum, considers that the cold matter nuclear (CNM) effects originate in the hard interaction of the nuclei at initial states of the collision. However, the particle production in those systems undergoes a hydro-dynamics evolution to freeze-out, which modifies the corresponding spectrum. In [19–21], it is argued that the p_T spectrum can be described by performing a temporal separation in the relaxation time approximation (RTA) of the Boltzmann transport equation [22]. The time separation corresponds to the hadrons produced in initial state hard collision and those produced in the equilibrium situation. Moreover, two-component models (thermal+hard) have been successful in describing experimental measurements [23–25]. In these approaches, the spectrum is decomposed into two parts: one related to the Boltzmann statistics and a second one characterized by the typical power-law behavior from pQCD. Nevertheless, the thermal nature of the spectrum has been posed in [26–28], where the two-component model takes into account a soft contribution coming from the longitudinal dissociation projectile-nucleus and a hard contribution due to the transverse production of jets. This seems to be enough to explain the data available without invoking collective flow. In a previous study [29], this approach has been considered to describe the spectra of produced pions in lead–lead (*PbPb*) collisions at the Large Hadron Collider (LHC). It was shown that the thermal parametrization can be substantially modified by taking into account the nuclear effects present in the gluon distribution function.

In this paper, thermal effects are investigated in the nuclear modification factor, R_{xA}, for neutral pion production, which has been measured by the PHENIX Collaboration at the Relativistic Heavy Ion Collider (RHIC) [30], in small system collisions. A salient feature of R_{xA} in these systems is that the Cronin peak tends to grow for smaller projectiles, which is in contradiction to the expected behavior from pQCD. In particular, the interface between the hard process described within the QCD k_T-factorization formalism and the thermal sector is investigated, which can play an important role in more central collisions. To this end, both the hard part of the spectrum and the multiplicity, dN/dy, determined by the soft part, are studied. One important issue regards the separation region where the collective effects are needed for the spectrum description. In [31], an analysis of the high multiplicity distribution of particles in pp collisions was performed, and it was argued that the bulk part of spectra ($p_T \lesssim 2.5$ GeV) can be described by a distribution-like blast-wave. In [32], an approach based on a blast-wave parametrization that incorporates Tsallis statistics was also considered in order to study the collective flow effects in pp collisions as a function of the center-of-mass energy, $\sqrt{s}$. Evidence of collective expansion in high energies was shown. This suggests a dependence of the radial flow as the multiplicity or collision energy increase. Such effects also become important with the increasing number of constituents of a projectile/nuclear target. Therefore, it is fundamental to analyze the produced spectrum for distinct projectiles in order to relate the emergence of thermal behavior in terms of $dN/dy, \sqrt{s}$ and the geometric parameters of the colliding system such N_{coll} and N_{part}, the number of participants.

This paper is organized as follows. In Section 2.1, the predictions of pQCD for the nuclear modification factor, R_{xA}, is discussed in different approaches. The main theoretical expressions for computing this observable are presented in the context of the pQCD k_T-factorization formalism. The main physical input is the nuclear unintegrated gluon distribution (nUGD). Following [29], the Moriggi–Peccini–Machado (MPM) analytical parametrization for the nuclear UGD is considered. It describes correctly the spectra of particles produced in pp collisions, as well as the nuclear modification factors in pPb collisions at the LHC. In Section 2.2, it is determined how the thermal corrections can be included in the spectrum is determined by using a blast–wave model. The focus is on pion production in $p + Al$, $p + Au$, $d + Au$, and $He + Au$ collisions. It is demonstrated that the Cronin peak decreases for larger projectiles, in opposition to what is expected, as only cold nuclear matter effects are considered. It is verified that R_{xA} presents a behavior almost independent of projectile species at $p_T \sim 10$ GeV. The same occurs for the thermal parameters of the system, such as the relaxation time, t_r, and temperature, T, which could suggest a correlation between the energy loss at large p_T and the production of a thermal system of particles. These results and corresponding discussions are presented in Section 3. Conclusions are summarized in Section 4.

2. Theoretical Framework and Main Predictions

2.1. Nuclear Effects in the Gluon Distribution in a Nucleus

In the collinear factorization formalism of pQCD, the XA cross-section can be expressed as the convolution of the parton distribution functions (PDFs) of the small projectile (labeled here as X) and the nucleus A with the hard parton level cross-section. In this context, the nuclear modification factor is described in terms of the modification of the parton distribution within a nucleus compared to the ones in a free nucleon. These modifications give rise to the shadowing/anti-shadowing effects observed in the nuclear structure functions probed in deep-inelastic-scattering (DIS) events [33–37]. Moreover, the multiple interactions among nucleons can be described in the Glauber model, where the nuclear cross-section is expected to scale with $N_{\rm coll}$. Corrections to the collinear factorization approach can take into account an intrinsic transverse momentum, $\langle k_T \rangle$, of partons at initial state. This effect increases with $N_{\rm coll}$, which leads to an enhancement of R_{xA} with respect to centrality [8,38,39].

An important aspect that appears in RHIC data [30] is that the Cronin peak tends to increase for smaller projectiles species in agreement with the ordering $R_{^3He+Au} < R_{d+Au} < R_{p+Au}$. However, the collinear approach with cold nuclear matter effects predicts the following: $R_{^3HeAu} > R_{d+Au} > R_{p+Au} > R_{p+Al}$. This prediction is in agreement with what is observed with respect to the nuclear structure functions. There, the shadowing/anti-shadowing contributions are intensified due to the atomic mass number, A. Another essential aspect presented by these data is that for more central collisions (0–5%), the factor $R_{^3He+Au}$ presents a suppression that is not seen for smaller projectiles. Moreover, it is worth mentioning that $R_{^3He+Au}$ increases for more peripheral collisions, while the factors $R_{p(d)A}$ decrease. In the multiple scattering model, it is expected that the average transverse momentum, $\langle k_T \rangle$, acquired by partons from the projectile as a result of interaction with the nuclear target is larger in more central collisions. Therefore, this effect raises the Cronin peak, and a decrease of R_{xA} is consequently anticipated in more peripheral reactions.

In the context of parton saturation approaches, like the color glass condensate (CGC) effective theory, the nuclear modification factor can be associated to the saturation of the nuclear UGD in the region of small p_T. This saturation is more intense in more central collisions and for large nuclei (dense color system). In those approaches, through the Glauber–Mueller formalism [40,41], the multiple interactions of colored partons in the projectile with the color field of the nuclear target modifies the transverse momentum of the gluons in target, and the Cronin peak is reproduced. One can find that the evolution in the rapidity of the nUGD and the nuclear geometry should influence the behavior of R_{pA} in terms of p_T [11,15,42]. The saturation approach has been used in different studies [43–48] in order to describe the p_T-spectrum and the ratio R_{pA} for dAu reaction at RHIC and pPb ones at the LHC. In AA collisions, the saturation approach has been utilized in [49–56].

In this study, we consider the k_T-factorization formalism in contrast to the collinear one. The main advantage is that the initial partons already present non-zero transverse momenta and the effects of parton saturation can be easily implemented. Now, the p_T-spectrum is given by the convolution of the unintegrated gluon distributions of the colliding nuclei, $\phi_{A,B}$, and the production cross-section at the parton level. The nUGD depends on the impact parameter, $\vec{b}$, of the reaction. In particular, the gluon production in $A + B$ collisions at very high energies is given by the following invariant cross section:

$$E\frac{d^3N(b)}{dp^3}^{AB \to g+X} = \frac{2\alpha_s}{C_F}\frac{1}{p_T^2}\int d^2\vec{s}\,d^2\vec{k}_T\,\phi_A(x_A, k_T^2, \vec{s})\phi_B(x_B, (\vec{p}_T - \vec{k}_T)^2, \vec{b} - \vec{s}), \quad (1)$$

where $x_{A,B} = (p_T/\sqrt{s})e^{\pm y}$ are the longitudinal momentum fraction of incoming partons in projectile A and target B, respectively. The quantity $\vec{s}$ is the transverse coordinate of the produced gluon. The strong coupling constant is α_s, and $C_F = (N_c^2 - 1)/2N_c$ is the QCD Casimir color factor with N_c being the color number. Here, some remarks are in order. The processes initiated by quarks, $q_f + g \to q_f$ and $g + q_f \to q_f$, are important in the fragmentation region and under large enough p_T [57,58]. Here, g and q_f denote the gluon and the fragmenting quark of flavour f, respectively. The data from PHENIX Collaboration [30] correspond to midrapidities ($\eta < 0.35$) and not so large transverse momentum. This is the reason for disregarding the quark contribution in the numerical calculations here.

The unintegrated gluon distribution depends on the transverse momentum k_T and longitudinal momentum fraction x. Here, in the numerical calculation, we use the MPM analytical parametrization [59] for the UGD. It is determined from the analysis of available data for light hadron production in $pp(\bar{p})$ collisions. It presents scaling on the variable $\tau = k_T^2 / Q_s^2(x)$, based on the geometric scaling property associated to the parton saturation formalism [60,61]. The characteristic momentum scale giving the transition between a dilute and a dense parton system is set by the saturation scale, Q_s. It is defined in terms of the longitudinal momentum fraction in the form $Q_s^2(x) = (x_0/x)^\lambda$. The MPM parametrization for a proton is written as follows:

$$\phi_p(x, k_T) = \phi_p(\tau) = \frac{3\sigma_0(1 + \delta n)}{4\pi^2 \alpha_s} \frac{\tau}{(1 + \tau)^{(2 + \delta n)}}, \tag{2}$$

where $\delta n = a\tau^b$, and $\lambda = 0.33$ is fixed [62,63]. The parameters σ_0, x_0, a, and b have been fitted from DESY-HERA (Deutsches Elektronen-Synchrotron, Hadron-Electron Ring Accelerator) data for proton structure function (see discussion in [59]). For central rapidity, $y \sim 0$, the x variable can be expressed as $x = p_T / \sqrt{s}$.

The incorporation of the nuclear effects in the gluon distribution is given by the Glauber–Gribov approach for multiple scattering. The main ingredients are the nuclear thickness function, $T_A(b)$, and the color dipole cross-section. The last quantity describes the multiple interactions of the leading Fock state of the projectile parton with the nucleons. The Woods–Saxon parametrization for the nuclear density has been considered for a large nucleus [64,65], and the thickness function has the normalization $\int d^2\vec{b}\, T_A(b) = A$. For deuteron, the Hulthén form was used [66], and for helium, the parametrization, presented in [67], is considered. Following the previous study [29], the nuclear UGD is given as follows:

$$\phi_A(x, k_T^2, b) = \frac{3}{4\pi^2 \alpha_s} k_T^2 \nabla_k^2 \mathcal{H}_0 \left\{ \frac{1 - S_{q\bar{q}A}(x, r, b)}{r^2} \right\}, \tag{3}$$

where $\mathcal{H}_0\{f(r)\} = \int r\,dr J_0(k_T r) f(r)$ is the Hankel transform of order 0. The quantity ∇_k^2 is the two-dimensional (2-D) Laplacian in momentum space. The key quantity is the dipole scattering matrix in configuration space, $S_{q\bar{q}A}(x, r, b)$. It can be determined from the cross-section for dipole scattering off a proton, $\sigma_{\rm dip}(x, r)$, in the following way:

$$S_{q\bar{q}A}(x, r, b) = \exp\left[-\frac{1}{2} T_A(b) \sigma_{\rm dip}(x, r) \right], \tag{4}$$

$$\sigma_{\rm dip}^{\rm MPM}(x, r) = \sigma_0 \left\{ 1 - \frac{2\left[\left(\frac{rQ_s(x)}{2} \right)^{1+\delta n} K_{1+\delta n}(rQ_s(x)) \right]}{\Gamma(1 + \delta n)} \right\}, \tag{5}$$

where, the last line gives the corresponding expression for the dipole-proton cross-section in the MPM model. In particular, it scales with $rQ_s(x)$, as is usual in geometric-scaling-based models. Here, K is the modified Bessel function of second kind and Γ is the Gamma function. The calculation is made straightforward by including any other phenomenological model for this quantity.

Considering only CNM effects, it is found that Equation (1) predicts only small changes in R_{xA} in collisions of gold nucleus Au with deuteron d or helium He. Essentially, the nuclear modification is driven by the large nucleus in a similar way as occurs in the collinear factorization formalism. Namely, the nuclear effects in the gluon distribution for small nuclei is tiny. In Section 3, the need for corrections to the initial distribution is discussed once the experimental measurements present a large difference as the projectile changes. The nuclear modification factor, R_{xA}, based on the discussion above, can be determined by using the Glauber model:

$$R_{xA} = \frac{\frac{d^3 N_{XA}}{dp^3}}{N_{\text{coll}} \frac{d^3 N_{pp}}{dp^3}}, \tag{6}$$

where $d^3 N_{XA}/dp^3$ is obtained by using Equations (1), (3), and (4).

The nuclear gluon distribution in Equation (3) is characterized by two main quantities: the saturation scale, $Q_s \sim x^{-0.33}$, which determines the increasing of the spectrum in terms of the collision energy, $\sqrt{s}$; and a power index δn that reproduces the power-like behavior on p_T in the large transverse momentum limit. In Ref. [59], the investigation into the spectrum was restricted to the region, where geometric scaling is expected. For RHIC data at 200 GeV, this is valid only at small p_T. In order to achieve a good description at large p_T, in the present paper, the parameter δn was modified as $\delta n = 1.2$, corresponding to the effective slope observed in the spectrum in pp collisions at RHIC. On the other hand, in order to keep the point of maximum in the UGD at the same place, the saturation scale has to be modified as $Q_s^2(x) \to (1 + \delta n)Q_s^2(x)$. The fragmentation process is described in a simplified way by taking the hadron transverse momentum in the approximate form $p_{Th} = \langle z \rangle p_{Tg}$. Here, p_{Tg} is the transverse momentum of the produced gluon, with $\langle z \rangle = 0.75$. The spectrum determination contains uncertainties, such as the appropriated description of the fragmentation process, the contribution of processes initiated by quarks at large p_T, and the determination of the mass of the gluon jet. However, in the ratio R_{xA}, these uncertainties are reasonably canceled, and the nuclear modification factor essentially provides the ratio of the UGDs in the nucleus and in the nucleon, $\phi_A(x, p_T, b)/\phi_p(x, p_T)$, at a given centrality.

Having determined the basic parameters and the formalism to compute the unintegrated gluon distribution in both nucleons and nuclei in the initial hard process, in the next Section, we present how the thermal effects are incorporated in the analysis.

2.2. Collective Expansion and the Blast-Wave Model

In [19–21], it has been argued that the p_T spectrum can be described by doing a time separation in the RTA of the Boltzmann transport equation [22]. In order to include the thermal corrections to the spectrum, predicted by Equation (1), We assume that the final state distribution f_{fin} can be expressed in the RTA,

$$f_{\text{fin}} = f_{\text{eq}} + (f_{\text{in}} - f_{\text{eq}})e^{-t_f/t_r}, \tag{7}$$

where t_f/t_r is the ratio of the freeze-out and relaxation times, with f_{in} given by Equation (1) and f_{eq} being the equilibrium distribution of the thermal system.

One way of investigating the collective properties on the spectrum is based in phenomenological models of blast-wave type [68], developed in order to capture the essential aspects of the thermal and hydrodynamic description of AA collisions. These models have been applied to heavy-ion collisions at RHIC [69–71], LHC [72–75] and CERN/SPS (European Organizationfor Nuclear Research/Super Proton Synchrotron) [76]. The velocity profile of the expanding medium is parametrized in the following way:

$$\rho = \tanh^{-1}(\beta_T) = \tanh^{-1}\left[\left(\frac{r}{R}\right)^m \beta_s\right], \tag{8}$$

where β_T is the transverse expansion velocity, m is the velocity profile's exponent, and β_s is the transverse expansion velocity at the surface. The average speed is $\langle \beta \rangle = \frac{2}{2+m}\beta_s$, r is the radial distance in the transverse plane from the centre of the fireball, and R is the fireballs radius. Here, a linear profile is considered, that is, $m = 1$. The spectrum of produced particles is given by the following:

$$f_{\text{eq}} \propto m_T \int_0^R r dr K_1 \left(\frac{m_T \cosh(\rho))}{T} \right) I_0 \left(\frac{p_T \sinh(\rho))}{T} \right), \tag{9}$$

where I_0 ard K_1 are the modified Bessel functions of the first and second kind, respectively, and T is the kinetic freeze-out temperature, T_{kin}, in the context of the Boltzmann-Gibbs blast-wave (BGBW) approach. In the RTA, considered here, $T = T_{\text{eq}}$ is the temperature characterizing the Boltzmann local equilibrium distribution, f_{eq}. It has been shown in [21] that for a given centrality, the temperature T_{eq} is larger than the kinetic one. This is consistent with the idea that the temperature decreases with the evolution of the system from the local equilibrium to the kinetic freeze-out. The spectrum is determined by the parameters $\langle \beta \rangle$ and T, which are adjusted from the experimental measurements.

As an exploratory study, the fitted parameters are presented in Table 1. A further study on possible constraints for the parameter ranges and discussion of data statistics is deserved. The expansion average velocity varies very little for the different systems. One obtains $\langle \beta \rangle \sim 0.55$, which is closer to the values determined in paper [29] for pions in *PbPb* collisions in $\sqrt{s} = 2.76$ TeV at the LHC. It is also noticed that in [31], the obtained values are $\langle \beta \rangle \sim 0.65$ in *pp* collisions, independently of the multiplicity. The temperature follows a trend similar to that observed for R_{xA} at large p_T (i.e., almost independent of the projectile species). This could suggest a relation between the energy loss in the large transverse momentum region and the equilibrium temperature of the system.

Table 1. Adjusted parameters' equilibrium temperature, $T = T_{\text{eq}}$, which characterizes the Boltzmann local equilibrium distribution; the average speed, $\langle \beta \rangle$; and the ratio of the freeze-out and relaxation times, t_f/t_r. Results are presented for three classes of centrality.

	(0–5)%			(0–20)%			
	$p + Al$	$p + Au$	$He + Au$	$p + Al$	$p + Au$	$d + Au$	$He + Au$
T (GeV)	0.054	0.055	0.041	0.043	0.046	0.045	0.035
$\langle \beta \rangle$	0.579	0.587	0.620	0.558	0.588	0.601	0.608
t_f/t_r	0.223	0.301	0.528	0.223	0.223	0.288	0.357

	(20–40)%			
	$p + Al$	$p + Au$	$d + Au$	$He + Au$
T (GeV)	0.032	0.028	0.045	0.034
$\langle \beta \rangle$	0.457	0.508	0.557	0.473
t_f/t_r	0.105	0.105	0.105	0.105

3. Results and Discussions

Our analysis is restricted to more central collisions (0–5, 0–20, 20–40)%, where the nuclear effects are more prominent. The average values of geometric parameters such as $\langle N_{\text{coll}} \rangle$ and $\langle N_{\text{part}} \rangle$ are calculated using a Glauber MC simulation, and they are explicitly presented in the PHENIX Collaboration papers (see Table II of [30] and page 6 of [77]). In Figure 1, the nuclear modification factor R_{xA} is shown for different systems, namely $p + Al$, $p + Au$, $d + Au$, and $He + Au$. For semi-central collisions such as (20–40)%, R_{xA} is close to unity at large p_T. This indicates the absence of relevant nuclear effects, which is independent of the projectile species. On the other hand, for the most central collisions, (0–5)%, there exists a suppression of R_{xA} at large p_T for all projectiles. This implies that the underlying mechanism of suppression does not depend on the colliding nucleus and has a strong dependence on centrality. In the figure, the lines represent the result obtained from Equations (6)–(9), including the thermal effects. In the relaxation time approximation,

$R_{xA} \sim e^{-t_f/t_r}$, the initial distribution scales with N_{coll}. This indicates that the relaxation time diminishes for more central collisions for any projectile. This can be understood supposing that the time should be inversely proportional to the energy density, which varies little for different colliding systems. For more peripheral reactions, the predicted Cronin peak is larger than that experimentally observed in the xAu collisions, which is related to the impact parameter dependence of the nuclear UGD. It decreases more slowly with the impact parameter than in case of the aluminium nucleus. In the small p_T region, it is observed a suppression due to the nuclear shadowing in the nUGD, followed by an enhancement near the maximum of the nUGD. In the initial distribution, an enhancement in the peak and a stronger shadowing effect are expected for the *He* nucleus. However, it was verified that the thermal contribution at small p_T contributes to diminishing these effects. A similar phenomenon occurs in heavy-ion collisions, as the produced particles in thermal equilibrium compensate the expected suppression due to strong nuclear shadowing from the nuclear UGD. In the case of small systems that are investigated here, the contribution of this piece, f_{eq}, to the spectrum is found to be small, $\sim 10\%$. This contribution is made clear if one looks at Figure 2, where the small p_T region of the spectrum is shown. There, the yield $dN_{xA}/d^2 p_T dy$ has been divided by N_{coll} and presented as a function of the transverse momentum. The dot-dashed lines represent the contribution of the particle produced in thermal equilibrium, and the solid line corresponds to the sum given by Equation (7). For less central collisions, where R_{xA} tends to unity at large p_T, such a contribution is small, and it is higher in more central collisions.

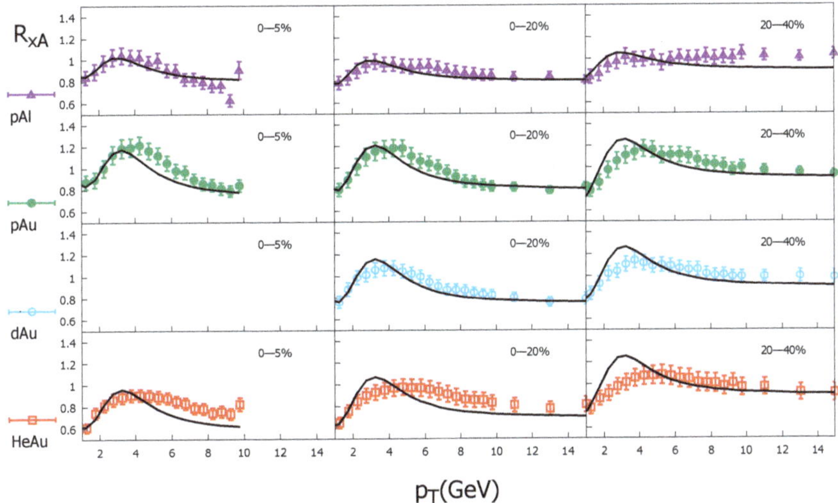

Figure 1. The nuclear modification factor, $R_{xA}(p_T)$, computed using Equations (6)–(9), with inclusion of thermal effects and comparison to data from PHENIX Collaboration [30]. Different projectile species and three more central classes of centrality are presented.

Each part of the spectrum has a distinct behavior in terms of the geometric parameter of the nuclear collision. The R_{xA} is analyzed in three different regions as a function of N_{coll}, as presented in Figure 3. At $p_T = 1.25$ GeV, where $R_{xA} < 1$, the dot-dashed lines correspond to the expected result by considering only f_{in}, which is determined by the nuclear shadowing present in the nuclear UGD. We draw attention to the fact that the contribution to R_{xA} from f_{in} presents fewer theoretical uncertainties compared to the absolute spectrum, as they are canceled in the ratio. In this limit, R_{xA} basically resembles the behavior of the ratio ϕ_A/ϕ_p at a given centrality class. For $N_{\text{coll}} \gtrsim 10$, data present the expected behavior, with a stronger suppression for increasing N_{coll}. However, in the case of $N_{\text{coll}} \lesssim 10$, the ratio R_{xA} is almost flat. For instance, in the case of $p + Al$, reaction R_{xA} increases for more central collisions compared to the peripheral ones. At $p_T = 4.25$ GeV, near the Cronin peak, the expected behavior is peak enhancement as N_{coll} increases. However, for $N_{coll} \gtrsim 10$, the observed behavior is the opposite; namely, the peak diminishes as the number of collisions increases. An interesting case is the nuclear modification factor at large p_T. The value $p_T = 9.75$ GeV is considered, where $R_{xA} \sim 1$ is expected due to the N_{coll}-scaling predicted by pQCD. However, the nuclear modification factor is substantially smaller than unity for the two more central classes of centrality. The most intriguing fact is that R_{xA} is independent of the projectile species and practically the same for each centrality. This suggests that R_{xA} in the large p_T region is testing observables, which have a poor dependence on the projectile or collision geometry, such as the multiplicity or transverse energy density. In [78], it was demonstrated that the obtained temperature in the Tsallis-blast-wave model scales with dN/η for pp, pA, and AA collisions but presents an intense reliance on the size of the colliding system.

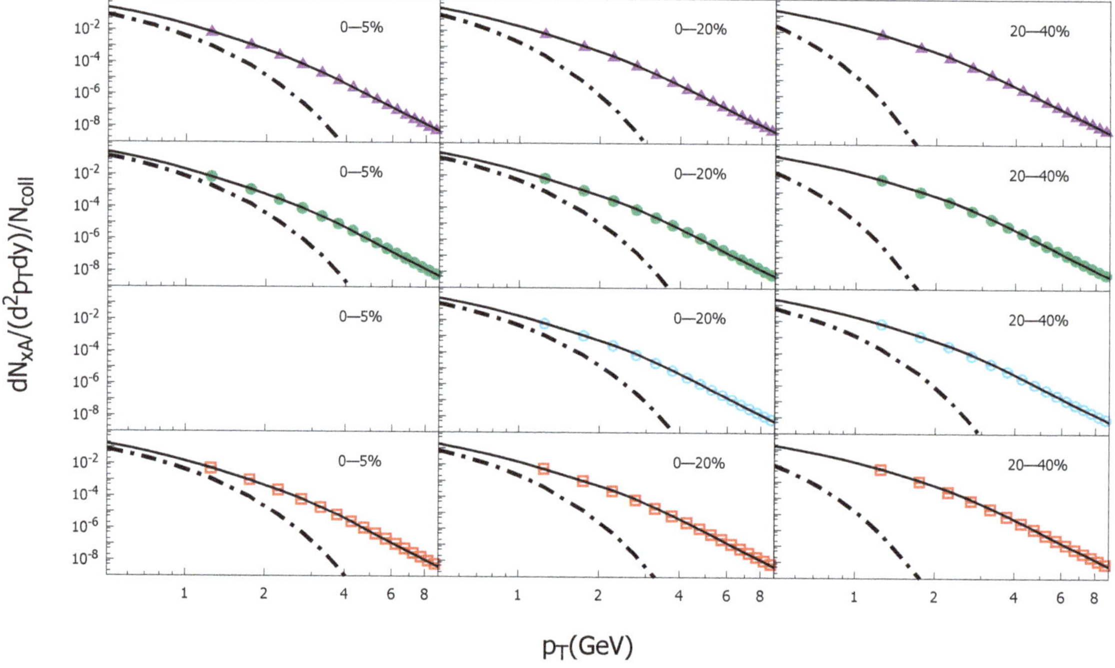

Figure 2. The yield $dN_{xA}/d^2 p_T dy$ for xA collisions divided by the number of binary collisions, N_{coll}, as a function of transverse momentum. Data from PHENIX Collaboration [30] at small transverse momentum region. Different projectile species and three more central classes of centrality are presented. The dot-dashed lines represent the contribution of the particle produced in thermal equilibrium, and the solid lines correspond to the sum given by Equation (7). See text for details.

Finally, the quantity $dN_{ch}/d\eta$ is studied, assuming that the charged hadron multiplicity (N_{ch}) is given predominantly by pions. Accordingly, the integrated pion spectra (after transformation $y \to \eta$) can be compared to the measured multiplicities. A comparison between predictions and the PHENIX data [77] is presented in Figure 4. In the figures, the black "×" symbols represent the results, and for better viewing, the dot-dashed lines correspond to their linear regression for each projectile species. For $N_{coll} \gtrsim 10$, the ratio $dN_{ch}/d\eta/N_{coll}$ follows the same pattern observed for R_{xA} at small p_T; namely, there is a scaling that is less steep on N_{coll}. In the intermediate region, $dN_{ch}/d\eta/N_{coll}$ is practically constant in terms of N_{coll}. In [79], an analysis of the energy loss parameter, δp_T, for different systems in AA collisions at RHIC and LHC (200 GeV) demonstrates that the energy loss imposed by the nuclear medium does not scale with the geometric parameters of the system (N_{coll}, N_{part}), but scales with quantities related to the system energy density and the multiplicity $dN_{ch}/d\eta$.

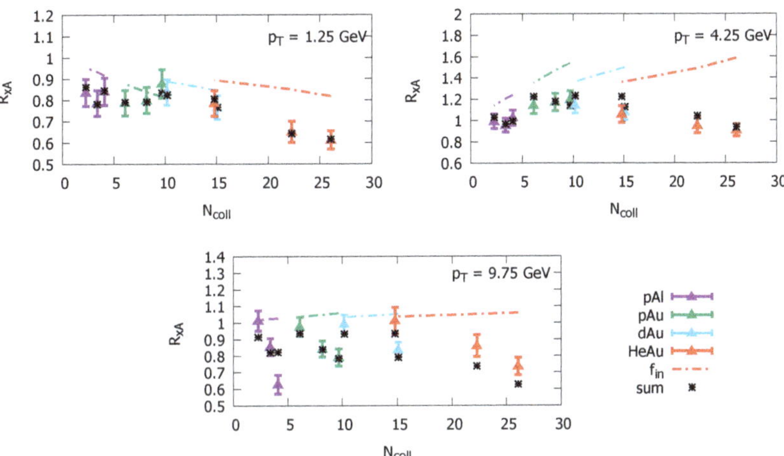

Figure 3. Nuclear modification factor as a function of the number of binary collisions for three values of transverse momentum, p_T. Dot-dashed lines represent the calculation without thermal effects by using only f_{in}. The black "×" symbols correspond to the full calculation given by Equations (6)–(9). Data from PHENIX Collaboration [30].

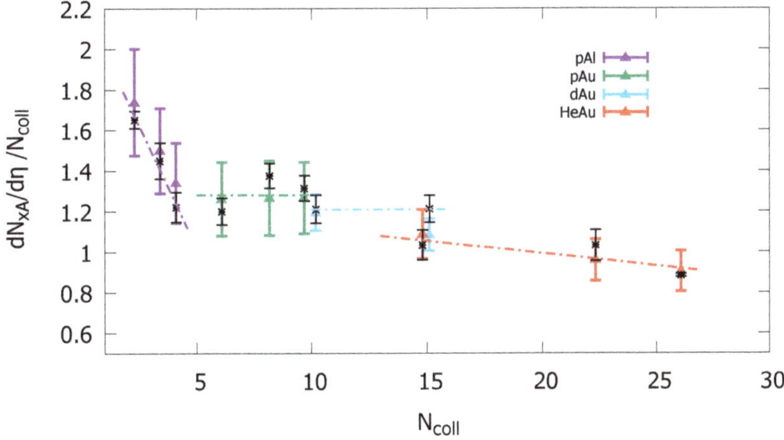

Figure 4. Multiplicity of charged hadrons measured by PHENIX Collaboration [77] as a function of N_{coll} for different projectile species and centrality classes. The black "×" symbols represent the numerical results, and for better viewing, the dot-dashed lines correspond to their linear regression for each projectile species.

4. Summary and Conclusions

In this paper, the relevance of the thermal effects is investigated in collisions of small systems based on the analysis of the transverse momentum spectra of neutral pions measured at RHIC (Relativistice Heavy Ion Collider. The Boltzmann equation in the relaxation time approximation has been considered. It was shown that the hard part contribution attributed to an initial production computed within the k_T-factorization formalism in pQCD (perturbative quantum chromodynamics) presents a different behavior from the one experimentally observed, even at large transverse momentum, p_T. The deviation can be understood in terms of enhancement/suppression of particles produced in the collective expansion of the thermal system. In particular, the Cronin peak tends to decrease in the case of larger-size projectiles, in opposition to what is expected, due to considering cold nuclear matter effects only. It is verified that the nuclear modification factor, R_{xA}, at $p_T \sim 10$ GeV does not depend on the projectile species. The same occurs for the thermal parameters of the system such as the relaxation time, t_r and temperature, T, which suggests a correlation between the energy loss at large p_T and the production of a thermal system of particles.

Author Contributions: L.M. and M.M. have contributed to the study equally, starting from the conceptualization of the problem, to the methodology, paper writing, and review. All authors have read and agreed to the published version of the manuscript.

Funding: This research was funded by the Brazilian National Council for Scientific and Technological Development (CNPq) under the contract number 306101/2018-1.

Data Availability Statement: The data used can be found in the corresponding references.

Conflicts of Interest: The authors declare no conflict of interest.

References

1. Khachatryan, V.; Sirunyan, A.M.; Tumasyan, A.; Adam, W.; Bergauer, T.; Dragicevic, M.; Erö, J.; Fabjan, C.; Friedl, M.; Frühwirth, R.; et al. Observation of long-range near-side angular correlations in proton-proton collisions at the LHC. *JHEP* **2010**, *9*, 91. [CrossRef]
2. Aad, G.; Abbott, B.; Abdallah, J.; Aben, R.; Abolins, M.; AbouZeid, O.S.; Abramowicz, H.; Abreu, H.; Abreu, R.; Abulaiti, Y.; et al. Observation of long-range elliptic azimuthal anisotropies in $\sqrt{s}$ =13 and 2.76 TeV *pp* collisions with the ATLAS Detector. *Phys. Rev. Lett.* **2016**, *116*, 172301. [CrossRef] [PubMed]
3. Khachatryan, V.; Sirunyan, A.M.; Tumasyan, A.; Adam, W.; Asilar, E.; Bergauer, T.; Brandstetter, J.; Brondolin, E.; Dragicevic, M.; Erö, J.; et al. Evidence for collectivity in *pp* collisions at the LHC. *Phys. Lett. B* **2017**, *765*, 193–220. [CrossRef]
4. Aaboud, M.; Aad, G.; Abbott, B.; Abeloos, B.; Abidi, S.H.; AbouZeid, O.S.; Abraham, N.L.; Abramowicz, H.; Abreu, H.; Abreu, R.; et al. Measurement of long-range multiparticle azimuthal correlations with the subevent cumulant method in *pp* and *p + Pb* collisions with the ATLAS detector at the CERN Large Hadron Collider. *Phys. Rev. C* **2018**, *97*, 024904. [CrossRef]
5. Adam, J.; Adamova, D.; Aggarwal, M.M.; Rinella, G.A.; Agnello, M.; Agrawal, N.; Ahammed, Z.; Ahmad, S.; Ahn, S.U.; Aiola, S.; Akindinov, A. Enhanced production of multi-strange hadrons in high-multiplicity proton-proton collisions. *Nat. Phys.* **2017**, *13*, 535–539. [CrossRef]
6. Acharya, S.; Adamová, D.; Adhya, S.P.; Adler, A.; Adolfsson, J.; Aggarwal, M.M.; Rinella, G.A.; Agnello, M.; Agrawal, N.; Ahammed, Z.; et al. Multiplicity dependence of (multi-)strange hadron production in proton-proton collisions at $\sqrt{s}$ = 13 TeV. *Eur. Phys. J. C* **2020**, *80*, 167. [CrossRef]
7. Cronin, J.; Frisch, H.J.; Shochet, M.; Boymond, J.; Mermod, R.; Piroue, P.; Sumner, R.L. Production of hadrons with large transverse momentum at 200, 300, and 400 GeV. *Phys. Rev. D* **1975**, *11*, 3105–3123. [CrossRef]
8. Wang, X.N. Systematic study of high p_T hadron spectra in *pp*, *pA* and AA collisions from SPS to RHIC energies. *Phys. Rev. C* **2000**, *61*, 064910. [CrossRef]
9. Kopeliovich, B.Z.; Nemchik, J.; Schafer, A.; Tarasov, A.V. Cronin effect in hadron production off nuclei. *Phys. Rev. Lett.* **2002**, *88*, 232303. [CrossRef]
10. Vitev, I.; Gyulassy, M. High p_T tomography of d + Au and Au + Au at SPS, RHIC, and LHC. *Phys. Rev. Lett.* **2002**, *89*, 252301. [CrossRef]
11. Kharzeev, D.; Kovchegov, Y.V.; Tuchin, K. Cronin effect and high-p_T suppression in *pA* collisions. *Phys. Rev.* **2003**, *D68*, 094013. [CrossRef]
12. Hwa, R.C.; Yang, C.B. Final state interaction as the origin of the Cronin effect. *Phys. Rev. Lett.* **2004**, *93*, 082302. [CrossRef]
13. Gelis, F.; Jalilian-Marian, J. From DIS to proton nucleus collisions in the color glass condensate model. *Phys. Rev. D* **2003**, *67*, 074019. [CrossRef]
14. Albacete, J.L.; Armesto, N.; Milhano, J.G.; Salgado, C.A.; Wiedemann, U.A. Numerical analysis of the Balitsky-Kovchegov equation with running coupling: Dependence of the saturation scale on nuclear size and rapidity. *Phys. Rev. D* **2005**, *71*, 014003. [CrossRef]

15. Baier, R.; Kovner, A.; Wiedemann, U.A. Saturation and parton level Cronin effect: Enhancement versus suppression of gluon production in *p-A* and *A-A* collisions. *Phys. Rev. D* **2003**, *68*, 054009. [CrossRef]

16. Blaizot, J.P.; Gelis, F.; Venugopalan, R. High-energy pA collisions in the color glass condensate approach. 1. Gluon production and the Cronin effect. *Nucl. Phys. A* **2004**, *743*, 13–56. [CrossRef]

17. Blaizot, J.P.; Iancu, E. The Quark gluon plasma: Collective dynamics and hard thermal loops. *Phys. Rep.* **2002**, *359*, 355–528. [CrossRef]

18. Jalilian-Marian, J.; Kovchegov, Y.V. Saturation physics and deuteron-gold collisions at RHIC. *Prog. Part. Nucl. Phys.* **2006**, *56*, 104–231. [CrossRef]

19. Tripathy, S.; Bhattacharyya, T.; Garg, P.; Kumar, P.; Sahoo, R.; Cleymans, J. Nuclear modification factor using Tsallis non-extensive statistics. *Eur. Phys. J. A* **2016**, *52*, 289. [CrossRef]

20. Tripathy, S.; Khuntia, A.; Tiwari, S.K.; Sahoo, R. Transverse momentum spectra and nuclear modification factor using Boltzmann transport equation with flow in Pb + Pb collisions at $\sqrt{s_{NN}}$ = 2.76 TeV. *Eur. Phys. J. A* **2017**, *53*, 99. [CrossRef]

21. Qiao, L.; Che, G.; Gu, J.; Zheng, H.; Zhang, W. Nuclear modification factor in Pb–Pb and p-Pb collisions using Boltzmann transport equation. *J. Phys. G Nucl. Part. Phys.* **2020**, *47*, 075101. [CrossRef]

22. Florkowski, W.; Ryblewski, R. Separation of elastic and inelastic processes in the relaxation-time approximation for the collision integral. *Phys. Rev. C* **2016**, *93*, 064903. [CrossRef]

23. Bylinkin, A.; Chernyavskaya, N.; Rostovtsev, A. Two components in charged particle production in heavy-ion collisions. *Nucl. Phys. B* **2016**, *903*, 204–210. [CrossRef]

24. Giannini, A.V.; Goncalves, V.P.; Silva, P.V.R.G. Thermal radiation and inclusive production in the running coupling k_T-factorization approach. *Eur. Phys. J. A* **2021** *57*, 43. [CrossRef]

25. Urmossy, K.; Barnaföldi, G.G.; Harangozó, S.; Biró, T.S.; Xu, Z. A 'soft + hard' model for heavy-ion collisions. *J. Phys. Conf. Ser.* **2017**, *805*, 012010. [CrossRef]

26. Adams, J.; Aggarwal, M.M.; Ahammed, Z.; Amonett, J.; Anderson, B.D.; Anderson, M.; Arkhipkin, D.; Averichev, G.S.; Bai, Y.; Balewski, J.; et al. The Multiplicity dependence of inclusive p_t spectra from *pp* collisions at $\sqrt{s}$ = 200-GeV. *Phys. Rev. D* **2006**, *74*, 032006. [CrossRef]

27. Trainor, T.A. Centrality evolution of p(t) and y(t) spectra from Au-Au collisions at s(NN)**(1/2) = 200-GeV. *Int. J. Mod. Phys. E* **2008**, *17*, 1499–1540. [CrossRef]

28. Trainor, T.A. A two-component model for identified-hadron $\mathbf{p_t}$ spectra from 5 TeV p-Pb collisions. *J. Phys. G* **2020**, *47*, 045104. [CrossRef]

29. Moriggi, L.S.; Peccini, G.M.; Machado, M.V.T. Role of nuclear gluon distribution on particle production in heavy ion collisions. *Phys. Rev. D* **2021**, *103*, 034025. [CrossRef]

30. Acharya, U.A.; Adare, A.; Aidala, C.; Ajitan, N.N.; Akiba, Y.; Al-Bataineh, H.; Alexander, J.; Alfred, M.; Andrieux, V.; Angerami, A.; et al. Systematic study of nuclear effects in p +Al, p +Au, d +Au, and ^{3}He+Au collisions at $\sqrt{s_{NN}}$ = 200 GeV using π^0 production. *Phys. Rev. C* **2022**, *105*, 064902. [CrossRef]

31. Rath, R.; Khuntia, A.; Sahoo, R.; Cleymans, J. Event multiplicity, transverse momentum and energy dependence of charged particle production, and system thermodynamics in *pp* collisions at the Large Hadron Collider. *J. Phys. G* **2020**, *47*, 055111. [CrossRef]

32. Jiang, K.; Zhu, Y.; Liu, W.; Chen, H.; Li, C.; Ruan, L.; Tang, Z.; Xu, Z.; Xu, Z. Onset of radial flow in p+p collisions. *Phys. Rev. C* **2015**, *91*, 024910. [CrossRef]

33. Eskola, K.J.; Honkanen, H. A Perturbative QCD analysis of charged particle distributions in hadronic and nuclear collisions. *Nucl. Phys.* **2003**, *A713*, 167–187. [CrossRef]

34. Helenius, I.; Eskola, K.J.; Paukkunen, H. Centrality dependence of inclusive prompt photon production in d+Au, Au+Au, p+Pb, and Pb+Pb collisions. *JHEP* **2013**, *5*, 30. [CrossRef]

35. Helenius, I.; Paukkunen, H.; Eskola, K.J. Nuclear PDF constraints from p+Pb collisions at the LHC. *arXiv* **2015**, arXiv:1509.02798. [CrossRef]

36. Kovarik, K.; Kusina, A.; Ježo, T.; Clark, D.B.; Keppel, C.; Lyonnet, F.; Morfín, J.G.; Olness, F.I.; Owens, J.F.; Schienbein, I.; et al. nCTEQ15—Global analysis of nuclear parton distributions with uncertainties in the CTEQ framework. *Phys. Rev. D* **2016**, *93*, 085037. [CrossRef]

37. Abdul Khalek, R.; Ethier, J.J.; Rojo, J. Nuclear parton distributions from lepton-nucleus scattering and the impact of an electron-ion collider. *Eur. Phys. J. C* **2019**, *79*, 471. [CrossRef]

38. Vitev, I.; Zhang, B.W. A Systematic study of direct photon production in heavy ion collisions. *Phys. Lett. B* **2008**, *669*, 337–344. [CrossRef]

39. Zhang, Y.; Fai, G.I.; Papp, G.; Barnafoldi, G.G.; Levai, P. High p_T pion and kaon production in relativistic nuclear collisions. *Phys. Rev. C* **2002**, *65*, 034903. [CrossRef]

40. Glauber, R. Cross-sections in deuterium at high-energies. *Phys. Rev.* **1955**, *100*, 242–248. [CrossRef]

41. Mueller, A.H. Small-x Behavior and Parton Saturation: A QCD Model. *Nucl. Phys. B* **1990**, *335*, 115–137. [CrossRef]

42. Albacete, J.L.; Armesto, N.; Kovner, A.; Salgado, C.A.; Wiedemann, U.A. Energy dependence of the Cronin effect from nonlinear QCD evolution. *Phys. Rev. Lett.* **2004**, *92*, 082001. [CrossRef]

43. Lappi, T.; Mäntysaari, H. Single inclusive particle production at high energy from HERA data to proton-nucleus collisions. *Phys. Rev. D* **2013**, *88*, 114020. [CrossRef]

44. Dumitru, A.; Hayashigaki, A.; Jalilian-Marian, J. Geometric scaling violations in the central rapidity region of d + Au collisions at RHIC. *Nucl. Phys.* **2006**, *A770*, 57–70. [CrossRef]

45. Goncalves, V.P.; Kugeratski, M.S.; Machado, M.V.T.; Navarra, F.S. Saturation physics at HERA and RHIC: An nified description. *Phys. Lett.* **2006**, *B643*, 273–278. [CrossRef]

46. Boer, D.; Utermann, A.; Wessels, E. Geometric Scaling at RHIC and LHC. *Phys. Rev.* **2008**, *D77*, 054014. [CrossRef]

47. Rezaeian, A.H. CGC predictions for p+A collisions at the LHC and signature of QCD saturation. *Phys. Lett. B* **2013**, *718*, 1058–1069. [CrossRef]

48. Kharzeev, D.; Kovchegov, Y.V.; Tuchin, K. Nuclear modification factor in d+Au collisions: Onset of suppression in the color glass condensate. *Phys. Lett. B* **2004**, *599*, 23–31. [CrossRef]

49. Armesto, N.; Salgado, C.A.; Wiedemann, U.A. Relating high-energy lepton-hadron, proton-nucleus and nucleus-nucleus collisions through geometric scaling. *Phys. Rev. Lett.* **2005**, *94*, 022002. [CrossRef]

50. Albacete, J.L.; Dumitru, A. A model for gluon production in heavy-ion collisions at the LHC with rcBK unintegrated gluon densities. *arXiv* **2010**, arXiv.1011.5161. [CrossRef]

51. Kharzeev, D.; Levin, E.; Nardi, M. Color glass condensate at the LHC: Hadron multiplicities in pp, pA and AA collisions. *Nucl. Phys.* **2005**, *A747*, 609–629. [CrossRef]

52. Tribedy, P.; Venugopalan, R. QCD saturation at the LHC: Comparisons of models to $p + p$ and $A + A$ data and predictions for $p + Pb$ collisions. *Phys. Lett. B* **2012**, *710*, 125–133. Erratum in *Phys. Lett. B* **2013**, *718*, 1154–1154. [CrossRef]

53. Lappi, T. Energy dependence of the saturation scale and the charged multiplicity in pp and AA collisions. *Eur. Phys. J. C* **2011**, *71*, 1699. [CrossRef]

54. Levin, E.; Rezaeian, A.H. Gluon saturation and energy dependence of hadron multiplicity in pp and AA collisions at the LHC. *Phys. Rev. D* **2011**, *83*, 114001. [CrossRef]

55. Albacete, J.L.; Marquet, C. Single inclusive hadron production at RHIC and the LHC from the color glass condensate. *Phys. Lett. B* **2010**, *687*, 174–179. [CrossRef]

56. Durães, F.; Giannini, A.; Goncalves, V.; Navarra, F. Testing the running coupling k_T-factorization formula for the inclusive gluon production. *Phys. Rev. D* **2016**, *94*, 054023. [CrossRef]

57. Czech, M.; Szczurek, A. Unintegrated CCFM parton distributions and pion production in proton-proton collisions at high energies. *Phys. Rev. C* **2005**, *72*, 015202. [CrossRef]

58. Czech, M.; Szczurek, A. Unintegrated parton distributions and pion production in pp collisions at RHIC's energies. *J. Phys. G* **2006**, *32*, 1253–1268. [CrossRef]

59. Moriggi, L.; Peccini, G.; Machado, M. Investigating the inclusive transverse spectra in high-energy pp collisions in the context of geometric scaling framework. *Phys. Rev. D* **2020**, *102*, 034016. [CrossRef]

60. McLerran, L.; Praszalowicz, M. Saturation and scaling of multiplicity, mean p_T and p_T distributions from $200\,\mathrm{GeV} \leq \sqrt{s} \leq 7\,\mathrm{TeV}$. *Acta Phys. Polon. B* **2010**, *41*, 1917–1926.

61. McLerran, L.; Praszalowicz, M. Saturation and scaling of multiplicity, mean p_T and p_T distributions from $200\,\mathrm{GeV} \leq \sqrt{s} \leq 7\,\mathrm{TeV}$—Addendum. *Acta Phys. Polon. B* **2011**, *42*, 99–103. [CrossRef]

62. Praszałowicz, M.; Francuz, A. Geometrical Scaling in Inelastic Inclusive Particle Production at the LHC. *Phys. Rev. D* **2015**, *92*, 074036. [CrossRef]

63. Staśto, A.M.; Golec-Biernat, K.J.; Kwiecinski, J. Geometric scaling for the total $\gamma^* p$ cross section in the low x region. *Phys. Rev. Lett.* **2001**, *86*, 596–599. [CrossRef] [PubMed]

64. De Jager, C.; De Vries, H.; De Vries, C. Nuclear charge and magnetization density distribution parameters from elastic electron scattering. *Atom. Data Nucl. Data Tabl.* **1974**, *14*, 479–508. Erratum in *Atom. Data Nucl. Data Tabl.* **1975**, *16*, 580–580. [CrossRef]

65. De Vries, H.; De Jager, C.; De Vries, C. Nuclear charge-density-distribution parameters from elastic electron scattering. *At. Data Nucl. Data Tables* **1987**, *36*, 495–536. [CrossRef]

66. Hulthén, L.; Sugawara, M. The two-nucleon problem. In *Structure of Atomic Nuclei/Bau der Atomkerne*; Springer: Berlin/Heidelberg, Germany, 1957; Volume 8/39, pp. 1–143. [CrossRef]

67. McCarthy, J.S.; Sick, I.; Whitney, R.R. Electromagnetic structure of the helium isotopes. *Phys. Rev. C* **1977**, *15*, 1396–1414. [CrossRef]

68. Schnedermann, E.; Sollfrank, J.; Heinz, U.W. Thermal phenomenology of hadrons from $200A$ GeV S+S collisions. *Phys. Rev. C* **1993**, *48*, 2462–2475. [CrossRef]

69. Abelev, B.I.; Aggarwal, M.M.; Ahammed, Z.; Anderson, B.D.; Arkhipkin, D.; Averichev, G.S.; Bai, Y.; Balewski, J.; Barannikova, O.; Barnby, L.S.; et al. Systematic measurements of identified particle spectra in pp, $d+$ Au and Au+Au collisions from STAR. *Phys. Rev. C* **2009**, *79*, 034909. [CrossRef]

70. Adams, J.; Adler, C.; Aggarwal, M.M.; Ahammed, Z.; Amonett, J.; Anderson, B.D.; Anderson, M.; Arkhipkin, D.; Averichev, G.S.; Badyal, S.K.; et al. Identified particle distributions in pp and Au + Au collisions at $\sqrt{s_{NN}} = 200$ GeV. *Phys. Rev. Lett.* **2004**, *92*, 112301. [CrossRef]

71. Adamczyk, L.; Adkins, J.K.; Agakishiev, G.; Aggarwal, M.M.; Ahammed, Z.; Ajitan, N.N.; Alekseev, I.; Anderson, D.M.; Aoyama, R.; Aparin, A.; et al. Bulk properties of the medium produced in relativistic heavy-ion collisions from the beam energy scan program. *Phys. Rev. C* **2017**, *96*, 044904. [CrossRef]
72. Abelev, B.; Adam, J.; Adamová, D.; Adare, A.M.; Aggarwal, M.M.; Rinella, G.A.; Agnello, M.; Agocs, A.G.; Agostinelli, A.; Ahammed, Z.; et al. Centrality dependence of π, K, p production in Pb-Pb collisions at $\sqrt{s_{NN}}$ = 2.76 TeV. *Phys. Rev. C* **2013**, *88*, 044910. [CrossRef]
73. Acharya, S.; Adamová, D.; Adhya, S.P.; Adler, A.; Adolfsson, J.; Aggarwal, M.M.; Rinella, G.A.; Agnello, M.; Agrawal, N.; Ahammed, Z.; et al. Production of charged pions, kaons, and (anti-)protons in Pb-Pb and inelastic pp collisions at $\sqrt{s_{NN}}$ = 5.02 TeV. *Phys. Rev. C* **2020**, *101*, 044907. [CrossRef]
74. Abelev, B.; Adam, J.; Adamová, D.; Adare, A.M.; Aggarwal, M.M.; Rinella, G.A.; Agnello, M.; Agocs, A.G.; Agostinelli, A.; Ahammed, Z.; et al. Multiplicity dependence of pion, kaon, proton and lambda production in p-Pb collisions at $\sqrt{s_{NN}}$ = 5.02 TeV. *Phys. Lett. B* **2014**, *728*, 25–38. [CrossRef]
75. Acharya, S.; Adamová, D.; Adler, A.; Adolfsson, J.; Aggarwal, M.M.; Rinella, G.A.; Agnello, M.; Agrawal, N.; Ahammed, Z.; Ahn, S.U.; et al. Multiplicity dependence of light-flavor hadron production in pp collisions at $\sqrt{s}$ = 7 TeV. *Phys. Rev. C* **2019**, *99*, 024906. [CrossRef]
76. Bearden, I.G.; Bøggild, H.; Boissevain, J.; Dodd, J.; Erazmus, B.; Esumi, S.; Fabjan, C.W.; Ferenc, D.; Fields, D.E.; Franz, A.; et al. Collective expansion in high-energy heavy ion collisions. *Phys. Rev. Lett.* **1997**, *78*, 2080–2083. [CrossRef]
77. Adare, A.; Aidala, C.; Ajitan, N.N.; Akiba, Y.; Alfred, M.; Andrieux, V.; Aoki, K.; Apadula, N.; Asano, H.; Ayuso, C.E.; et al. Pseudorapidity dependence of particle production and elliptic flow in asymmetric nuclear collisions of $p + Al$, $p + Au$, $d + Au$, and $^3He + Au$ at $\sqrt{s_{NN}}$ = 200 GeV. *Phys. Rev. Lett.* **2018**, *121*, 222301. [CrossRef]
78. Gu, J.B.; Li, C.Y.; Wang, Q.; Zhang, W.C.; Zheng, H. Collective expansion in pp collisions using the Tsallis statistics. *arXiv* **2022**, arXiv:2201.02091. [CrossRef]
79. Adare, A.; Afanasiev, S.; Aidala, C.; Ajitan, N.N.; Akiba, Y.; Akimoto, R.; Al-Bataineh, H.; Alexander, J.; Alfred, M.; Al-Ta'ani, H.; et al. Scaling properties of fractional momentum loss of high-p_T hadrons in nucleus-nucleus collisions at $\sqrt{s_{NN}}$ from 62.4 GeV to 2.76 TeV. *Phys. Rev. C* **2016**, *93*, 024911. [CrossRef]

Article

Analytical Calculations of the Quantum Tsallis Thermodynamic Variables

Ayman Hussein [1] and Trambak Bhattacharyya [2,*]

[1] Zewail City of Science, Technology and Innovation, Giza 12578, Egypt; s-ayman.mh@zewailcity.edu.eg
[2] Bogoliubov Laboratory of Theoretical Physics, Joint Institute for Nuclear Research, Dubna 141980, Russia
* Correspondence: bhattacharyya@theor.jinr.ru

Abstract: In this paper, we provide an account of analytical results related to the Tsallis thermodynamics that have been the subject matter of a lot of studies in the field of high-energy collisions. After reviewing the results for the classical case in the massless limit and for arbitrarily massive classical particles, we compute the quantum thermodynamic variables. For the first time, the analytical formula for the pressure of a Tsallis-like gas of massive bosons has been obtained. The study serves both as a brief review of the knowledge gathered in this area, and as original research that forwards the existing scholarship. The results of the present paper will be important in a plethora of studies in the field of high-energy collisions including the propagation of non-linear waves generated by the traversal of high-energy particles inside the quark-gluon plasma medium showing the features of non-extensivity.

Keywords: Tsallis statistics; Tsallis thermodynamics; thermodynamic variables; integral representation

Citation: Hussein, A.; Bhattacharyya, T. Analytical Calculations of the Quantum Tsallis Thermodynamic Variables. *Physics* **2022**, *4*, 800–811. https://doi.org/10.3390/physics4030051

Received: 27 April 2022
Accepted: 17 June 2022
Published: 19 July 2022

Publisher's Note: MDPI stays neutral with regard to jurisdictional claims in published maps and institutional affiliations.

1. Introduction

Power-law distributions have been routinely used to describe particle yields in high-energy collision physics. It has been observed that the pions, kaons, protons (and other hadrons) originated in these collision events follow a power-law distribution in the transverse momentum (p_T) space. The power-law formula, utilized by experiments such as STAR [1], PHENIX [2], ALICE [3] and CMS [4], use the following form of a power-law transverse momentum distribution,

$$\frac{d^2N}{dp_T dy} = p_T \frac{dN}{dy} \frac{(n-1)(n-2)}{nC(nC + m_0(n-2))} \left(1 + \frac{m_T - m_0}{nC} \right)^{-n}, \tag{1}$$

which has some correspondence with the form of the Tsallis transverse momentum distribution, proposed by Cleymans and Worku in 2012 [5,6],

$$\frac{d^2N}{dp_T dy} = \frac{gV}{(2\pi)^2} p_T m_T \cosh y \left(1 + (q-1)\frac{m_T \cosh y - \mu}{T} \right)^{-\frac{q}{q-1}}. \tag{2}$$

In Equations (1) and (2), n, C, m_0, V (volume), q (Tsallis parameter), T (Tsallis temperature) and μ (chemical potential) are fit parameters, g is degeneracy, $m_T = \sqrt{p_T^2 + m^2}$ is the transverse mass of a particle with the mass m and y is rapidity. These distributions can be associated with the Tsallis statistical mechanics, developed by C. Tsallis in 1988 [7], a statistics that has long been used to tackle a medium with fluctuation, long-range correlation [8–11], small system size [12] and fractal structure [13]. It has been shown that the Tsallis-like distribution, proposed in Refs. [5,6], obeys thermodynamic relations. Later on, from the definition of the Tsallis entropy, the distribution (2) has been shown to be the zeroth-order approximation of the exact Tsallis-like transverse momentum distribution in the Tsallis-2 prescription [14] that was shown to be rather useful for the Large Hadron

Collider (LHC) phenomenology [15–19]. The same distribution can also be obtained from the q-dual statistics, proposed in Ref. [20]. In the papers that spearheaded the study of Tsallis thermodynamics, there were comparisons between the Tsallis-like classical and quantum distributions and their Boltzmann–Gibbs (BG) counterparts (see, e.g., [5]) which is achieved once the q-parameter approaches unity.

A reader may also have felt an implicit necessity to be able to compare the thermodynamic variables in the two formulations—Tsallis and BG. One might always take a numerical approach to provide an answer, as the analytical formulae of the Tsallis thermodynamic variables were not widely available, as opposed to their Boltzmann–Gibbs counterparts, which are expressible in terms of the modified Bessel functions. However, this does not mean that there were no attempts to find analytical results. Lavagno already in 2002 provided these expressions in terms of the q-modified Bessel functions of the second kind [21]. However, the properties of this group of q-modified special functions is not widely available and known to physicists. Hence, it was necessary to explore this question further. Thermodynamic variables are important as their relationships form the equation of state that is an important input to study, for example, the evolution of the quark-gluon plasma (QGP) medium [22–28], propagation of non-linear waves in the QGP using the hydrodynamic equation [29–35].

There was a renewed interest in this attempt during Professor Jean Cleymans' visit to India in 2015. This attempt was based on the observation that the Tsallis-like distributions can be written using the Taylor's series expansion in the increasing order of $(q-1)^n$, $n \in \mathbb{Z}^{\geq}$ ($\mathbb{Z}^{\geq}$ represents the set of non-negative integers). In a joint paper [36], the Tsallis thermodynamic variables for an ideal gas of massive particles were explored. However, eventually it was realized that the results were restricted by the fact that the Tsallis distribution was truncated at $\mathcal{O}(q-1)^2$. Such an early truncation led to restrictions in the phase space dictated by the ratios involving q, T and the single-particle energy, E_p; see also [10]. Therefore, the question of finding an unapproximated analytical expression of the Tsallis thermodynamic variables was still open. In the meantime, one of the authors (T.B.) joined the group of Jean Cleymans in Cape Town, and some progress ensued. Cleymans proved that the calculations for the massless case can be performed analytically, and that was a breakthrough. This was the inspiration behind another paper in collaboration with him [37] that elaborated a method to analytically calculate the Tsallis thermodynamic variables for the massive particles without an approximation (like the Taylor's series expansion, or considering the massless case) using the Mellin–Barnes contour integral representation of the Tsallis distribution. It was found that the interesting features (like poles) that were missed in the Taylor approximated calculations are intact in the massless as well as the massive case. In this paper, we extend the existing knowledge to the quantum domain, and propose a method to calculate analytical formulae of the quantum Tsallis thermodynamic variables. In the present study, isotropic momentum distributions are considered. So, the results can also be useful for the quark-gluon plasma medium formed in the early universe [38] or other branches of physics that use isotropic distributions [39–43]. For a more realistic scenario to treat high-energy collision physics, anisotropy may be considered but we reserve that for future studies.

Being involved in such a journey with Jean Cleymans as a friend, philosopher and guide is truly rewarding. The present paper serves as our tribute to the memory and inspiring scientific curiosity of Jean Cleymans.

2. Review: Tsallis Thermodynamics: $m = 0, \mu = 0$

The Tsallis thermodynamic variables can be written in terms of the Tsallis single-particle distribution. The single-particle distributions can be obtained following three different averaging schemes [44], which we name Tsallis-1, 2 and 3. These schemes differ in the definition of the mean values (e.g., the mean energy), utilized for the constrained maximization of the Tsallis entropy. In the first scheme, the mean is defined as $\langle O \rangle = \sum_i p_i O_i$, in the second scheme $\langle O \rangle = \sum_i p_i^q O_i$ and in the third scheme

$\langle O \rangle = \sum_i p_i^q O_i / \sum_i p_i^q$, where $\{p_i\}$ are the probabilities of micro-states. It is worthwhile to mention that following the previous studies [5,6], we follow the second averaging scheme. The present paper focuses entirely on the analytical method to calculate the Tsallis thermodynamic variables, and based on this prescription, it is relatively straightforward to extend the calculations for other forms of the single-particle distributions. With this understanding, we consider the following isotropic quantum single-particle distributions (positive sign for fermions (f), negative sign for bosons (b)):

$$n_{b/f} = \frac{1}{\left(1 + (q-1)\frac{E_p - \mu}{T}\right)^{\frac{q}{q-1}} \pm 1}, \tag{3}$$

where $E_p = \sqrt{p^2 + m^2}$ is the single-particle energy of a particle with the mass, m, and the three-momentum, p (with magnitude p). This distribution is not entirely phenomenological because it can be obtained (after certain approximations) from the constrained maximization of the Tsallis entropy, as shown in [45]. The distribution (3) is similar to (but not exactly the same as) the one, proposed in [46,47]. For the sake of completeness, we also quote the popular classical (Maxwell–Boltzmann, MB) Tsallis-like single-particle distribution that gives rise to Equation (2):

$$n_{MB} = \left(1 + (q-1)\frac{E_p - \mu}{T}\right)^{-\frac{q}{q-1}}. \tag{4}$$

With the help of the single-particle distributions (n_s; s = b, f, MB), the thermodynamic variables such as the pressure, P, the mean energy, E, and the mean number of particles, N, can be expressed as follows:

$$P = g \int \frac{d^3 p}{(2\pi)^3} \frac{p^2}{3\,E_p} n_s; \quad E = gV \int \frac{d^3 p}{(2\pi)^3} E_p\, n_s; \quad N = gV \int \frac{d^3 p}{(2\pi)^3} n_s. \tag{5}$$

2.1. Classical Case

In this Section we tabulate the results for the classical (Tsallis Maxwell–Boltzmann) case in the massless approximation. The pressure, P, the energy density, $\epsilon = E/V$, and the number density, $\rho = N/V$, with $\mu = 0$ are given by [37]

$$P = \frac{gT^4}{6\pi^2} \frac{1}{(2-q)(3/2-q)(4/3-q)} \tag{6}$$

$$\epsilon = \frac{gT^4}{2\pi^2} \frac{1}{(2-q)(3/2-q)(4/3-q)} = 3P \tag{7}$$

$$\rho = \frac{gT^3}{2\pi^2} \frac{1}{(2-q)(3/2-q)}. \tag{8}$$

Interestingly, from the above expressions (e.g., for P), the first pole of q appears at $q = 4/3$, which is close to $1.\overline{3}$ [48]. This puts an upper-bound on the q value that is a parameter to be determined from the experimental data. Experimental observations indeed show that q values do obey this upper-bound which is imposed because of the finite values of the thermodynamic variables. However, other considerations may further shrink the range [49]. It is also noteworthy that the upper-bound of q (denoted as $q_{max}^{(D)}$), obtained from the thermodynamic considerations, changes with the dimension of the system as $q_{max}^{(D)} < 1 + 1/(D-1)$. Hence, for $D = 4$, the value of $q_{max}^{(D)}$ is $4/3$.

2.2. Quantum Case

Tsallis quantum thermodynamic variables in the massless limit are given by the following closed analytic formulae; see Ref. [34] for deatils.

2.2.1. Bosons

$$P_b = \frac{gT^4}{6\pi^2(q-1)^3 q}\left[3\psi^{(0)}\left(\frac{3}{q}-2\right) + \psi^{(0)}\left(\frac{1}{q}\right) - 3\psi^{(0)}\left(\frac{2}{q}-1\right) - \psi^{(0)}\left(\frac{4}{q}-3\right)\right], \quad (9)$$

$$\epsilon_b = \frac{gT^4}{2\pi^2(q-1)^3 q}\left[3\psi^{(0)}\left(\frac{3}{q}-2\right) + \psi^{(0)}\left(\frac{1}{q}\right) - 3\psi^{(0)}\left(\frac{2}{q}-1\right) - \psi^{(0)}\left(\frac{4}{q}-3\right)\right], \quad (10)$$

$$\rho_b = \frac{gT^3}{2\pi^2(q-1)^2 q}\left[2\psi^{(0)}\left(\frac{2}{q}-1\right) - \psi^{(0)}\left(\frac{3}{q}-2\right) - \psi^{(0)}\left(\frac{1}{q}\right)\right]. \quad (11)$$

2.2.2. Fermions

$$P_f = \frac{gT^4}{6\pi^2(q-1)^3 q}\left[3\Phi\left(-1,1,\frac{2}{q}-1\right) - 3\Phi\left(-1,1,\frac{3}{q}-2\right) + \Phi\left(-1,1,\frac{4}{q}-3\right)\right.$$
$$\left. - \Phi\left(-1,1,\frac{1}{q}\right)\right], \quad (12)$$

$$\epsilon_f = \frac{gT^4}{2\pi^2(q-1)^3 q}\left[3\Phi\left(-1,1,\frac{2}{q}-1\right) - 3\Phi\left(-1,1,\frac{3}{q}-2\right) + \Phi\left(-1,1,\frac{4}{q}-3\right)\right.$$
$$\left. - \Phi\left(-1,1,\frac{1}{q}\right)\right], \quad (13)$$

$$\rho_f = \frac{gT^3}{2\pi^2(q-1)^2 q}\left[-2\Phi\left(-1,1,\frac{2}{q}-1\right) + \Phi\left(-1,1,\frac{3}{q}-2\right) + \Phi\left(-1,1,\frac{1}{q}\right)\right]. \quad (14)$$

Here $\psi^{(0)}(z)$ is the digamma function, and $\Phi(a,b,z)$ is Lerch's transcendent [50], both having the poles at $z = 0$. One observes that, similar to the classical case, the first pole in the thermodynamic variables (for example, pressure) appears at $q = 4/3$. Hence, the upper-bound $q < 4/3$ is still relevant. Before moving to the next Section, let us comment that the above results can be extended to treat very light particles, as discussed in Ref. [34]. It is possible to obtain the $\mathcal{O}(m^2 T^2)$ correction to the above approximated results, which may also work for the light quarks such as up and down.

3. Review and New Results: Tsallis Thermodynamics: $m \neq 0, \mu = 0$

In this Secton, we quote the closed analytical formula of the pressure in a gas of massive classical and quantum particles without utilizing any approximation. The classical case has already been considered earlier [37]. However, we are not aware of any other results for the quantum case with arbitrarily massive particles (albeit results are available for slightly massive particles). In this Secton, we only quote the obtained results for classical and quantum (boson) particles. Detailed mathematical steps for obtaining quantum results are described in Section 4.

3.1. Classical Case (Review)

In this Section, we tabulate analytical results of classical Tsallis thermodynamics for arbitrarily massive particles. The calculation involves an integral representation of a power-law function that appears in the Tsallis statistics. Due to the nature of the integrals involved, the convergence conditions lead to two separate formulae for the thermodynamic variables for two regions $q > 1 + T/m$, which we call the "upper region" and $q \leq 1 + T/m$, which we call the "lower region". The origin of these regions are explained in Section 4, where the quantum case is considered.

3.1.1. Upper Region: $q > 1 + T/m$

The analytical expression valid for the upper region is:

$$
\begin{aligned}
P_{\mathrm{U}} = \frac{g\,m^4}{16\pi^{\frac{3}{2}}} \left(\frac{T}{(q-1)\,m}\right)^{\frac{q}{q-1}} &\left[\frac{\Gamma\left(\frac{4-3q}{2(q-1)}\right)}{\Gamma\left(\frac{2q-1}{2(q-1)}\right)}\; {}_2F_1\left(\frac{q}{2(q-1)},\frac{4-3q}{2(q-1)};\frac{1}{2};\frac{T^2}{(q-1)^2 m^2}\right)\right. \\
&\left. -\frac{2T}{(q-1)m}\times\frac{\Gamma\left(\frac{3-2q}{2(q-1)}\right)}{\Gamma\left(\frac{q}{2(q-1)}\right)}\; {}_2F_1\left(\frac{2q-1}{2(q-1)},\frac{3-2q}{2(q-1)};\frac{3}{2};\frac{T^2}{(q-1)^2 m^2}\right)\right],
\end{aligned}
\tag{15}
$$

where $\Gamma(z)$ is the Gamma function, and ${}_2F_1(a,b,c;z)$ is the hypergeometric function [50].

3.1.2. Lower Region: $q \le 1 + T/m$

The analytical expression valid for the lower region is:

$$
P_{\mathrm{L}} = \frac{g\,m^4}{2^{\frac{q}{q-1}}\pi^{\frac{3}{2}}} \left[\frac{(q-1)^2\,(3-q)\,\Gamma\left(\frac{1}{q-1}\right)}{(4-3q)\,(3-2q)\,(2-q)\Gamma\left(\frac{1+q}{2(q-1)}\right)}\right]
$$
$$
\times\; {}_2F_1\left(\frac{2q-1}{2(q-1)},\frac{q}{2(q-1)},\frac{3-q}{2(q-1)},1-\frac{(q-1)^2\,m^2}{T^2}\right).
\tag{16}
$$

It is worth noticing that both the expressions in general require $q < 4/3$ for the consistency of the framework apart from the (upper or lower) limits, based on the convergence criterion.

3.2. Quantum Case: Bosons (New Results)

In this Section, we tabulate the newly found analytical results of the Tsallis thermodynamics for arbitrarily massive bosons. In comparison with the classical case, there is an extra step in the quantum calculations that entails expressing the quantum single-particle distributions as a superposition of an infinite number of classical distributions. However, all the other procedures are the same as those in the classical case. In the quantum case also, two analytical formulae for the upper and the lower regions are obtained.

3.2.1. Upper Region: $q > 1 + T/m$

$$
\begin{aligned}
P_{\mathrm{U,b}} = \sum_{s=1}^{s_0}\frac{g m^4}{32\pi^2\Gamma\left(\frac{qs}{q-1}\right)} &\left(\frac{2T}{m(q-1)}\right)^{\frac{qs}{q-1}} \left[\Gamma\left(\frac{qs}{2(q-1)}\right)\Gamma\left(\frac{qs}{2(q-1)}-2\right)\right. \\
&\times\; {}_2F_1\left(\frac{qs}{2(q-1)},\frac{qs}{2(q-1)}-2;\frac{1}{2};\frac{T^2}{m^2(q-1)^2}\right)-\frac{2T}{m(q-1)}\Gamma\left(\frac{qs-3q+3}{2(q-1)}\right) \\
&\left.\times\;\Gamma\left(\frac{qs+q-1}{2(q-1)}\right)\; {}_2F_1\left(\frac{qs-3q+3}{2(q-1)},\frac{qs+q-1}{2(q-1)};\frac{3}{2};\frac{T^2}{m^2(q-1)^2}\right)\right].
\end{aligned}
\tag{17}
$$

3.2.2. Lower Region: $q \le 1 + T/m$

$$
\begin{aligned}
P_{\mathrm{L,b}} = \frac{g T^4}{16(q-1)^4}\sum_{s=1}^{s_0} {}_2\tilde{F}_1 &\left(\frac{qs}{2(q-1)}-2,\frac{qs}{2(q-1)}-\frac{3}{2};\frac{qs}{q-1}-\frac{3}{2};1-\frac{m^2(q-1)^2}{T^2}\right) \\
&\times \sec\left(\frac{\pi qs}{q-1}\right)\left[\frac{\Gamma\left(\frac{qs-4q+4}{2(q-1)}\right)}{\Gamma\left(\frac{q(s-5)+5}{2(1-q)}\right)\Gamma\left(\frac{qs}{2(1-q)}+\frac{1}{2}\right)\Gamma\left(\frac{qs}{2(q-1)}+\frac{1}{2}\right)}\right. \\
&\left. -\frac{\Gamma\left(\frac{qs}{2(q-1)}-\frac{3}{2}\right)}{\Gamma\left(\frac{qs}{2(q-1)}\right)\Gamma\left(\frac{qs}{2(1-q)}+1\right)\Gamma\left(\frac{qs}{2(1-q)}+3\right)}\right].
\end{aligned}
\tag{18}
$$

Here, the regularized hypergeometric function ${}_2\tilde{F}_1(a,b;c;z) = {}_2F_1(a,b;c;z)/\Gamma(c)$. Ideally, s_0 is a very large number, depending on the desired agreement between the numerical and analytical results. Equations (17) and (18) (also repeated in Equations (30) and (31) below) are the main results of the paper.

4. Methodology: The Pressure of a Gas of Bosons Following the Tsallis Distribution

From Equation (5), the pressure for the bosons is:

$$P_b = g \int \frac{d^3 p}{(2\pi)^3}\, \frac{p^2}{3\, E_p}\, n_b, \tag{19}$$

where n_b is the Tsallis Bose–Einstein single-particle distribution given by Equation (3). The spherical symmetry of the integrand implies that:

$$P_b = \frac{g}{6\pi^2} \int_0^\infty \frac{p^4}{\sqrt{m^2+p^2}}\, \frac{1}{\left[1+(q-1)\frac{\sqrt{m^2+p^2}}{T}\right]^{\frac{q}{q-1}}-1}\, dp, \tag{20}$$

where $\mu = 0$ is set. Now, we describe the steps to obtain Equations (17) and (18).

4.1. Rescaling the Integration Variable

To simplify the calculations, the $k = p/m$ is defined, so that the pressure reads:

$$P_b = \frac{gm^4}{6\pi^2} \int_0^\infty \frac{k^4}{\sqrt{1+k^2}}\, \frac{1}{\left[1+\frac{m(q-1)}{T}\sqrt{1+k^2}\right]^{\frac{q}{q-1}}-1}\, dk. \tag{21}$$

4.2. Infinite Summation

Now, one observes that, similar to the Boltzmann–Gibbs case, the Tsallis quantum distributions can be expressed in terms of an infinite summation of the Tsallis MB distributions:

$$\frac{1}{\left[1+\frac{m(q-1)}{T}\sqrt{1+k^2}\right]^{\frac{q}{q-1}}\pm 1} = \sum_{s=1}^\infty (-1)^{a(s+1)}\left(1+\frac{m(q-1)}{T}\sqrt{1+k^2}\right)^{-\frac{qs}{q-1}}, \tag{22}$$

where $a=0$ ($a=1$) yields the bosonic (fermionic) distribution. This step allows us to write down the Tsallis pressure in a bosonic gas in a form similar to its classical counterpart, except for a summation sign in front and a power index s in the denominator. Hence, one obtains:

$$P_b = \frac{gm^4}{6\pi^2} \sum_{s=1}^\infty \int_0^\infty \frac{k^4}{\sqrt{1+k^2}}\, \frac{1}{\left(1+\frac{m(q-1)}{T}\sqrt{1+k^2}\right)^{\frac{qs}{q-1}}}\, dk. \tag{23}$$

4.3. Contour Integral Representation

Next, we use the Mellin–Barnes contour representation [51–53] of the power-law function appearing in the integrand in Equation (23), and follow the procedure, described in Ref. [37]. A power-law function can be written as a Mellin–Barnes contour integration:

$$\frac{1}{(X+Y)^\lambda} = \frac{1}{2\pi i} \int_{\epsilon-i\infty}^{\epsilon+i\infty} \frac{\Gamma(-z)\Gamma(z+\lambda)}{\Gamma(\lambda)}\, \frac{Y^z}{X^{\lambda+z}}\, dz, \tag{24}$$

where $\text{Re}(\lambda) > 0$ and $\text{Re}(\epsilon) \in (-\text{Re}(\lambda), 0)$, which is the case here since $\lambda = qs/(q-1) > 0 \Leftrightarrow q, s \geq 1$. Moreover, one can take $X = m(q-1)\sqrt{1+k^2}/T$ and $Y = 1$, or other combinations, all of which yield the following expression:

$$P_{\rm b} = \frac{gm^4}{12\pi^3 i} \sum_{s=1}^{\infty} \left(\frac{T}{m(q-1)}\right)^{\frac{qs}{q-1}} \int_{\epsilon-i\infty}^{\epsilon+i\infty} \frac{\Gamma(-z)\,\Gamma\left(z+\frac{qs}{q-1}\right)}{\Gamma\left(\frac{qs}{q-1}\right)} \left(\frac{T}{m(q-1)}\right)^{z} dz$$

$$\times \int_0^{\infty} k^4 (1+k^2)^{-\frac{z}{2}-\frac{1}{2}-\frac{qs}{2(q-1)}}\, dk. \tag{25}$$

After performing the k-integration, the pressure reads:

$$P_{\rm b} = \frac{gm^4}{32\pi^{\frac{5}{2}} i} \sum_{s=1}^{\infty} \left(\frac{T}{m(q-1)}\right)^{\frac{qs}{q-1}} \int_{\epsilon-i\infty}^{\epsilon+i\infty} \frac{\Gamma(-z)\Gamma\left(z+\frac{qs}{q-1}\right)\Gamma\left(\frac{qs}{2(q-1)}+\frac{z}{2}-2\right)}{\Gamma\left(\frac{qs}{q-1}\right)\Gamma\left(\frac{qs}{2(q-1)}+\frac{z}{2}+\frac{1}{2}\right)}$$

$$\times \left(\frac{T}{m(q-1)}\right)^{z} dz. \tag{26}$$

The convergence of the scaled momentum integration requires $\text{Re}(z) \geq 0$.

4.4. Wrapping Contour Clockwise: $q > 1 + T/m$

To identify the poles, in order to obtain the residues of the integrand, the transformation $z \to 2z$ is made along with the use of the Legendre's duplication formula [54] and Cauchy's residue formula [55], so that the pressure now reads:

$$P_{\rm b} = \frac{gm^4}{64\pi^{\frac{7}{2}} i} \sum_{s=1}^{\infty} \frac{2^{\frac{qs}{q-1}}}{\Gamma\left(\frac{qs}{q-1}\right)} \left(\frac{T}{m(q-1)}\right)^{\frac{qs}{q-1}} \int_{\epsilon-i\infty}^{\epsilon+i\infty} \Gamma(-z)\Gamma\left(-z+\frac{1}{2}\right)$$

$$\times \Gamma\left(z+\frac{qs}{2(q-1)}\right)\Gamma\left(\frac{qs}{2(q-1)}+z-2\right)\left(\frac{T}{m(q-1)}\right)^{2z} dz$$

$$= (-2\pi i) \times \frac{gm^4}{64\pi^{\frac{7}{2}} i} \sum_{s=1}^{\infty} \frac{2^{\frac{qs}{q-1}}}{\Gamma\left(\frac{qs}{q-1}\right)} \left(\frac{T}{m(q-1)}\right)^{\frac{qs}{q-1}}$$

$$\times \sum_{\ell=0}^{\infty} \left\{ \text{Res}^{(1)}[f(z), z=\ell] + \text{Res}^{(2)}\left[f(z), z=\ell+\frac{1}{2}\right] \right\}, \tag{27}$$

where $f(z)$ is defined as:

$$f(z) \equiv \Gamma(-z)\Gamma\left(\frac{1}{2}-z\right)\Gamma\left(\frac{qs}{2(q-1)}+z\right)\Gamma\left(\frac{qs}{2(q-1)}+z-2\right)\left(\frac{T}{m(q-1)}\right)^{2z},$$

and the contour is wrapped clockwise so that residues receive a contribution only from the poles of $\Gamma(-z)$ at the positive integers ($\text{Res}^{(1)}$) including zero, and the poles of $\Gamma(-z+1/2)$

at the positive half-integers ($\mathrm{Res}^{(2)}$). This clockwise wrapping of contour imposes the convergence condition $q > 1 + T/m$ when $z \to \infty$. $\mathrm{Res}^{(1)}$ and $\mathrm{Res}^{(2)}$ are defined as follows:

$$
\begin{aligned}
\mathrm{Res}^{(1)} &= \mathrm{Res}\left[f(z), \{z = \ell \ni \ell \in \mathbb{Z}^{\geq}\}\right] \\
&= \frac{(-1)^{\ell+1}\left(\frac{T}{m(q-1)}\right)^{2\ell}}{\ell!} \Gamma\left(\frac{1}{2} - \ell\right)\Gamma\left(\ell + \frac{qs}{2(q-1)} - 2\right) \\
&\quad \times \Gamma\left(\ell + \frac{qs}{2(q-1)}\right)
\end{aligned}
\tag{28}
$$

$$
\begin{aligned}
\mathrm{Res}^{(2)} &= \mathrm{Res}\left[f(z), \{z = \ell + \frac{1}{2} \ni \ell \in \mathbb{Z}^{\geq}\}\right] \\
&= \frac{(-1)^{\ell+1}\left(\frac{T}{m(q-1)}\right)^{2\ell+1}}{\ell!} \Gamma\left(-\ell - \frac{1}{2}\right)\Gamma\left(\ell + \frac{qs}{2(q-1)} - \frac{3}{2}\right) \\
&\quad \times \Gamma\left(\ell + \frac{qs}{2(q-1)} + \frac{1}{2}\right).
\end{aligned}
\tag{29}
$$

In substituting Equations (28) and (29) into Equation (27), the infinite summation over ℓ can be expressed in terms of the hypergeometric function $_2F_1$ [50], and the pressure in the region $q > 1 + T/m$, given by Equation (17), reads:

$$
\begin{aligned}
P_{\mathrm{U,b}} =& \sum_{s=1}^{s_0} \frac{gm^4}{32\pi^2\Gamma\left(\frac{qs}{q-1}\right)}\left(\frac{2T}{m(q-1)}\right)^{\frac{qs}{q-1}}\left[\Gamma\left(\frac{qs}{2(q-1)}\right)\Gamma\left(\frac{qs}{2(q-1)} - 2\right)\right. \\
&\times {_2F_1}\left(\frac{qs}{2(q-1)}, \frac{qs}{2(q-1)} - 2; \frac{1}{2}; \frac{T^2}{m^2(q-1)^2}\right) - \frac{2T}{m(q-1)}\Gamma\left(\frac{qs-3q+3}{2(q-1)}\right) \\
&\left.\times \Gamma\left(\frac{qs+q-1}{2(q-1)}\right){_2F_1}\left(\frac{qs-3q+3}{2(q-1)}, \frac{qs+q-1}{2(q-1)}; \frac{3}{2}; \frac{T^2}{m^2(q-1)^2}\right)\right],
\end{aligned}
\tag{30}
$$

when the infinite summation is truncated at $s = s_0$.

4.5. Analytic Continuation: $q \leq 1 + T/m$

Instead of keeping the dimension of the momentum space arbitrary and analytically continuing the integrand prior to wrapping (since it does not lead to a closed form), the result, obtained in Equation (30) using Ref. [50], is analytically continued, and one obtains the following result in the complementary (lower) region:

$$
\begin{aligned}
P_{\mathrm{L,b}} =& \frac{8T^4}{16(q-1)^4}\sum_{s=1}^{s_0} {_2\tilde{F}_1}\left(\frac{qs}{2(q-1)} - 2, \frac{qs}{2(q-1)} - \frac{3}{2}; \frac{qs}{q-1} - \frac{3}{2}; 1 - \frac{m^2(q-1)^2}{T^2}\right) \\
&\times \sec\left(\frac{\pi qs}{q-1}\right)\left[\frac{\Gamma\left(\frac{qs-4q+4}{2(q-1)}\right)}{\Gamma\left(\frac{q(s-5)+5}{2(1-q)}\right)\Gamma\left(\frac{qs}{2(1-q)} + \frac{1}{2}\right)\Gamma\left(\frac{qs}{2(q-1)} + \frac{1}{2}\right)}\right. \\
&\left. - \frac{\Gamma\left(\frac{qs}{2(q-1)} - \frac{3}{2}\right)}{\Gamma\left(\frac{qs}{2(q-1)}\right)\Gamma\left(\frac{qs}{2(1-q)} + 1\right)\Gamma\left(\frac{qs}{2(1-q)} + 3\right)}\right],
\end{aligned}
\tag{31}
$$

where the regularized hypergeometric function, $_2\tilde{F}_1(a, b; c; z)$, is defined as for Equation (18).

5. Numerical Results and Discussion

Here, some comments about the comparison of numerical results with the results, obtained from the analytical formulae, are in order. Let us check the massless limit first. We notice that the final result works rather well for the case of massless particles ($m = 0$), as it should. To check this we substitute numerical values in Equation (18) and compare the result with the numerical value of the integral in Equation (20). We take, for example, $q = 1.2$, $g = 1$, $T = 0.08$ GeV, $m = 10^{-7}$ GeV, such that $q = 1.2 \ll 1 + T/m$. This condition implies that Equation (31) was used. For $s_0 = 20$, both the numerical and the analytical results (obtained also from Equation (9)) agree up to eleven significant digits and the value of pressure is 2.20098×10^{-5} GeV^4.

Next, let us consider light particles like the positively-charged pions (of the mass of 0.140 GeV), produced in proton-proton (p-p) collisions at the LHC. We observe that for $q = 1.154$, $g = 1$ and $T = 0.0682$ GeV (values taken from [5]), the value of s_0 significantly differs from the massless case when we consider the pions. For the pions ($q = 1.154 < 1 + T/m = 1.487$), a similar agreement between the analytical and numerical results can be reached for $s_0 = 5$ and the pressure turns out to be 5.4318×10^{-6} GeV^4.

We also consider more massive particles like the protons (of the mass of 0.938 GeV), produced in p-p collisions at the LHC. For $q = 1.107$, $g = 2$ and $T = 0.073$ GeV (the values are taken from [5]), a similar agreement between the numerical and analytical results, both of which are 3.5597×10^{-7} GeV^4 , can be obtained including just two terms, i.e., $s_0 = 2$. It is noteworthy that the protons are fermions, and the corresponding formula for pressure can be obtained by multiplying $(-1)^{s+1}$ with each term of, for example, Equation (30), as $q = 1.107 > 1 + T/m \approx 1.078$. In these examples, the heavier the particle, the faster the infinite summation convergence. This trend is repeated when we change only mass, while keeping q and T values unaltered.

For a gas of positively charged pions produced at the RHIC (Relativistic Heavy Ion Collider) (center-of-mass energy, $\sqrt{s_{NN}} = 200$ GeV, Au-Au collisions, $q = 1.090$ and $T = 0.117$ GeV [56]), the pressure is found to be 3.0089×10^{-5} GeV^4. Considering all the charged particles produced at the LHC ($\sqrt{s_{NN}} = 2760$ GeV Pb-Pb collisions, $q = 1.135$ and $T = 0.096$ GeV [19]), the pressure is found to be 6.5733×10^{-5} GeV^4

We conclude from the comparison of numerical and analytical results that the latter works considerably well. We hope that the main results, reported in this paper, will sufficiently reduce the overall computation time. We have checked that for some of the above examples, computation time is almost ten times reduced when the analytical formulae are used.

6. Summary, Conclusions and Outlook

In summary, we presented a brief review of the studies, related to the Tsallis thermodynamics, that may be important in the further studies of the quark-gluon plasma and many other systems that display fluctuation and long-range correlation. We also presented a detailed description of how to extend those existing findings to the quantum domain (Equations (30) and (31)). We used the contour integral representation of the power-law function and followed the ritual proposed in [37], after expressing the quantum distributions in terms of an infinite summation of classical MB distributions. We elaborated the analytical computation of the pressure of a bosonic gas following the Tsallis statistics, and the final result can be expressed as a summation that appears from the superposition of classical distributions. However, we noticed that in the examples discussed, only a finite number of terms are needed, and the number of required terms for convergence decreases with mass (when q, and T, are kept unaltered). The integral representation, also known as the Mellin–Barnes representation, has extensively been used in the studies involving loop calculations in quantum field theory [53]. Hence, in a way, this is one of the examples where techniques established in one field of research benefit another. Although not mentioned in the paper, extension to the fermionic case is straightforward. The only difference in summation comes owing to a factor $(-1)^{s+1}$ appearing with each term. In this paper, we provided the results only of the pressure of a Tsallis-like bosonic gas. Other thermodynamic variables of such a system can be calculated by appropriately differentiating pressure. Extension to the $\mu \neq 0$ case can be performed with a proper identification of the variables

X and Y in Equation (24). Moreover, in this case, the convergence condition for clockwise wrapping is modified [37].

There may be many different applications of the present study but we would like to mention a particular field that has caught some recent interest. Of late, there have been studies [34,35] reporting the propagation of non-linear waves in the quark-gluon plasma fluid (both ideal and viscous) in which constituents follow the Tsallis-like distributions. In studies [34,35], a Tsallis-like MIT bag equation of state, considering massless (or very light) particles, was used. It will be interesting to modify the equation of state incorporating the present findings. It will also be interesting to extend the study for hadronic gases. It has been shown [14] that the exact Tsallis single-particle distribution is expressed in terms of a series summation, and the distributions, used in Equation (3) are only the approximations. For low-energy collisions (e.g., in the future experiments at the NICA (Nuclotron-based Ion Collider fAcility) and FAIR (Facility for Antiproton and Ion Research) facilities, terms beyond the one, used in the present paper, may be important. It will be worthwhile to investigate how those additional terms would affect the present results, and hence, the studies utilizing them.

Author Contributions: Conceptualization, T.B.; methodology, T.B. and A.H.; software, T.B. and A.H.; validation, T.B. and A.H.; formal analysis, T.B. and A.H.; investigation, T.B. and A.H.; writing—original draft preparation, T.B.; writing—review and editing, A.H.; supervision, T.B. All authors have read and agreed to this version of the manuscript.

Funding: This research was partially funded by the joint project between the JINR (Joint Institute for Nuclear Research, Dubna, Russia) and IFIN-HH (Horia Hulubei National Institute for R&D in Physics and Nuclear Engineering, Ilfov, Romania).

Data Availability Statement: No new data have been generated.

Acknowledgments: A.H. acknowledges all-round support from Alia Dawood during this work and stimulating discussions with Mohamed Elekhtiar and Mohamed Al Begaowe. T.B. gratefully acknowledges discussions with Sylvain Mogliacci regarding the intricacies of the Mellin–Barnes representation used in the paper as well as generous support from the University of Cape Town where the foundation of this work was prepared. Authors thank Rajendra Nath Patra for providing fit parameter values of RHIC data.

Conflicts of Interest: The authors declare no conflict of interest.

References

1. Abelev, B.I.; et al. [STAR Collaboration]. Strange particle production in $p + p$ collisions at $\sqrt{s} = 200$ Gev. *Phys. Rev. C* **2007**, *75*, 064901. [CrossRef]
2. Adare, A.; et al. [PHENIX Collaboration]. Identified charged hadron production in $p + p$ collisions at $\sqrt{s} = 200$ Gev and 62.4 Gev. *Phys. Rev. C* **2011**, *83*, 064903. [CrossRef]
3. Aamodt, K.; et al. [The ALICE Collaboration]. Production of pions, kaons and protons in pp collisions at $\sqrt{s} = 900$ GeV with ALICE at the LHC. *Eur. Phys. J. C* **2011**, *71*, 1655. [CrossRef]
4. Khachatryan, V.; et al. [CMS Collaboration]. Strange particle production in pp collisions at $\sqrt{s} = 0.9$ and 7 TeV. *J. High Energy Phys.* **2011**, *5*, 64. [CrossRef]
5. Cleymans, J.; Worku, D. The Tsallis distribution in proton-proton collisions at $\sqrt{s} = 0.9$ TeV at the LHC. *J. Phys. G Nucl. Part. Phys.* **2012**, *39*, 025006. [CrossRef]
6. Cleymans, J.; Worku, D. Relativistic thermodynamics: Transverse momentum distributions in high-energy physics. *Eur. Phys. J. A* **2012**, *48*, 160. [CrossRef]
7. Tsallis, C. Possible generalization of Boltzmann-Gibbs statistics. *J. Stat. Phys.* **1988**, *52*, 479–487. [CrossRef]
8. Wilk, G.; Włodarczyk, Z. Interpretation of the Nonextensivity Parameter q in Some Applications of Tsallis Statistics and Lévy Distributions. *Phys. Rev. Lett.* **2000**, *84*, 2770. [CrossRef]
9. Wilk, G.; Włodarczyk, Z. Multiplicity fluctuations due to the temperature fluctuations in high-energy nuclear collisions. *Phys. Rev. C* **2009**, *79*, 054903. [CrossRef]
10. Osada, T.; Wilk, G. Nonextensive hydrodynamics for relativistic heavy-ion collisions. *Phys. Rev. C* **2009**, *77*, 044903. [CrossRef]
11. Biro, T.S.; Molnar, E. Fluid dynamical equations and transport coefficients of relativistic gases with non-extensive statistics. *Phys. Rev. C* **2012**, *85*, 024905. [CrossRef]
12. Birö, T.S.; Barnaföldi, G.G.; Van, P. Quark-gluon plasma connected to finite heat bath. *Eur. Phys. J. A* **2013**, *49*, 110. [CrossRef]

13. Deppman, A. Thermodynamics with fractal structure, Tsallis statistics, and hadrons. *Phys. Rev. D* **2016**, *93*, 054001. [CrossRef]
14. Parvan, A.S.; Bhattacharyya, T. Hadron transverse momentum distributions of the Tsallis normalized and unnormalized statistics. *Eur. Phys. J. A* **2020**, *56*, 72. [CrossRef]
15. Cleymans, J.; Lykasov, G.I.; Parvan, A.S.; Sorin, A.S.; Teryaev, O.V.; Worku, D. Systematic properties of the Tsallis Distribution: Energy Dependence of Parameters in High-Energy $p - p$ Collisions. *Phys. Lett. B* **2013**, *723*, 351. [CrossRef]
16. Marques, L.; Cleymans, J.; Deppman, A. Description of high-energy pp collisions using Tsallis thermodynamics: Transverse momentum and rapidity distributions. *Phys. Rev. D* **2015**, *91*, 054025. [CrossRef]
17. Tripathy, S.; Tiwari, S.K.; Younus, M.; Sahoo, R. Elliptic flow in Pb+Pb collisions at $\sqrt{s_{NN}} = 2.76$ TeV at the LHC using Boltzmann transport equation with non-extensive statistics. *Eur. Phys. J. A* **2018**, *54*, 38. [CrossRef]
18. Acharya, S.; et al. [ALICE Collaboration]. Production of deuterons, tritons, ^{3}He nuclei, and their antinuclei in pp collisions at $\sqrt{s} = 0.9, 2.76$, and 7 TeV. *Phys. Rev. C* **2018**, *97*, 024615. [CrossRef]
19. Azmi, M.D.; Bhattacharyya, T.; Cleymans, J.; Paradza, M.W. Energy density at kinetic freeze-out in Pb-Pb collisions at the LHC using the Tsallis distribution. *J. Phys. G* **2020**, *47*, 045001. [CrossRef]
20. Parvan, A.S. Equivalence of the phenomenological Tsallis distribution to the transverse momentum distribution of q-dual statistics. *Eur. Phys. J. A* **2020**, *56*, 106. [CrossRef]
21. Lavagno, A. Relativistic nonextensive thermodynamics. *Phys. Lett. A* **2002**, *301*, 13–18. [CrossRef]
22. Gyulassy, M.; Matsui, T. Quark gluon plasma evolution in scaling hydrodynamics. *Phys. Rev. D* **1984**, *29*, 419–425. [CrossRef]
23. Akase, Y.; Mizutani, M.; Muroya, S.; Namiki, M.; Yasuda, M. Hydrodynamical evolution of QGP fluid with phase transition and particle distribution in high-energy nuclear collisions. *Prog. Theor. Phys.* **1991**, *85*, 305–320. [CrossRef]
24. Csernai, L.P.; Anderlik, C.; Keranen, A.; Magas, V.K.; Manninen, J.; Strottman, D.D. QGP hydrodynamics for RHIC energies. *Acta Phys. Hung. A* **2003**, *17*, 271–280. [CrossRef]
25. Song, H.; Heinz, U.W. Extracting the QGP viscosity from RHIC data—A Status report from viscous hydrodynamics. *J. Phys. G* **2009**, *36*, 064033. [CrossRef]
26. Teaney, D.A. Viscous hydrodynamics and the quark gluon plasma. In *Quark Gluon Plasma 4*; Hwa, R.C., Wang, X.-N., Eds.; World Scientific: Singapore, 2010; pp. 207–266. [CrossRef]
27. Chaudhuri, A.K. Knudsen number, ideal hydrodynamic limit for elliptic flow and QGP viscosity in $\sqrt{s} = 62$ and 200 GeV Cu+Cu/Au+Au collisions. *Phys. Rev. C* **2010**, *82*, 047901. [CrossRef]
28. Jaiswal, A.; Roy, V. Relativistic hydrodynamics in heavy-ion collisions: General aspects and recent developments. *Adv. High Energy Phys.* **2016**, *2016*, 9623034. [CrossRef]
29. Fowler, G.N.; Raha, S.; Stelte, N.; Weiner, R.M. Solitons in nucleus-nucleus collisions near the speed of sound. *Phys. Lett. B* **1982**, *115*, 286–290. [CrossRef]
30. Fogaça, D.A.; Filho, L.G.F.; Navarra, F.S. Nonlinear waves in a quark gluon plasma. *Phys. Rev. C* **2010**, *81*, 055211. [CrossRef]
31. Fogaça, D.A.; Navarra, F.S.; Filho, L.G.F. Korteveg-de Vries solitons in a cold quark-gluon plasma. *Phys. Rev. D* **2010**, *84*, 054011. [CrossRef]
32. Fogaça, D.A.; Navarra, F.S. Gluon condensates in a cold quark-gluon plasma. *Phys. Lett. B* **2011**, *700*, 236–242. [CrossRef]
33. Fogaça, D.A.; Navarra, F.S.; Filho, L.G.F. On the radial expansion of tubular structures in a quark-gluon plasma. *Nucl. Phys. A* **2012**, *887*, 22–41. [CrossRef]
34. Bhattacharyya, T.; Mukherjee, A. Propagation of non-linear waves in hot, ideal, and non-extensive quark-gluon plasma. *Eur. Phys. Jour. C* **2020**, *80*, 656. [CrossRef]
35. Sarwar, G.; Hasanujjaman, M.; Bhattacharyya, T.; Rahaman, M.; Bhattacharyya, A.; Alam, J.E. Nonlinear waves in a hot, viscous and non-extensive quark-gluon plasma. *Eur. Phys. J. C* **2022**, *82*, 189. [CrossRef]
36. Bhattacharyya, T.; Cleymans, J.; Khuntia, A.; Pareek, P.; Sahoo, R. Radial flow in non-extensive thermodynamics and study of particle spectra at LHC in the limit of small (q – 1). *Eur. Phys. J. A* **2016**, *52*, 30. [CrossRef]
37. Bhattacharyya, T.; Cleymans, J.; Mogliacci, S. Analytic results for the Tsallis thermodynamic variables. *Phys. Rev. D* **2018**, *94*, 094026. [CrossRef]
38. Sanches, M.S., Jr.; Navarra, F.S.; Fogaça, D.A. The quark gluon plasma equation of state and the expansion of the early Universe. *Nucl. Phys. A* **2015**, *937*, 1–16. [CrossRef]
39. Bhatia, A.B. Vibration spectra and specific heats of cubic metals. I. Theory and application to sodium. *Phys. Rev.* **1955**, *97*, 363. [CrossRef]
40. Taylor, C.D.; Lookman, T.; Scott, L.R. Ab initio calculations of the uranium-hydrogen system: Thermodynamics, hydrogen saturation of a-U and phase-transformation to UH3. *Acta Mater.* **2010**, *58*, 1045. [CrossRef]
41. Aguiar, J.C.; Mitnik, D.; DiRocco, H.O. Electron momentum density and Compton profile by a semi-empirical approach. *J. Phys. Chem. Solids* **2015**, *83*, 64–69. [CrossRef]
42. Sharma, G.; Joshi, K.B.; Mishra, M.C.; Kothari, R.K.; Sharma, Y.C.; Vyas, Y.C.V.; Sharma, B.K. Electronic structure of AlAs: A Compton profile study. *J. Alloys Compd.* **2009**, *485*, 682–686. [CrossRef]
43. Kawasuso, A.; Maekawa, M.; Fukaya, Y.; Yabuuchi, A.; Mochizuki, I. Polarized positron annihilation measurements of poly-crystalline Fe, Co, Ni, and Gd based on Doppler broadening of annihilation radiation. *Phys. Rev. B* **2011**, *83*, 100406(R). [CrossRef]

44. Tsallis, C.; Mendes, R.S.; Plastino, A.R. The role of constraints within generalized nonextensive statistics. *Phys. A* **1998**, *261*, 534–554. [CrossRef]
45. Bhattacharyya, T.; Parvan, A.S. Analytical results for the classical and quantum Tsallis hadron transverse momentum spectra: The zeroth order approximation and beyond. *Eur. Phys. J. A* **2021**, *57*, 206. [CrossRef]
46. Conroy, J.M.; Miller, H.G.; Plastino, A.R. Thermodynamic consistency of the q-deformed Fermi-Dirac distribution in nonextensive thermostatics. *Phys. Lett. A* **2010**, *374*, 4581–4584. [CrossRef]
47. Büyükkiliç, F.; Demirhan, D. A fractal approach to entropy and distribution functions. *Phys. Lett. A* **1993**, *181*, 24–28. [CrossRef]
48. A Repeated Digit Is Represented by a Bar. Available online: https://en.wikipedia.org/wiki/Repeating_decimal (accessed on 19 April 2022).
49. Bhattacharyya, T.; Cleymans, J.; Mogliacci, S.; Parvan, A.S.; Sorin, A.S.; Teryaev, O.V. Non-extensivity of the QCD p_{T}-spectra. *Eur. Phys. J. A* **2018**, *54*, 222. [CrossRef]
50. Erdélyi, A.; Magnus, W.; Oberhettinger, F.; Tricomi, F.G. *Higher Transcendental Functions*; Krieger: New York, NY, USA, 1981; Volume 1.
51. Davydychev, A.I.; Tausk, J.B. Two-loop self-energy diagrams with different masses and the momentum expansion. *Nucl. Phys. B* **1993**, *397*, 123–142. [CrossRef]
52. Boos, É.É.; Davydychev, A.I. A method of calculating massive Feynman integrals. *Theor. Math. Phys.* **1991**, *89*, 1052–1064. [CrossRef]
53. Smirnov, V.A. *Evaluating Feynman Integrals*; Springer: Berlin/Heidelberg, Germany, 2004. [CrossRef]
54. Wolfram MathWorld. Available online: https://mathworld.wolfram.com/LegendreDuplicationFormula.html (accessed on 19 April 2022).
55. Weber, H.J.; Arfken, G.B. *Essential Mathematical Methods for Physicists*; Academic Press: San Diego, CA, USA, 2004.
56. Patra, R.N.; Mohanty, B.; Nayak, T.K. Centrality, transverse momentum and collision energy dependence of the Tsallis parameters in relativistic heavy-ion collisions. *Eur. Phys. J. Plus* **2021**, *136*, 702. [CrossRef]

physics

Review

Stochastic Variational Method for Viscous Hydrodynamics [†]

Takeshi Kodama * and Tomoi Koide

Instituto de Física, Universidade Federal do Rio de Janeiro, Rio de Janeiro 21941-972, Brazil; tomoikoide@gmail.com or koide@if.ufrj.br
* Correspondence: kodama.takeshi@gmail.com or tkodama@if.ufrj.br
† This article is dedicated to the memory of the late Professor Jean Cleymans.

Abstract: In this short review, we focus on some of the subjects, related to J. Cleymans' pioneering contribution of statistical approaches to the particle production process in heavy-ion collisions. We discuss these perspectives from the effects of stochastic processes in collective variables of hydrodynamic description, which is described by a stochastic variational method. In this connection, we stress also the necessity of the inclusion of surface and quantum effects in the study of relativistic heavy-ion reactions.

Keywords: stochastic variation; viscous hydrodynamics; surface and quantum effects

Citation: Kodama, T.; Koide, T. Stochastic Variational Method for Viscous Hydrodynamics. *Physics* **2022**, *4*, 847–864. https://doi.org/10.3390/physics4030054

Received: 23 June 2022
Accepted: 11 July 2022
Published: 26 July 2022

1. Introduction

The history of hydrodynamic description goes back to even before the 18th Century, and as we know, it has been applied widely to the studies of continuum media in many different scenarios. In fact, the hydrodynamic approach works not only for the description of a variety of phenomena around us, but also for the systems in completely different scales, from elementary particles to the Universe. As a microscopic example, the relativistic hydrodynamic approach is an important tool for the study of the properties of the matter produced in relativistic heavy-ion collisions. There, the properties of the matter created in these processes are reflected in the behaviors of collective flow variables [1–6]. On the other hand, as a macroscopic example, hydrodynamic modeling has been an indispensable tool in the studies of astrophysical phenomena. More recently, in particular, a full general relativistic hydrodynamics is required in the investigation of the so-called multi-messenger astrophysics such as neutron star or black hole mergers; see, e.g., [7].

The above wide range of the applicability of the hydrodynamic description comes from the fact that its basic kinematic structure is nothing but a set of local conservation laws, such as energy, momentum, charges, etc., and expressed in the forms of the equations of continuity. Of course, the proper definitions of the local conserved densities and the continuum assumption depend on systems and are not always trivial. These may enter some subtle aspects, as we discuss below. However, once the system is represented as a continuum medium in the form of local conserved densities and its flows, the corresponding equations of continuity are applicable independently of the properties and size of the matter. This is the principal reason why the hydrodynamic approach can have a wide applicability irrespective of the scale of the system.

On the other hand, it is evident that the set of these equations of continuity alone does not form a closed system to describe dynamics. They are simply kinematic constraints for the dynamical evolution of continuum media. To describe the time evolution, we have to introduce "forces" in a closed form: these forces should be specified as functions (or functionals) of conserved densities. For example, in the case of an ideal fluid description, any infinitesimal part of the fluid is assumed to stay in thermodynamic equilibrium adiabatically during the whole time evolution. In such a case, the force is simply given by the pressure, which is a function of other local densities.

As sketched above, the hydrodynamic description requires several assumptions and also depends on the external input, such as the equation of state (EoS). Thus, the hydrodynamic description is basically phenomenological. Even its most fundamental assumption of the continuum medium for the matter is not always justified in a quantitative manner. This becomes pronounced in the applications to microscopic systems such as relativistic heavy-ion collisions. As described below, the applicability of hydrodynamics is intimately related to the concept of coarse-graining procedure through which we introduce the local densities of the medium and corresponding thermodynamic properties.

To clarify the concept of the coarse-graining procedure, let us consider, as an example, the statistical model for the particle productions in heavy-ion collisions. In this model, we consider the whole ensemble of produced particles in a huge number of collisional events, within a certain kinematic domain. By assuming the grand-canonical ensemble, this system is characterized by a few thermodynamic parameters, such as temperature and chemical potentials. These parameters can be adjusted to reproduce the relative abundance of different species. This statistical approach in heavy-ion collisions indicates new mechanisms of particle productions, such as Hagerdorn's prediction of the limiting temperature in hadronic gas, J/ψ suppression, strangeness enhancement, and thermal photon and lepton pair signals [8–20]. These observations converged to the idea of the formation of quark–gluon plasma (QGP), in the early 2000s.

The relativistic hydrodynamic description of the dynamics of QGP is nothing but a more microscopic and dynamical version of the statistical model. In fact, this is literally the most microscopic system of fluid ever known [21] (see English translation in [22]). There, instead of identifying a unique hot QGP fireball as the average over many collisional events, we consider that such a hot and dense state of the QGP stage is created initially in the small domain at the instant of the collision and develops as a fluid in spacetime. The time evolution of the hot QGP depends on the different initial condition defined on an event-by-event (EbE) basis. This is a natural step toward the spacetime description of the QGP, starting from the statistical model.

In the case of relativistic heavy-ion collisions, however, the proper system is microscopic so that the spacetime evolution of the QGP fluid is not observable directly in experiments. For a given experimental event, we only know a few pieces of global information, which characterizes the initial condition of two colliding nuclei and the momentum distribution of the produced particles with their identification in the final state. Thus, to apply the hydrodynamic description under such a restrictive condition, we introduce three distinct steps in practice: (1) preparation of hydrodynamic initial conditions from the particle system, (2) hydrodynamic evolution, and (3) particlization from the fluid final state. In the first step, we usually rely on the use of a some well-established event generator based on nucleon–nucleon high-energy collisions. Then, the initial distributions of the energy and the momentum are constructed for a given collision geometry of the colliding nucleus, which is characterized by, for example, the impact parameter (see also the discussion below). This is done by a smoothing procedure over the energy, the momentum, and the location of primordial particles produced in each event generator. In this paper, we refer to such a procedure as geometric coarse graining.

Once the initial condition is determined, in the second step, we follow its hydrodynamic evolution until the assumptions of the hydrodynamic description are violated. For example, if the density of the fluid becomes less than a certain critical value, the fluid description is considered to be invalid. The hypersurface in the four-dimensional spacetime where the fluid attains the critical density is called the freeze-out surface. In the third step, the final fluid state on the freeze-out surface is mapped into the corresponding momentum distribution of the produced particles, in terms of the so-called freeze-out process. Technically speaking, the freeze-out mechanism involves complicated physical mechanisms, such as the chemical and thermal freeze-out, after-burner, etc, but we do not enter into the details here.

The above steps are repeated to generate an ensemble of many "hydrodynamic collisional events" for a specific collision geometry. In a nucleus–nucleus collision, the basic parameters for a collision geometry are the impact parameter (collision centrality) and the corresponding collisional plane. We identify the simulated event ensembles as that of the corresponding experimental events classified by the same collisional geometry in a statistical sense. From this identification, we can calculate statistical correlations of observables in both ensembles and compare them. Such correlations can characterize non-trivial signals of the presence of collective motions as dynamic responses of the QGP matter. Here, we stress that the EbE analysis does not mean the one-to-one correspondence between a hydrodynamic simulation and experimental data in an individual collisional event. We rather suppose that the ensemble of the hydrodynamic events is statistically equivalent as a whole to the corresponding experimental events. This is a kind of coarse-graining procedure in the identification of observables. For simplicity, we refer to this procedure as the statistical coarse-graining. This EbE hydrodynamic analysis was first introduced in [23].

As described above, a hydrodynamic approach in the heavy-ion collisions focuses on the collective flow of the system, introducing some coarse-graining procedures of different natures. The collective features in collisional events are reflected in the characteristic correlations among detected particles. Some observables are known to be related directly to the global initial condition geometry (such as impact parameter and collision plane) and others to those EbE fluctuations, e.g., the local inhomogeneities. The EbE analysis has revealed that the fluctuations in the initial conditions with the same class of the global geometry manifest themselves in the collective flow observables, such as v_2, v_3, establishing hydrodynamic natures.

Here, we stress again that the determination of correlations among observables requires the ensemble average over different events with a specified initial global geometry. Therefore, strictly speaking, the success of the hydrodynamic description does not necessarily mean that all assumptions behind it are justified literally as they are. In the hydrodynamic description, we assume that the QGP is formed as a continuum medium represented by an ensemble of small domains, called fluid elements. They should be specified by a set of local conserved densities in thermodynamic equilibrium, at least approximately. For this to happen, the size of a fluid element must be orders of magnitude smaller than that of the whole fluid, but at the same time, it should be sufficiently dense so that it contains a huge number of the QGP particles. In addition, the local thermal relaxation time must be sufficiently shorter than the time scale of the hydrodynamic evolution to keep the fluid element in a state close to the thermal equilibrium during its evolution. However, as mentioned, we are not considering one-to-one correspondence between the two ensembles, hydrodynamic evolutions, and experimental events, and hence, the observables contain the statistical coarse-graining. In this sense, the hydrodynamic description for the rather global collective flow parameters, such as v_2, v_3, are insensitive to more local dynamics of the system. To investigate the more detailed behavior of the QGP dynamics, nonlinear correlations should be analyzed since there are different initial conditions that reproduce the same collective flow parameters. For example, the distinction of finite v_3 flows between a global triangular anisotropy and the presence of a hot spot in the peripheral region is not clear [24,25]. The measurements of higher-order correlations in observables will further refine these questions with respect to the coarse-graining scale, and such efforts are being made by studying the non-linear response of the collective parameters [26,27].

Up to now, we have not considered the presence of viscosity, which is of course an important factor for the hydrodynamic description. In ideal hydrodynamics, the fluid elements are thermally equilibrated during the time evolution and smoothly connected to their neighbors. Their time evolution should be adiabatic, and the total entropy is conserved. In such a case, it is well known that the hydrodynamic equation of the system is equivalent to the set of classical equations of motion of each fluid element. The so-called Lagrange coordinate system faithfully reflects such an image. Then, it is rather straight to see that we can apply the variational method to derive the hydrodynamic equations, defining the action of the system simply introducing the internal energy of each fluid element. As is known, the variational approach has many advantages in formulating the dynamical problem in an elegant and transparent way to deal with, e.g., the relation between the conservation laws and symmetries. As an example, ideal relativistic hydrodynamic equations are derived from the same action, just written in a covariant form [28].

From the variational point of view, ideal hydrodynamics describes the optimal trajectory of the fluid elements, where they interact with the neighbors exchanging the internal energy in an adiabatic way. Then, if we follow this image to deal with dissipative processes, we need to include the non-adiabatic processes in the interactions among the fluid elements. This is not trivial in a self-contained variational scheme without introducing an artificial modification such as Rayleigh's dissipation function.

One possibility is to extend the domain of dynamical variables as classical deterministic ones by introducing stochastic variables in the definition of the action. This is a natural way to include the fluctuations associated with the coarse-graining mechanism. Such a generalized variational method is known as the stochastic variational method (SVM) [29–32]. In this review, we focus our attention on some known systems where the SVM approach is applicable. The action is defined as the average over the whole ensemble of stochastic events. The variation is taken with respect to these stochastic dynamical variables, whose initial and final conditions are specified in terms of their distributions.

The original motivation of the SVM was to derive the Schrödinger equation in terms of noises around the classical trajectory [29], but the SVM approach is shown to be valid to derive viscous hydrodynamics, as well. It is shown that the Navier–Stokes and Gross–Pitaevskii, in addition to the generalized diffusion equations are derived within the framework of the SVM [33]. There, the quantum uncertainties and uncertainties in hydrodynamic descriptions due to the coarse-graining procedure are dealt with by the same framework [32,34–36]. The crucial point of the SVM approach compared to the ordinary variation method is that we deal with the non-differentiability and the time-reversed stochastic trajectories.

This paper is organized as follows. In Section 2, we describe the essential structure of the stochastic variational approach and discuss several known results, such as the derivation of the Schrödinger equation. In Section 3, we apply the SVM to the system of fluids and derive generalized viscous hydrodynamics. In Section 4, we discuss the roles of the new term, which does not appear in standard viscous hydrodynamics. This term, on the one hand, plays a crucial role for quantum-mechanical systems and, on the other hand, generates the surface tension in generalized viscous hydrodynamics. In Section 5, we discuss the uncertainty relation in hydrodynamics. The application to curved geometric systems is discussed in Section 6. Section 7 is devoted to concluding remarks.

2. Stochastic Variation Method

Let us consider a single-particle system with mass m. In the standard variational principle of classical mechanics, it is known that the physical particle trajectory $\mathbf{x}(t)$ is given by the optimized path of the action:

$$I[\mathbf{x}(t)] = \int_{t_i}^{t_f} dt\, L(\dot{\mathbf{x}}(t), \mathbf{x}(t)),$$ (1)

where t_i and t_f are the initial and final times and dot denotes the time derivative. The Lagrangian, L, is defined by

$$L(\dot{\mathbf{x}}(t), \mathbf{x}(t)) = \frac{m}{2}\dot{\mathbf{x}}(t)^2 - V(\mathbf{x}(t)),$$ (2)

where the potential energy is denoted by $V(\mathbf{x}(t))$. There, it is implicitly assumed that the virtual trajectories for variation are differentiable. Then, as known, the optimized path of the action is given by the solution of the Newton equation. If we permit considering, however, non-differential virtual trajectories in the variational procedure, the optimized dynamics could be modified from the Newton equation. The SVM is one such generalized variational method.

The typical example of non-differentiable trajectories is found in the Brownian motion. In a manner analogous to this, in the SVM, the particle trajectory is assumed to be given by the following forward stochastic differential equation (SDE):

$$d\mathbf{x}(t) = \mathbf{u}_+(\mathbf{x}(t), t)dt + \sqrt{2\nu}d\mathbf{W}_+(t) \quad (dt > 0),$$ (3)

where $\mathbf{u}_+$ is a time-dependent vector field to be determined, and $\mathbf{W}_+(t)$ is given by the standard Wiener process, satisfying

$$E[d\mathbf{W}_+(t)] = 0, \quad E[dW_+^i(t)dW_+^j(t')] = |dt|\,\delta_{t\,t'}\,\delta_{ij},$$ (4)

where δ symbol denotes the Kronecker delta, and $i, j = 1, 2, 3$, stay for the component index of space-like vectors. The ensemble average for the Wiener process is denoted by $E[\ ]$. The intensity of the stochasticity is characterized by the parameter ν.

The standard definition of velocity in classical mechanics is not applicable in stochastic trajectories because the left-hand and right-hand limits of a zigzag path do not agree. To accommodate this ambiguity, we consider also the backward time evolution of the trajectory described by the backward SDE:

$$d\mathbf{x}(t) = \mathbf{u}_-(\mathbf{x}(t), t)dt + \sqrt{2\nu}d\mathbf{W}_-(t) \quad (dt < 0),$$ (5)

where $\mathbf{W}_-(t)$ represents another standard Wiener process. The vector function $\mathbf{u}_-(\mathbf{x}, t)$ is determined from $\mathbf{u}_+(\mathbf{x}, t)$ using the consistency condition, discussed below.

Nelson introduced two different time derivatives [37]: one is the mean forward derivative D_+ and the other the mean backward derivative D_-, which are defined by

$$D_{\pm}\mathbf{x}(t) = \lim_{dt \to 0\pm} E\left[\frac{\mathbf{x}(t+dt) - \mathbf{x}(t)}{dt}\Big|\mathbf{x}(t)\right] = \mathbf{u}_{\pm}(\mathbf{x}(t), t).$$ (6)

Here, the expectation value is the conditional average for fixing $\mathbf{x}(t)$, and we used the fact that $\mathbf{x}(t)$ is Markovian.

The two unknown vector functions $\mathbf{u}_{\pm}(\mathbf{x}, t)$ are not independent. To see this, let us introduce the particle distribution, which is defined by

$$\rho(\mathbf{x}, t) = \int d^3\mathbf{R}_i\, \rho_0(\mathbf{R}_i)E[\delta(\mathbf{x} - \mathbf{x}(t))],$$

where $\mathbf{R}_i = \mathbf{x}(t_i)$ denotes the initial position of the stochastic particle, and its distribution is characterized by $\rho_0(\mathbf{R}_i)$. Applying the forward and backward SDEs to this definition, two Fokker–Planck equations are obtained ($\partial_t \equiv \partial/\partial t$):

$$\partial_t\rho(\mathbf{x}, t) = -\nabla \cdot \{\mathbf{u}_+(\mathbf{x}, t)\rho(\mathbf{x}, t)\} + \nu\nabla^2\rho(\mathbf{x}, t),$$ (7)
$$\partial_t\rho(\mathbf{x}, t) = -\nabla \cdot \{\mathbf{u}_-(\mathbf{x}, t)\rho(\mathbf{x}, t)\} - \nu\nabla^2\rho(\mathbf{x}, t).$$ (8)

To conform Equation (8) to Equation (7), $\mathbf{u}_-(\mathbf{x}, t)$ should be chosen to satisfy the consistency condition:

$$\mathbf{u}_+(\mathbf{x}, t) = \mathbf{u}_-(\mathbf{x}, t) + 2\nu\nabla \ln \rho(\mathbf{x}, t).$$ (9)

A more general representation of the consistency condition is shown in Equation (25) of [32]. It is also noteworthy that these Fokker–Planck equations are reduced to the same equation of continuity:

$$\partial_t\rho(\mathbf{x}, t) = -\nabla \cdot (\rho(\mathbf{x}, t)\mathbf{v}(\mathbf{x}, t)),$$ (10)

where the mean velocity field is defined by

$$\mathbf{v}(\mathbf{x}, t) = \frac{\mathbf{u}_+(\mathbf{x}, t) + \mathbf{u}_-(\mathbf{x}, t)}{2}.$$ (11)

To apply the stochastic variation to the single-particle Lagrangian (2), one needs to replace d/dt with D_+ and D_- of Equation (6). Suppose that the kinetic term of the Lagrangian is replaced with the average of the two time derivatives. Then, the single-particle stochastic Lagrangian is defined by

$$L_{\mathrm{sto}}(\widehat{\mathbf{x}}, D_+\widehat{\mathbf{x}}, D_-\widehat{\mathbf{x}}) = \frac{m}{2} \frac{(D_+\mathbf{x}(t))^2 + (D_-\mathbf{x}(t))^2}{2} - V(\mathbf{x}(t)).$$ (12)

Let us define the variation of the stochastic trajectory by

$$\mathbf{x}(t) \longrightarrow \mathbf{x}'(t) = \mathbf{x}(t) + \delta f(\mathbf{x}(t), t),$$ (13)

where $\delta f(\mathbf{x}, t)$ is an infinitesimal function satisfying

$$\delta f(\mathbf{x}, t_i) = \delta f(\mathbf{x}, t_f) = 0.$$ (14)

The stochastic variation of the stochastic Lagrangian (12) leads to the stochastic Euler–Lagrange equation:

$$\left[D_- \frac{\partial L_{\mathrm{sto}}}{\partial (D_+\mathbf{x}(t))} + D_+ \frac{\partial L_{\mathrm{sto}}}{\partial (D_-\mathbf{x}(t))} - \frac{\partial L_{\mathrm{sto}}}{\partial \mathbf{x}(t)} \right]_{\mathbf{x}(t)=\mathbf{x}} = 0.$$ (15)

Substituting Equation (12), one finds:

$$(\partial_t + \mathbf{v} \cdot \nabla)\mathbf{v} = -\frac{1}{m}\nabla V + 2v^2 \nabla \frac{\nabla^2 \sqrt{\rho}}{\sqrt{\rho}}.$$ (16)

Note that $\rho(\mathbf{x}, \mathbf{t})$ is described by the equation of continuity (10).

The above result, Equation (16), of the stochastic variation can be cast into a more familiar form. Let us introduce a complex function, defined by

$$\psi(\mathbf{x}, t) = \sqrt{\rho(\mathbf{x}, t)} e^{i\theta(\mathbf{x}, t)},$$ (17)

where the phase, θ, is defined by

$$\mathbf{v}(\mathbf{x}, t) = 2v\nabla\theta(\mathbf{x}, t).$$ (18)

The evolution equation of $\theta(\mathbf{x}, t)$ is obtained from Equation (16), and then, $\psi(\mathbf{x}, t)$ satisfies

$$i\partial_t \psi(\mathbf{x}, t) = \left[-v\nabla^2 + \frac{1}{2vm} V(\mathbf{x}) \right] \psi(\mathbf{x}, t).$$ (19)

When

$$v = \frac{\hbar}{2m},$$ (20)

is chosen, Equation (19) becomes the Schrödinger equation; here, $\hbar$ is the reduced Planck's constant. Then, actually, Equation (16) coincides with the so-called Madelung's hydrodynamic representation of the Schrödinger equation.

3. Generalized Viscous Hydrodynamics

In Section 2 above, it is shown that quantization is the stochastic optimization of the classical action. That is, when we observe a single-particle system with a macroscopic scale where the non-differentiability of the particle trajectory becomes negligibly small, we can apply the classical variation to the single-particle action, and the Newton equation is obtained. However, if the system has a microscopic scale, one has to optimize the corresponding action with the stochastic variation, and the Schrödinger equation is derived. In this sense, we can understand that classical mechanics emerges as the result of the coarse-graining caused by large differences in scales of observables.

Let us apply this idea to another example of coarse-graining in hydrodynamics. The behavior of the ideal fluid is described by the Euler equation, while that of the viscous (Newtonian) fluid is described by the Navier–Stokes–Fourier (NSF) equation. In an ideal fluid, the local thermal equilibrium is perfectly satisfied for each fluid element during their time evolution. Therefore, the thermodynamic property of the fluid elements changes in a quasi-static manner. In other words, the time scale of the collective dynamics of an ideal fluid is much longer than the local relaxation time for thermal equilibrium. On the other hand, when the fluid changes in a relatively smaller scale, there exist small deviations from the thermal equilibrium in the internal states. Since this deviation is not represented by a function of thermodynamic quantities, the fluid obtains extra acceleration mechanisms attributed to the non-equilibrium nature of the fluid elements. Therefore, one can consider that the difference between the Euler and NSF equations comes from that of the coarse-graining in the spacetime scale in the hydrodynamic behavior. In such a smaller scale, the trajectories of the fluid elements are not always smooth, as is the case of the Schrödinger equation. If this idea is correct, one can obtain the NSF equation from the Euler equation by using the stochastic variation.

As mentioned in the Introduction, the motion of a fluid can be represented by the ensemble of the motions of fluid elements. We consider a simple fluid composed of N constituent particles with mass m. The Lagrangian of the system is given by the sum of the contributions from each fluid element as

$$L = \int d^3\mathbf{R}_i \rho_0(\mathbf{R}_i) \left[\frac{m}{2} \left(\frac{d\mathbf{x}(t)}{dt} \right)^2 - E_{\mathbf{x}(t)} \right],$$
(21)

where $\mathbf{x}(t)$ is the Lagrange coordinate of the fluid element. Each fluid element involves a fixed number of constituent particles, which is conserved in the time evolution. Then, the initial distribution of the fluid elements is characterized by $\rho_0(\mathbf{R}_i)$, normalized by N,

$$\int d^3\mathbf{R}_i \, \rho_0(\mathbf{R}_i) = N.$$
(22)

The internal energy per particle in the fluid element $\mathbf{R}_i$ at time t is denoted by $E_{\mathbf{x}(t)}$ and given as $E_{\mathbf{x}(t)} = \varepsilon/\rho$. Here, $\varepsilon = \varepsilon(\rho)$ is the internal energy density and $\rho = \rho(\mathbf{x}(t), t)$ is the number density of the constituent particles of the fluid element at time t for a given initial position $\mathbf{R}_i$.

Applying the classical variation to this Lagrangian, one obtains the Euler equation, which describes the motion of the ideal fluid:

$$(\partial_t + \mathbf{v} \cdot \nabla)\mathbf{v} = -\frac{1}{m\rho}\nabla P,$$
(23)

where

$$P = -\frac{d}{d\rho^{-1}} \frac{\varepsilon(\rho)}{\rho}.$$
(24)

In the Lagrange coordinates moving with the fluid elements of the ideal fluid, the specific entropy of each fluid element is constant. Thus, the quantity (24) can be identified with the thermodynamic pressure,

$$P = -\left(\frac{\partial}{\partial \rho^{-1}} \frac{\varepsilon}{\rho} \right)_s,$$
(25)

where s is the specific entropy of the fluid element. The detailed mathematical derivation of the pressure term in the variation is shown in [32].

Let us apply the stochastic variation to the ideal classical Lagrangian (21). As was mentioned in the derivation of the Schrödinger equation, there is an ambiguity for the replacement in the kinetic term of the stochastic Lagrangian, and we simply assumed that the stochastic kinetic term is given by the average of the contributions of D_+ and D_-. Here, we consider a more general situation. Suppose that the kinetic term is given by the most general quadratic form of $D_\pm$, defined in Equation (6),

$$\frac{m}{2}\left[B_+\{A_+(D_+\mathbf{x}(t))^2 + A_-(D_-\mathbf{x}(t))^2\} + B_-(D_+\mathbf{x}(t))(D_-\mathbf{x}(t))\right], \tag{26}$$

where

$$A_\pm = \frac{1}{2} \pm \alpha_A, \tag{27}$$

$$B_\pm = \frac{1}{2} \pm \alpha_B. \tag{28}$$

and α_A and α_B are real parameters. In the vanishing limit of the noise, $\nu \to 0$, the two mean derivatives are reduced to the standard time derivative, and thus, the kinetic term (26) coincides with the classical kinetic term,

$$\lim_{\nu \to 0} \frac{m}{2}\left[B_+\{A_+(D_+\mathbf{x}(t))^2 + A_-(D_-\mathbf{x}(t))^2\} + B_-(D_+\mathbf{x}(t))(D_-\mathbf{x}(t))\right] = \frac{m}{2}\left(\frac{d\mathbf{x}(t)}{dt}\right)^2.$$

Using this, the general form of the stochastic Lagrangian, corresponding to Equation (21), is given by

$$L = \int d^3\mathbf{R}_i \rho_0(\mathbf{R}_i) E\left[\frac{m}{2}\left\{B_+A_+(D_+\mathbf{x}(t))^2 + B_+A_-(D_-\mathbf{x}(t))^2 + B_-(D_+\mathbf{x}(t))(D_-\mathbf{x}(t))\right\}\right.$$
$$\left. - \frac{\varepsilon(\rho(\mathbf{x}(t),t))}{\rho(\mathbf{x}(t),t)}\right]. \tag{29}$$

The stochastic variation finally leads to the following equation,

$$m(\partial_t + \mathbf{v}\cdot\nabla)v^i = 2\kappa\partial_i\frac{\nabla^2\sqrt{\rho}}{\sqrt{\rho}} - \frac{1}{\rho}\partial_i\{P - \zeta(\nabla\cdot\mathbf{v})\} + \frac{1}{\rho}\sum_{j=1}^{D}\partial_j\left(\eta E^{ij}\right), \tag{30}$$

where the traceless symmetric stress tensor is defined by

$$E^{ij} = \frac{1}{2}\left(\partial_i v^j + \partial_j v^i\right) - \frac{1}{3}(\nabla\cdot\mathbf{v})\delta_{ij}.$$

The shear viscosity η and the bulk viscosity ζ are defined by

$$\eta = 2m\alpha_A(1 + 2\alpha_B)\nu\rho, \tag{31}$$

$$\zeta = \mu + \frac{\eta}{3}. \tag{32}$$

The second coefficient of viscosity, μ, emerges from the variation of ε through the changes of the associated specific entropy, as suggested in [33]. One can find that Equation (30) is identical to the NSF equation,

$$m(\partial_t + \mathbf{v}\cdot\nabla)v^i = -\frac{1}{\rho}\partial_i\{P - \zeta(\nabla\cdot\mathbf{v})\} + \frac{1}{\rho}\sum_{j=1}^{D}\partial_j\left(\eta E^{ij}\right), \tag{33}$$

except for the term that contains κ, which is discussed in Section 4 below.

Here, the stochastic optimization of the averaged behavior of fluid dynamics was considered in terms of the fluctuating motion of the fluid element. It is however possible to consider models, where the noise terms are added directly to hydrodynamics; for recent studies, see [38,39] (and references therein).

4. Quantum Effects and Surface Energy

The coefficient κ is defined by

$$\kappa = 2\alpha_B \nu^2 / m .$$

This does not appear in the NSF equation (33), but one can immediately notice that it turns out to be the quantum potential term by choosing $\kappa = \hbar^2 / 4m$. Indeed, Equation (30) is sometimes used as a model of a quantum viscous fluid [40,41].

The appearance of the κ term in generalized hydrodynamics has been discussed for a long period of time. For example, Brenner pointed out that, since the velocity of a tracer particle of a fluid is not necessarily parallel to the mass velocity, the existence of these two velocities should be taken into account in the formulation of hydrodynamics. This theory is called bivelocity hydrodynamics [32,34,42–44], and Equation (30) is understood to be one of the variants.

Another example is related to the diffuse-interface models of hydrodynamics [45]. The properties of the interface between two fluids has been studied since the 19th Century by Young, Gauss, Maxwell, Gibbs, Rayleigh, van der Waals, and others. In particular, Korteweg [46] considered that the behavior of liquid–vapor fluids near phase transitions is described by the Navier–Stokes–Korteweg (NSK) equation, and our Equation (30) is its special case. Then, the κ term describes the capillary action; for more details, see review [45].

As a matter of fact, the κ term is related intimately to the surface tension and can be incorporated into the classical Lagrangian. Modifying the internal energy term of the Lagrangian Equation (21) as

$$L \to L = \int d^3\mathbf{R}_i \rho_0(\mathbf{R}_i) \left[\frac{m}{2}\left(\frac{d\mathbf{x}(t)}{dt}\right)^2 - \frac{\varepsilon}{\rho} - \frac{\kappa}{4}\left(\frac{\nabla\rho}{\rho}\right)^2 \right], \tag{34}$$

the classical variational approach leads to the κ term in the equation of motion Equation (30) without the viscous term ($\zeta, \eta = 0$). This type of internal energy was introduced in the Thomas–Fermi model for the nuclear density distribution, showing the relation between the κ term and the surface energy for a saturating system, such as nuclear matter [47,48].

To illustrate the role of the noise in the SVM formulation as the surface energy, let us consider the hydrostatic equilibrium of a fluid with such a κ term in the Lagrangian. The density distribution at the hydrostatic equilibrium for a given total number of constituent particles is given by

$$\delta\{E_{ToT}[\rho] - \lambda N[\rho]\} = 0, \ \forall \delta\rho,$$

where

$$E_{ToT}[\rho] = \int d^3\mathbf{x}\,\rho(\mathbf{x}) \left\{ \frac{\varepsilon(\rho)}{\rho} + \frac{\kappa}{4}\left(\frac{\nabla\rho}{\rho}\right)^2 \right\}, \tag{35}$$

and

$$N[\rho] = \int d^3\mathbf{x}\,\rho(\mathbf{x}) \tag{36}$$

is the total number of particles. Here, the number density ρ should be non-negative and λ is the Lagrangian multiplier (chemical potential). Applying the usual variational procedure with respect to the density ρ, we readily obtain the following differential equation for the spherical symmetric case,

$$\frac{d^2\phi}{dr^2} = \frac{1}{\kappa}\left(\frac{d\varepsilon}{d\rho} - \lambda\right)\phi, \tag{37}$$

where r is the radial coordinate and

$$\phi(r) = r\sqrt{\rho(r)}.$$

In order to satisfy the boundary condition for ϕ at $r \to \infty$, i.e., $\lim_{r\to\infty}\rho(r) \to 0$, Equation (37) should be solved as an eigenvalue problem of λ for a given value of central density $\rho|_{r=0}$.

For a system such as nuclear matter, the internal energy par particle as a function of ρ is characterized by the following properties:

$$\frac{\varepsilon}{\rho} \to 0, \ \rho \to 0,$$

$$\left.\frac{\varepsilon}{\rho}\right|_{\rho_{eq}} < 0,$$

$$\left.\frac{d}{d\rho}\left(\frac{\varepsilon}{\rho}\right)\right|_{\rho=\rho_{eq}} = 0, \tag{38}$$

$$\frac{\varepsilon}{\rho} \to \infty, \ \rho \to \rho_S,$$

where ρ_{eq} is the equilibrium density for an infinite matter and $\rho_S\,(> \rho_{eq})$ is the maximum density due to the strong repulsive forces at close distances between particles (hard core). As an example, let us consider the following function:

$$\frac{\varepsilon}{\rho} = \epsilon\left(-\frac{1}{4} + \frac{1}{(\rho_S/\rho_{eq} - \rho/\rho_{eq})^2} - 2\frac{1}{(\rho_S/\rho_{eq} - 1)^3}\frac{\rho}{\rho_{eq}}\right), \tag{39}$$

where ϵ is a positive constant with the dimension of energy.

Figure 1 shows the radial dependences of the number density, $\bar{\rho} = \rho/\rho_{eq}$, which are given by the eigenfunctions of Equation (37) by setting $\rho_S = 2\rho_{eq}$. The result is plotted by using an adimensional radial variable $x = \sqrt{\epsilon/\kappa}\,r$. The solid, dotted, and dashed lines correspond to $\bar{\rho}(0) = 1.18, 1.19$, and 1.20, respectively. The line for $\bar{\rho} = 1$ corresponds to the infinite homogeneous matter in its hydrostatic equilibrium. These density distributions show the plateau in the central bulk domain and the exponentially decreasing surface. Thus, these distributions are reminiscent of the known Wood–Saxon distribution for the nuclear density. The surface thicknesses of the three lines are almost constant and scaled by the value of κ. Indeed, because of the definition of x, one can easily see that the larger κ, the wider the surface thickness. Due to the practically vanishing radial derivative except for the surfaces, the positive contribution of the κ term in Equation (35) comes basically from the surfaces that have almost the same thickness for the three lines. Therefore, to minimize the energy fixing the total N, the area of the surface should be reduced. This is the reason why the central density increases for a smaller N.

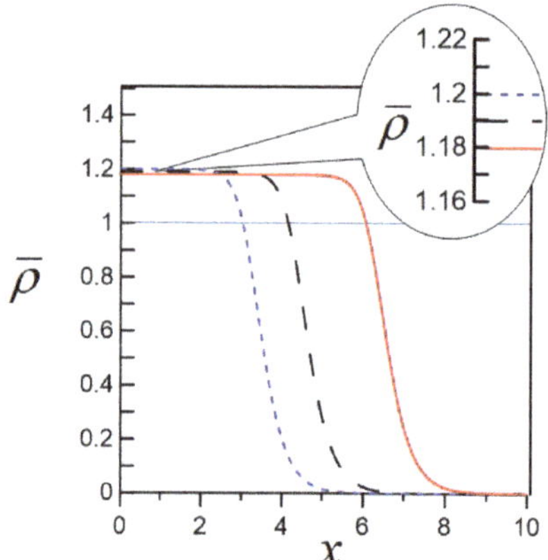

Figure 1. The radial dependences of the number density, obtained by solving Equation (37). Here, $\bar{\rho} = \rho/\rho_{\mathrm{eq}}$, $x = \sqrt{\epsilon/\kappa}\,r$, and $\rho_S = 2\rho_{\mathrm{eq}}$ is set. The solid, dotted, and dashed lines correspond to $\bar{\rho}(0) = 1.18, 1.19$, and 1.20, respectively. The line $\bar{\rho} = 1$ corresponds to the infinite homogeneous matter in its hydrostatic equilibrium. See text for details.

5. Uncertainty Relation in Hydrodynamics

The uncertainty relations characterize an important feature in quantum physics, and its comprehension requires unceasing improvement. Recently, the present authors proposed a new formulation of the uncertainty relations, which reproduce quantum mechanical results and are also applicable to general stochastic dynamics within the framework of SVM [34]. Using this, one can derive the uncertainty relations for the motion of the fluid element of a viscous (Newtonian) fluid.

To illustrate the method, let us consider the single-particle stochastic Lagrangian (12), which is applied to obtain the Schrödinger equation. In this Lagrangian, the two time-derivatives are introduced and are attributed to the non-differentiability of the stochastic trajectories. Then, two momenta are introduced through the Legendre transformation of the stochastic Lagrangian:

$$\mathbf{p}_{\pm}(\mathbf{x}, t) = 2 \left. \frac{\partial L_{\mathrm{sto}}}{\partial D_{\pm}\mathbf{x}(t)} \right|_{\mathbf{x}(t)=\mathbf{x}}. \tag{40}$$

Here, the factor 2 in the definitions is introduced for a convention to reproduce the classical result in the vanishing limit of ν.

The standard deviation of position of a quantum particle is defined by

$$\sigma_{x^i}^{(2)} = \lceil (\delta x^i)^2 \rfloor ,$$

where $\delta f = f(\mathbf{x}, t) - \lceil f \rfloor$, the following expectation value is introduced:

$$\lceil f \rfloor = \frac{1}{N} \int d^D \mathbf{x}\, \rho(\mathbf{x}, t) f(\mathbf{x}, t) , \tag{41}$$

where N is the normalization factor of ρ analogous to Equation (22) and is a unity, $N = 1$, for quantum mechanics of single particle system, and D denotes the number of the spatial dimension. The standard deviation of momentum is defined by the average of the two standard deviations:

$$\sigma_{p^i}^{(2)} = \frac{\lceil (\delta p_+^i)^2 \rfloor + \lceil (\delta p_-^i)^2 \rfloor}{2} ,$$

Using these definitions and the Cauchy–Schwarz inequality, we can show that the product of $\sigma^{(2)}_{x^i}$ and $\sigma^{(2)}_{p^j}$ satisfies the inequality [34]:

$$\sigma^{(2)}_{x^i}\sigma^{(2)}_{p^j} \geq \frac{\hbar^2}{4}\delta_{ij} + (m\lceil \delta x^i \delta v^j \rfloor)^2 , \tag{42}$$

where $\mathbf{v}(\mathbf{x}, t)$ is defined by Equation (11). The second term on the right-hand side of Equation (42) is expressed in terms of the quantum-mechanical notation:

$$m\lceil \delta x^i \delta v^j \rfloor = \mathrm{Re}[\langle (x^i_{\mathrm{op}} - \langle x^i_{\mathrm{op}}\rangle)(p^j_{\mathrm{op}} - \langle p^j_{\mathrm{op}}\rangle)\rangle] , \tag{43}$$

where $\mathbf{x}_{\mathrm{op}}$ and $\mathbf{p}_{\mathrm{op}}$ are the position and momentum operators, respectively, and $\langle \ \rangle$ denotes the expectation value with a wave function. With this identification, one finds that Equation (42) is the Robertson–Schrödinger inequality. When the second term on the right-hand side is ignored, this becomes the Kennard inequality.

The advantage of the present approach compared to the standard canonical formulation is that it is easily extended to the generalized coordinates systems. Let us consider the generalized coordinate, q^i, and the corresponding canonical momentum, p_j. Then, the uncertainty relation in the generalized coordinates system is given by [35]

$$\sigma^{(2)}_{q^i}\sigma^{(2)}_{p_j} \geq \frac{\hbar^2}{4}\left| \delta^i_j - \int d^D q \partial_j \{J\rho(q^i - E[q^i(t)])\} + \int J d^D q\, \rho(q^i - E[q^i(t)])\Gamma^k_{jk}\right|^2$$
$$+ |E[(\delta q^i(t))(\delta p_j(t))]|^2 , \tag{44}$$

where J is the Jacobian of the generalized coordinates and Γ^i_{jk} is the Christoffel symbol. The term next to δ^i_j on the right-hand side gives a finite contribution when q^i is a periodic variable. Actually, for the angle variable θ in the polar coordinates, this is reduced to

$$\sigma^{(2)}_{\theta}\sigma^{(2)}_{L} \geq \frac{\hbar^2}{4}\left| 1 - 2\pi \int_0^{\infty} r dr \rho(r, 2\pi, t)\right|^2 + |E[(\delta\theta(t))(\delta L(t))]|^2 , \tag{45}$$

where L is the angular momentum. For the eigenstate of the angular momentum, the right-hand side vanishes, $\sigma^{(2)}_{\theta}\sigma^{(2)}_{p_\theta} \geq 0$. The above inequality resolves the famous problem in the angular uncertainty relation.

In the above calculations, the lower bound of the inequality is due to the consistency condition (9), which comes from the consistency between the forward and backward SDEs. That is, the finite minimum uncertainty is attributed to the non-differentiability of trajectories. Therefore, the similar inequality should be satisfied for the motion of the fluid element in viscous hydrodynamics. We now apply the same procedure to the Lagrangian of generalized viscous hydrodynamics (29). Different from the quantum-mechanical case, we consider the motion of the fluid element. Thus, the uncertainty relation in this application reads the restriction for the fluctuating motion of the fluid element. Then, the uncertainty relation for the fluid element of generalized viscous hydrodynamics [32,34] is given by

$$\sigma^{(2)}_{x^i}\sigma^{(2)}_{p^j} \geq m^2 \frac{(\tilde{\zeta}^2 - \kappa)^2}{\nu^2 + \tilde{\zeta}^2}\delta_{ij} , \tag{46}$$

where m is the mass of the constituent particle of a simple fluid and $\tilde{\zeta}$ is the kinematic viscosity

$$\tilde{\zeta} = \frac{\eta}{2m\rho} . \tag{47}$$

Different from the quantum-mechanical case, the minimum uncertainty on the right-hand side of the inequality (46) is not a constant, but a function of thermodynamic variables through viscosity. For the case of the NSF equation (33), where $\kappa = 0$, one can find that the uncertainty will be enhanced for larger viscosity.

In the same inequality (46), the right hand side shows that the minimum uncertainty of the inviscid fluid is modified by the effect of viscosity. It is natural to assume that the minimum uncertainty of the inviscid fluid is larger than the quantum-mechanical minimum, $\hbar/2$. Suppose that the viscous effect does not improve uncertainty beyond the inviscid (and thus quantum-mechanical) minimum. To satisfy this condition, the kinematic viscosity has the following lower bound [35]:

$$\zeta \geq \frac{\sqrt{3}}{2}\frac{\hbar}{m}.\tag{48}$$

This lower bound is the same order of magnitude as the Kovtun–Son–Starients (KSS) bound of the shear viscosity [49]. This may suggest the close relation between the KSS bound and the minimum uncertainty relation. See also [50].

In quantum mechanics, it is known that the minimum uncertainty state is given by the coherent state. Even in viscous hydrodynamics, one can write down the minimum uncertainty state, which is given by the generalized coherent state; for details, see [35].

6. Quantum Mechanics in Curved Spacetime

The framework of the SVM is applicable to particle systems described by generalized coordinates in curved geometries. For this application, the modification of the noise term in the SDEs is crucial. For example, suppose that the forward SDE for the radial component in the polar coordinates is expressed by the direct generalization of Equation (3):

$$dr(t) = u_+^r(r(t),\theta(t),t)dt + \sqrt{2\nu}dW_+(t) \quad (dt) > 0,\tag{49}$$

where $W_+(t)$ is the Wiener process. This equation however does not function because $dW_+(t)$ can take any real value at random, and thus, the positivity $r(t) > 0$ can be violated.

In the following, we discuss a non-relativistic SVM system in the curved geometry following the formulation developed in [51,52]. Let us consider a curved spacetime geometry characterized by the metric $g_{\mu\nu}$ where g_{00} is a function only of time, and the indices denoted by the Greek letters stay for the time (0) and space (1,2,3) components. Moreover, there is no mixture terms of the spacetime components, $g_{0i} = 0$. The position in this generalized coordinates is denoted by q^μ. On the other hand, one can find local Minkowskian coordinates around q^μ, which is denoted by y^a. The Minkowskian metric is $\eta_{ab} = \mathrm{diag}(-1,1,1,1)$. The Latin indices a,b,c etc. stay for the local Minkowskian coordinates, i,j,k, etc. are reserved to denote the spatial components of q^μ. Then, one can introduce the tetrads, defined by

$$e_a^\mu(Q) = \left.\frac{\partial q^\mu}{\partial y^a}\right|_{q=Q},\tag{50}$$

$$\bar{e}_\mu^a(Q) = \left.\frac{\partial y^a}{\partial q^\mu}\right|_{q=Q}.\tag{51}$$

These tetrads satisfy

$$g_{\mu\nu}(Q)e_a^\mu(Q)e_b^\nu(Q) = \eta_{ab},\tag{52}$$

$$\eta_{ab}\bar{e}_\mu^a(Q)\bar{e}_\nu^b(Q) = g_{\mu\nu}(Q).\tag{53}$$

where the convention of taking summation for repeated indices is used. The forward SDE in generalized coordinates is given by

$$dq^0(t) = c\,dt\,, \tag{54}$$

$$dq^i(t) = u^i_+(q(t))dt + \sqrt{2\nu}\underline{e}^i_a(q(t)) \circ_s dW^a_+(t) \quad (dt) > 0\,, \tag{55}$$

where c denotes the speed of light, $q(t) = (q^0(t), \vec{q}(t))$ and the Stratonovich definition of the product is introduced:

$$f(q_t) \circ_s dW^a(t) = f(q_{t+dt/2})dW^a(t)\,, \tag{56}$$

for an arbitrary smooth function $f(x)$. Since $dW^a_+(t)$ is defined in the local Minkowskian coordinates, one can use the same Wiener process been applied so far. The backward SDE is defined in a similar way; see [51,52] for details.

Let us consider the single-particle system of mass m, which is described by the following stochastic Lagrangian:

$$L_{\text{sto}} = \frac{m}{4}\{(D_+y^a)\,\eta_{ab}\,(D_+y^b) + (D_-y^a)\,\eta_{ab}\,(D_-y^b)\} - V\,, \tag{57}$$

where $y^a = y(q(t))$ and V is a potential energy. Here, the kinetic term is expressed by the average of the two contributions, D_+ and D_-. For the case of more general replacement, one obtains the viscous term [52].

The stochastic variation then leads to

$$v^\mu \nabla_\mu v^i + \frac{g^{ij}}{m}\partial_j V = 2\nu^2 g^{ij}\left(\partial_j \frac{1}{\sqrt{\rho}}\nabla_{LB}\sqrt{\rho} - \frac{1}{2}R^k_j \partial_k \ln\rho\right)\,, \tag{58}$$

where ∇_μ, ∇_{LB}, and $R_j{}^k$ are the covariant derivative, the Laplace–Beltrami operator, and the Ricci tensor $R^k_j = g^{kl}R_{jl} = g^{kl}R^\mu{}_{jl\mu}$, respectively. The four-velocity field is defined by

$$v^\mu(Q) = (v^0, \vec{v}(Q)) = \left(c, \frac{u^i_+(Q) + u^i_+(Q)}{2}\right)\,. \tag{59}$$

The probability density ρ satisfies the equation of continuity,

$$\nabla_\mu(\rho v^\mu) = 0\,. \tag{60}$$

The remarkable feature of Equation (58) is the last term on the right-hand side, which is induced by the interplay between quantum fluctuation and spacetime curvature. We call this the quantum-curvature (QC) term. To investigate the role of this term, $\nu = \hbar/(2m)$ is set. Then, taking the limit of the flat spacetime, Equation (58) is reduced to Madelung's hydrodynamic representation of the Scrödinger Equation (16). To express the above result in terms of the wave function, we have to obtain the equation of the phase of the wave function from Equation (58). This is impossible in general due to the $q(t)$-dependence of $R_j{}^k$. That is, no wave function can be introduced to express quantum dynamics in general geometries. It should be emphasized that, normally, the existence of the Hilbert space is required from the beginning of quantization, but this is not trivial a priori.

However, if the Ricci tensor is a constant, then one can still introduce the wave function, while the corresponding Schrödinger equation becomes non-linear. For example, let us set

$$R_j{}^k = -2\gamma\delta^k_j\,, \tag{61}$$

$$v^i = \frac{\hbar}{m}g^{ij}\partial_j\theta\,. \tag{62}$$

24. Wen, D.; Castilho, W.M.; Lin, K.; Qian, W.L.; Hama, Y.; Kodama, T. On the peripheral tube description of the two-particle correlations in nuclear collisions. *J. Phys. G Nucl. Part. Phys.* **2019**, *46*, 035103. [CrossRef]
25. Hama, Y.; Kodama, T.; Qian, W.L. Two-particle correlations at high-energy nuclear collisions, peripheral-tube model revisited. *J. Phys. G Nucl. Part. Phys.* **2020**, *48*, 015104. [CrossRef]
26. Bhalerao, R.S.; Ollitrault, J.-Y.; Pal, S. Event-plane correlators. *Phys. Rev. C* **2013**, *88*, 024909. [CrossRef]
27. Bhalerao, R.S.; Ollitrault, J.-Y.; Pal, S. Characterizing flow fluctuations with moments. *Phys. Lett. B* **2015**, *742*, 94–98. [CrossRef]
28. Gaspar Elsas, J.H.; Koide, T.; Kodama, T. Noether's theorem of relativistic–electromagnetic Ideal hydrodynamics. *Braz. J. Phys.* **2015**, *45*, 334–339. [CrossRef]
29. Yasue, K. Stochastic calculus of variation. *J. Funct. Anal.* **1981**, *41*, 327–340. [CrossRef]
30. Zambrini, J.C. Stochastic dynamics: A review of stochastic calculus of variations. *Int. J. Theor. Phys.* **1985**, *24*, 277–327. [CrossRef]
31. Koide, T.; Kodama, T.; Tsushima, T. Unified description of classical and quantum behaviours in a variational principle. *J. Phys. Conf. Ser.* **2015**, *626*, 012055. [CrossRef]
32. de Matos, G.G.; Kodama, T.; Koide, T. Uncertainty relations in hydrodynamics. *Water* **2020**, *12*, 3263. [CrossRef]
33. Koide, T.; Kodama, T. Navier-Stokes, Gross-Pitaevskii and generalized diffusion equations using the stochasticvariational method. *J. Phys. A Math. Gen.* **2012**, *45*, 255204. [CrossRef]
34. Koide, T.; Kodama, T. Generalization of uncertainty relation for quantum and stochastic systems. *Phys. Lett. A* **2018**, *382*, 1472–1480. [CrossRef]
35. Gazeau, J.-P.; Koide, T. Uncertainty relation for angle from a quantum-hydrodynamical perspective. *Ann. Phys.* **2020**, *416*, 168159. [CrossRef]
36. Koide, T. Viscous control of minimum uncertainty state in hydrodynamics. *J. Stat. Mech. Theo. Exp.* **2022**, *2022*, 023210. [CrossRef]
37. Nelson, E. Derivation of the Schrödinger equation from Newtonian mechanics. *Phys. Rev.* **1966**, *150*, 1079–1085. [CrossRef]
38. De, A.; Shen, C.; Kapusta, J.I. Stochastic hydrodynamics meets hydro-kinetics. *arXiv* **2022**, arXiv:2203.02134. [CrossRef]
39. Simmons, S.A.; Pillay, J.C.; Kheruntsyan, K.V. Phase-space stochastic quantum hydrodynamics for interacting Bose gases. *arXiv* **2022**, arXiv:2202.10609. [CrossRef]
40. Brull, S.; Méhats, F. Derivation of viscous correction terms for the isothermal quantum Euler mode. *Z. Angew. Math. Mech. [J. Appl. Math. Mech.]* **2010**, *90*, 219–230. [CrossRef]
41. Bresch, D.; Gisclon, M.; Lacroix-Violet, I. On Navier-Stokes-Korteweg and Euler-Korteweg systems: Application to quantum fluids models. *Arch. Ration. Mech. Anal.* **2019**, *233*, 975–1025. [CrossRef]
42. Brenner, H. Is the tracer velocity of a fluid continuum equal to its mass velocity? *Phys. Rev. E* **2004**, *70*, 061201. [CrossRef]
43. Koide, T.; Ramos, R.O.; Vicente, G.S. Bivelocity picture in the nonrelativistic limit of relativistic hydrodynamics. *Braz. J. Phys.* **2015**, *45*, 102–111. [CrossRef]
44. Reddy, M.L.; Dadzie, S.K.; Ocone, R.; Borg, M.K.; Reese, J.M. Recasting Navier-Stokes equations. *J. Phys. Commun.* **2019**, *3*, 105009. [CrossRef]
45. Anderson, D.M.; McFadden, G.B.; Wheeler, A.A. Diffuse-interface methods in fluid mechanics. *Annu. Rev. Fluid Mech.* **1998**, *30*, 139–165. [CrossRef]
46. Korteweg, D.J. Sur la forme que prennent les équations du mouvement si l'on tient compte de forces capillaires causées par les variations de densité considérables mais continues et sur la théorie de la capillarité dans l'hypothèse d'une variation continue de la densité. *Arch. Néerl. Sci. Exact. Natur.* **1901**, *6*, 1—24.
47. Berg, R.A.; Willets, L. Nuclear surface effects. *Phys. Rev.* **1956**, *101*, 201–204. [CrossRef]
48. Willets, L. Theories of the nuclear surface. *Rev. Mod. Phys.* **1958**, *30*, 542–549. [CrossRef]
49. Kovtun, P.K.; Son, D.T.; Starinets, A.O. Viscosity in strongly interacting quantum field theories from black hole physics. *Phys. Rev. Lett.* **2005**, *94*, 111601. [CrossRef]
50. Danielewicz, P.; Gyulassy, M. Dissipative phenomena in quark-gluon plasmas. *Phys. Rev. D* **1985**, *31*, 53–62. [CrossRef]
51. Koide, T.; Kodama, T. Novel effect induced by spacetime curvature in quantum hydrodynamics. *Phys. Lett. A* **2019**, *383*, 2713–2718. [CrossRef]
52. Koide, T.; Kodama, T. Variational formulation of compressible hydrodynamics in curved spacetime and symmetry of stress tensor. *J. Phys. A Math. Theor.* **2020**, *53*, 215701. [CrossRef]
53. Unruh, W.G. Experimental black-hole evaporation? *Phys. Rev. Lett.* **1981**, *46*, 1351–1353. [CrossRef]
54. Lahav, O.; Itah, A.; Blumkin, A.; Gordon, C.; Rinott, S.; Zayats, A.; Steinhauer, J. Realization of a sonic black hole analog in a Bose-Einstein con-densate. *Phys. Rev. Lett.* **2010**, *105*, 240401. [CrossRef] [PubMed]
55. Steinhauer, J. Observation of quantum Hawking radiation and its entanglement in an analogue black hole. *Nat. Phys.* **2016**, *12*, 959–965. [CrossRef]
56. Bose, S.; Mazumdar, A.; Morley, G.W.; Ulbricht, H.; Toroš, M.; Paternostro, M.; Geraci, A.A.; Barker, P.F.; Kim, M.S.; Milburn, G. Spin entanglement witness for quantum gravity. *Phys. Rev. Lett.* **2017**, *119*, 240401. [CrossRef] [PubMed]
57. Marletto, C.; Vedral, V. Gravitationally induced entanglement between two massive particles is sufficient evidence of quantum effect in gravity. *Phys. Rev. Lett.* **2017**, *119*, 240402. [CrossRef]
58. Christodoulou, M.; Rovelli, C. On the possibility of laboratory evidence for quantum superposition of geometries. *Phys. Lett. B* **2019**, *792*, 64–68. [CrossRef]

59. Dankel, T.G., Jr. Mechanics on manifolds and the incorporation of spin into Nelson's stochastic mechanics. *Arch. Ration. Mech. Anal.* **1970**, *37*, 192–221. [CrossRef]
60. Dohrn, D.; Guerra, F. Geodic correction to stochastic parallel displacement of tensor. In *Stochastic Behavior in Classical and Quantum Hamiltonian Systems*; Casati, G., Ford, J., Eds.; Springer: Berlin/Heidelberg, Germany, 1979; pp. 241–249. [CrossRef]
61. Dohrn, D.; Guerra, F. Nelson's stochastic mechanics on Riemannian manifolds. *Lett. Nuovo Cimento* **1978**, *22*, 121–127. [CrossRef]
62. Koide, T.; Kodama, T. Stochastic variational method as quantization scheme: Field quantization of the complex Klein–Gordon equation. *Prog. Theor. Exp. Phys.* **2015**, *2015*, 093A03. [CrossRef]
63. de Matos, G.G.; Kodama, T.; Koide, T. Possible enhancement of collective flow anisotropy induced by uncertainty relation for fluid element. 2022, *in preparation*.
64. Koide, T.; Kodma, T. Scalar ideal hydrodynamic incorporating quantum-field-theoretical fluctuation. 2022, *in preparation*.

Article

Jean Cleymans, Stringy Thermal Model, Tsallis Quantum Statistics

Tamás S. Biró [1,2]

1   Wigner Research Centre for Physics, Konkoly-Thege Miklós út 29-33, 1121 Budapest, Hungary; biro.tamas@wigner.hu; Tel.: +36-20-435-1283
2   Faculty of Physics, Babeş-Bolyai University (UBB), 1 Mihail Kogălniceanu Street, 400084 Cluj-Napoca, Romania

Abstract: My memories on Jean Cleymans and a brief advocation of the stringy thermal model, describing massless constituents with the energy-per-particle and temperature relation, $E/N = 6T = 1$ GeV, are presented. Another topic, the Kubo–Martin–Schwinger (KMS) relation applied to the Tsallis distribution in quantum statistics is also sketched, which was triggered by our discussions with Jean.

Keywords: string; quark-gluon plasma; thermal model; KMS relation; Tsallis distribution

1. My Meetings with Jean

I briefly recapitulate my meetings with Jean Cleymans; the knowledge of his unexpected recent death is still shocking. He was a real demiurge, an active spirit in organizing a number of initiatives for the better sake of the high energy physics community. Probably the most known is his advocation of establishing the cooperation of the South African Republic with CERN (European Organization for Nuclear Research, Geneva, Switzerland) and JINR (Joint Institute for Nuclear Research, Dubna, Russia).

I met Jean in Cape Town, in 2004, when he was a main organizer of the *Strangeness in Quark Matter* meeting; and then later, at the meeting held in Stellenbosch, a nearby small town famous for wine making. Several visits had been organized for the sequel, during which I met some of his younger collaborators too. Most memorable to me were Azwinndini Muronga, and in later years Andre Peshier, whom I already knew from Giessen, Germany.

Jean also reciprocated quite a few visits in Budapest. In 2007 he took part in the *Zimányi'75* memorial workshop: he chaired a session with Greco, Hamar, Petreczky and Mócsy. He delivered a talk, entitled *Transverse energy and charged hadron production from GSI to RHIC*. A few years later, in 2011, he talked again at the *Zimányi School*, in December, on *The thermal model at the LHC*. This indicates that one of his favorite topics must have been the "thermal model" of hadrons—a minimalistic theory applied to a great number of experimental results ever since.

I remember Jean walking in the winter fare in midtown Budapest, watching the typical European activity in the pre-Christmas time. He looked over the heads in the crowd.

We have discussed with Jean several physics questions during those years. In this paper, I pick up two of the topics because they are characteristic to Jean's interests as a physicist and because these we discussed a lot and I had the feeling to have succeeded to convince him on the actuality and perspective of these. One topic was to include the presence of strings, connecting massless particles in a first-principles thermodynamical treatment, as the quark-gluon plasma (QGP) counterpart of the hadronic thermal model [1]. Another point was to lure Jean into the non-extensive thermodynamics perspective, with all its complications when using cut power-law type energy distributions instead of exponentials and treating their thermodynamical consequences [2,3]. Here, a particular question—the relation between fermion and anti-fermion quantum statistics when based on the q-generalization of the exponential function—had grown from our discussions. Jean had chosen the "cut-and-paste" solution [4], we with Gergely Barnaföldi and Keming Shen

Citation: Biró, T.S. Jean Cleymans, Stringy Thermal Model, Tsallis Quantum Statistics. *Physics* **2022**, *4*, 873–879. https://doi.org/10.3390/physics4030056

Received: 6 July 2022
Accepted: 25 July 2022
Published: 2 August 2022

Publisher's Note: MDPI stays neutral with regard to jurisdictional claims in published maps and institutional affiliations.

in Budapest adventured around to try "deeper" approaches, paying tribute to the particle–hole symmetry, in particular to the Kubo–Martin–Schwinger (KMS) relation [5].

In the present paper, the fundamentals of these two topics are presented. Even if this is not a first appearance of these topics, perhaps, the thoughts we shared with Jean Cleymans are worth revisiting.

2. Stringy Thermal Models

One of the most intriguing features taught to us by the thermal model of hadronization [6,7] in heavy-ion collisions is the scaling of experimental points on a single curve in the temperature vs. baryochemical potential, T–μ, plane. Among a few interpretation possibilities, the most popular was that this curve represents a constant energy per particle, $E/N \approx 1$ GeV [3,8,9].

Concentrating on the low baryon number region, typical for RHIC (Relativistic Heavy Ion Collider) and LHC (Large Hadron Collider) experiments, the E/N ratio is surpirsingly high for a thermal ensemble of massless quarks and gluon partons. On the hadron side, there is no such problem [10].

For a massive, non-relativistic ideal gas of monoatomic constituents, one expects

$$\frac{E}{N} = m + \frac{3}{2}T, \tag{1}$$

using units, where the Boltzmann constant and the speed of light values are set to unity; as well the Planck's constant is set to unity in what follows; here, m is the particle mass. In this case, the measured energy-per-particle value and the conjectured temperature, also fitted to parts of transverse momentum spectra, i.e., $T \approx 167$ MeV, lead to a conclusion of $m \approx 750$ MeV. Indeed, this average hadron mass, close to the ρ-meson mass, is not unreasonable.

On the other hand, for extreme relativistic pointlike particle plasmas without interaction in the Boltzmann limit, acute at high temperatures, one obtains:

$$\frac{E}{N} = 3T, \tag{2}$$

which is modified by only ten percent when assuming Bose–Einstein distribution. Meantime, one can see that this relation does not satisfy the experimentally fitted values, cited above. Conclusively, the QGP side at around the color deconfinement, known from lattice QCD (quantum chromodynamics) simulations, cannot consist of an ideal, non-interacting plasma of massless partons.

Motivated by this, an interacting model of massless QGP particles and the consequent thermodynamics were considered [1,11]. The interaction energy was modeled as strings with individual contributions of $\varepsilon_{\text{int}} = \sigma\langle \ell \rangle$. The average length, $\langle \ell \rangle$, of such strings is a function of density, n, for straight strings, which are optimal: $\langle \ell \rangle \sim n^{-1/3}$. This gives rise to a free energy density as follows:

$$f(n,T) = f_{\text{id}}(n,T) + An^{2/3}, \tag{3}$$

where the subscript "id" denotes the ideal gas formula and A is proportional to the string tension.

The thermodynamical consequences of such a term are multiple. Due to the homogeneity assumption, one has the known relation, connecting the energy density, ε, the pressure, p, the entropy density, s and the chemical potential, μ:

$$f = \varepsilon - Ts = \mu n - p, \tag{4}$$

with the partial derivatives,

$$\mu = \frac{\partial f}{\partial n}, \qquad s = -\frac{\partial f}{\partial T}. \tag{5}$$

In the stringy QGP, defined by Equation (3), one obtains:

$$s = s_{\mathrm{id}}, \qquad \mu = \mu_{\mathrm{id}} + \frac{2}{3} A n^{-1/3},$$

$$\varepsilon = \varepsilon_{\mathrm{id}} + A n^{2/3}, \qquad p = p_{\mathrm{id}} - \frac{1}{3} A n^{2/3}. \tag{6}$$

The negative contribution to the pressure indicates that strings pull, not push. The entropy density has only an ideal gas contribution, while the energy density receives the same correction term as the free energy density. It is a particular feature that there is a combination without interaction correction:

$$\varepsilon + 3p = \varepsilon_{\mathrm{id}} + 3p_{\mathrm{id}} = 6p_{\mathrm{id}}. \tag{7}$$

It is noteworthy that the correction to the chemical potential is decreasing with increasing density [12]. The ideal part, in the Boltzmann approximation proportional to the logarithm of the density, is increasing on the other hand. The common effect of these two terms is a minimum at some n. Moreover at given temperatures, when this minimum is negative, the $\mu(n)$ curve crosses the zero axis, signalling changes in the chemical behavior of strings. Below such temperatures the minimum of $\mu(n)$ is positive and the density of string sources will be diminished indefinitely. The critical temperature, interfacing these two cases, proved to be proportional to $\sqrt{A}$. This agrees with the early lattice gauge theory calculation results [13,14].

More problematic is the border of mechanical stability. According to Equation (6), the pressure may become negative. The stringy interaction term, $A n^{2/3}$, for massless sources with densities of $n \sim T^3$, represents an AT^2 order correction to the free QGP pressure. This is again in accordance with the lattice QCD findings, most noticeable in the studies of the interaction measure, $\Delta = (\varepsilon - 3p)/T^4$. We have analyzed such corrections among others in Refs. [1,12]. Assuming a thermal massless density of string sources, $n = \gamma T^3$, one concludes that $\Delta = 2\gamma A/T^2$. This is indeed an observed behaviour at above the color deconfinement temperature, $T > T_c$, in lattice QCD equation of state studies. This quantity drops to zero below this temperature, so the stringy interaction must be converted into masses in the hadron resonance gas.

The $p = 0$ mechanical stability limit line is an assumed point of rapid hadronization. Any interaction term in the equation of state which reduces the pressure while increasing the energy density modifies the expected E/N ratio.

$$\left. \frac{E}{N} \right|_{\mathrm{hadronization}} = \left. \frac{\varepsilon}{n} \right|_{p=0} \tag{8}$$

can be expressed observing that for ideal gases, $n = p_{\mathrm{id}}/T$ and $p = p_{\mathrm{id}} - p_{\mathrm{int}}$ along with $\varepsilon = \varepsilon_{\mathrm{id}} + \varepsilon_{\mathrm{int}}$. For a bag model type approach, $p_{\mathrm{int}} = \varepsilon_{\mathrm{int}} = B$, where B defines the bag pressure, and one obtains at $p = 0$:

$$\left. \frac{E}{N} \right|_{\mathrm{bag}} = \frac{\varepsilon_{\mathrm{id}} + p_{\mathrm{id}}}{p_{\mathrm{id}}/T} = 4T. \tag{9}$$

In view of the experimental results, this is indeed insufficient: a MIT (Massachusetts Institute of Technology) bag model equation of state for a quark-gluon plasma cannot match the measurements.

To the contrary, the stringy model just has the correct interaction terms. There, $\varepsilon_{\mathrm{int}} = 3p_{\mathrm{int}} = A n^{2/3}$, and one arrives at the estimate (cf. Equation (7)):

$$\left. \frac{E}{N} \right|_{\mathrm{string}} = \frac{\varepsilon_{\mathrm{id}} + 3p_{\mathrm{id}}}{p_{\mathrm{id}}/T} = 6T. \tag{10}$$

This straight and remarkable result encouraged us for further studies. We have studied the mechanical instability border line, $p = 0$, also at a finite baryochemical potential [1]. Then the very formulas are more involved; however, the main message is the same: $E/N = 6T \approx 1$ GeV curve describes all other data as well, even that taken at higher baryon densities. Citing Berndt Müller [13]: "These results which are generally true for systems composed of massless particles, also remain valid when interactions are included." This is said about the relation $p_{\mathrm{id}} = \varepsilon_{\mathrm{id}}/3$. By that the $E/N = 1$ GeV value remains the same for all chemical potentials.

3. Tsallis-Fermi Problem

Another of our common projects with Jean was to explore the consequences ofTsallis-distributed hadrons and eventually quarks and gluons for the thermodynamics and, therefore, for the thermal model predictions too [15,16]. At a first glance a transverse momentum, p_T, distribution, which is not exponential in its tail, was predicted by perturbative QCD calculations quite early. On the other hand, Rolf Hagedorn had suggested to interpolate towards a Boltzmann exponential, typical in thermal equilibrium situations, by the so-called "cut power-law" distribution [17]. It turned out in the 1990s and, with higher momentum, after 2000 that such functions of $(1 + ax)^{-b}$ type can be viewed as a mathematical generalization of the exponential function and can be derived as canonical distributions from an altered entropy formula. Since then it is tagged as "Tsallis distribution", as the canonical energy distribution, associated with the q-entropy formula, promoted by Constantino Tsallis since 1988 [18,19].

The core of the use of Tsallis distribution is to replace the exponential function, $\exp(x)$, in statistical formulas by

$$e_q(x) \equiv (1 + (1-q)x)^{\frac{1}{1-q}}, \tag{11}$$

where q is the Tsallis (non-extensivity) parameter.

This approach has provided good agreement to the measured spectra in the Boltzmann approximation [20]. However, dealing with quarks and gluons on the one side or mesons and baryons on the other side, one is tempted to consider quantum statistics. In case of fermions it is even unavoidable at low temperatures, near the Fermi energy.

There are several ways to approach the quantum statistical pendants based on Tsallis distribution [4,5]. In particular, there is a symmetry between particles and holes (negative energy states) in the statistical approach to quantum field theory; referred to as the Kubo–Martin–Schwinger (KMS) relation, which explores the symmetry between negative and positive frequency waves in thermal equilibrium according to elementary commutation relations. In the case of the Bose–Einstein and Fermi–Dirac distribution, in the $q = 1$ case, the relation,

$$n(x) + n(-x) = \mp 1, \tag{12}$$

needs to be fulfilled for the Bose–Einstein distribution (with minus sign) and for the Fermi–Dirac distribution (with plus sign), respectively [12,14,16]. The resolution of this constraint leads to the form:

$$n(x) = \frac{1}{\exp(x) \mp 1}. \tag{13}$$

This fulfillment is based on the elementary identity,

$$\exp(x) \cdot \exp(-x) = 1. \tag{14}$$

Now, replacing the exponential function with another one, $e_q(x)$, brings a difficulty. For the Tsallis distribution, this product is not unity:

$$e_q(x) \cdot e_q(-x) \neq 1. \tag{15}$$

Instead, the basic formula reads:

$$e_q(x) \cdot e_{2-q}(-x) = 1. \tag{16}$$

So, either one shall not use the Tsallis' $e_q(x)$ (11) in the Bose–Einstein and Fermi–Dirac distributions instead of the exponential function or a more sophisticated relation is needed.

Both ways are possible. One may generalize the exponential function in another way, e.g., using the deformed exponential promoted by Giorgio Kaniadakis [21]:

$$e_k(x) = \left(\sqrt{1+k^2x^2} + kx \right)^{1/k}. \tag{17}$$

In this case,

$$e_k(x) \cdot e_k(-x) = 1. \tag{18}$$

Generalizing further, an even function, $b(x^2)$, can also be used, and by that extension, $e_k(x)$ can be related to a ratio of Tsallis expressions [5]. An expression like

$$e_b(x) = \left(\sqrt{1+k^2x^2b^2(x^2)} + kxb(x^2) \right)^{1/k}, \tag{19}$$

indeed satisfies $e_b(x)e_b(-x) = 1$. On the other hand,

$$f_q(x) = \frac{e_q(x/2)}{e_q(-x/2)} \tag{20}$$

demonstrates this feature too. One can then equate:

$$e_b(x) = f_q(x)^{1-q} = t, \tag{21}$$

and conclude that

$$b(x^2) = \frac{1}{2kx} \left[f_q(x)^{1-q} - f_q(-x)^{1-q} \right]. \tag{22}$$

In addition, as a further alternative, a combination of the naïve Bose–Einstein and Fermi–Dirac distributions,

$$n_q(x) = \frac{1}{e_q(x) \mp 1}, \tag{23}$$

can also be utilized to fulfill the KMS relation. It turns out that the linear combination,

$$n(x) = \frac{1}{2} \left[n_q(x) + n_{2-q}(x) \right], \tag{24}$$

also satisfies $n(x) + n(-x) = \mp 1$.

4. Conclusions

In conclusion, the phenomenology of heavy-ion collisions and the search for signals of quark-gluon plasma (QGP) formation sometimes meet with fundamental concepts. These concepts sometimes can be and indeed were handled in terms of simple enough models and elementary considerations. One of these models was the thermal model, which played an important role in the career of Jean Cleymans. It intrigued me to take steps towards understanding why and how the energy-per-particle ratio interrelation with temperature, $E/N = 6T$, is possible, in particular, with massless constituents, as high-T QCD (quantum chromodynamics) expected it.

Another vast streamline of the development of concepts is connected with statistical physics. While at the beginning, in the 1960s and 1970s the application of thermodynamics was almost unthinkable to high energy physics, the attitude had changed dramatically in the 1980s. Analogous to this, in the 2000s and 2010s the use of non-extensive statistics modified and extended the early Boltzmannian based concepts. Here, Jean could be convinced and joined to this enterprise to check the statistical consequences of an altered entropy formula together with an altered form of canonical distributions. Experiments namely supported this view much more strongly than the early fits of exponentials to a narrow window of spectra available to the date.

In relation to this advanced statistical approach, with Jean we have discussed what to do with the quantum statistics. He had chosen a cut-and-paste approach as a fast and practical cure to the particle–hole problem, inherent in field theory due to the Kubo–Martin–Schwinger (KMS) relation. We have investigated a more general class of possible solutions, all smooth at the Fermi surface. I believe that even not being a co-author, he deserves to be acknowledged and late but clearly mentioned connection with this issue too.

Funding: This research was funded by the Hungarian National Research Development & Innovation Office, Hungarian Scientific Research Fund (NKFIH OTKA), grant number K123815, and by the Romanian research project PN-III-P4-ID-PCE-2020-0647, hosted at the Babeş-Bolyai University (UBB) Cluj, Romania.

Data Availability Statement: Not applicable.

Acknowledgments: The author T.S.B. thanks Raghunath Sahoo for insisting on a contribution to Jean Cleymans memorial Special Issue.

Conflicts of Interest: The author declares no conflict of interest.

Abbreviations

The following abbreviations are used in this manuscript:

CERN	Conseil (Organisation) Européenne pour la Recherche Nucléaire (European Organization for Nuclear Research)
GSI	Gesellschaft für Schwerionenforschung (German Society for Heavy Ion Research)
JINR	Joint Institute for Nuclear Research
QGP	quark-gluon plasma
QCD	quantum chromodynamics
KMS	Kubo–Martin–Schwinger
LHC	Large Hadron Collider
MIT	Massachusetts Institute of Technology
RHIC	Relativistic Heavy Ion Collider

References

1. Biró, T.S.; Cleymans, J. Hadronization line in the stringy matter. *Phys. Rev. C* **2008**, *78*, 034902. [CrossRef]
2. Cleymans, J.; Lykasov, G.I.; Parvan, A.S.; Sorin, A.S., Teryaev, O.V., Worku, D. Systematic properties of the Tsallis distribution: Energy dependence of parameters in high energy p–p collisions. *Phys. Lett. B* **2013**, *723*, 351–354. [CrossRef]
3. Cleymans, J.; Paradza, M.W. Tsallis statistics in high energy physics: Chemical and thermal freeze-outs. *Physics* **2020**, *2*, 654–664. [CrossRef]
4. Cleymans, J.; Worku, D. The Tsallis distribution in proton–proton collisions at $\sqrt{s} = 0.9\ TeV$. *J. Phys. G* **2012**, *39*, 025006. [CrossRef]
5. Biró, T.S.; Shen, K.M.; Zhang, B.W. Non-extensive quantum sttaistics with particle–hole symmetry. *Phys. A Stat. Mech Appl.* **2015**, *428*, 410–415. [CrossRef]
6. Cleymans, J. The thermal-statistical model for particle production. *EPJ Web. Conf.* **2010**, *7*, 01001. [CrossRef]
7. Cleymans, J.; Redlich, K. Unified description of freeze-out parameters in relativistic heavy ion collisions. *Phys. Rev. Lett.* **1998**, *81*, 5284–5286. [CrossRef]
8. Cleymans, J.; Oeschler, H.; Redlich, K.; Wheaton, S. Transition from baryonic to mesonic freeze-out. *Phys. Lett. B* **2005**, *615*, 50–54. [CrossRef]
9. Cleymans, J.; Oeschler, H.; Redlich, K.; Wheaton, S. Comparison of chemical freeze-out criteria in heavy-ion collisions. *Phys. Rev. C* **2006**, *73*, 034905. [CrossRef]
10. Sharma, N.; Cleymans, J.; Hyppolite, B.; Paradza, M. Comparison of p-p, p-Pb, and Pb-Pb collisions in the thermal model: Multiplicity dependence of thermal parameters. *Phys. Rev. C* **2019**, *99*, 044914. [CrossRef]
11. Biró, T.S.; Ürmössy, K. Transverse hadron spectra from a stringy quark matter. *J. Phys. G* **2009**, *36*, 064044. [CrossRef]
12. Biró, T.S.; Jakovác, A.; Schram, Z. Nuclear and quark matter at high temperature. *Eur. Phys. J. A* **2017**, *53*, 52. [CrossRef]
13. Müller, B. *The Physics of Quark–Qluon Plasma*; Springer: Berlin/Heidelberg, Germany, 1985. [CrossRef]

14. Biró, T.S.; Jakovác, A. QCD above T_c: Hadrons, partons, and the continuum. *Phys. Rev. D* **2014**, *90*, 094029. [CrossRef]
15. Biró, T.S.; Barnaföldi, G.G.; Ván, P. New entropy formula with fluctuating reservoir. *Phys. A Stat. Mech. Appl.* **2013**, *417*, 215–220. [CrossRef]
16. Biro, T.S. *Is There a Temperature? Conceptual Challenges at High Energy, Acceleration and Complexity*; Springer Science+Business Media, LLC: New York, NY, USA, 2011. [CrossRef]
17. Hagedorn, R. Statistical thermodynamics of strong interactions at high energies. *Nuovo Cim. Suppl.* **1965**, *3*, 147–186; Reprinted in *Quark-Gluon Plasma: Theoretical Foundations*; Kapusta, J., Müller, B., Rafelski, J., Eds.; Elsevier B.V.: Amsterdam, The Netherlands, 2003; pp. 24–63. Available online: https://cds.cern.ch/record/346206/ (accessed on 29 July 2022).
18. Tsallis, C. Possible generalization of Boltzmann–Gibbs statistics. *J. Stat. Phys.* **1988**, *52*, 479–487. [CrossRef]
19. Tsallis, C. *Introduction to Nonextensive Statistical Mechanics: Approaching a Complex World*; Springer: New York, NY, USA, 2009. [CrossRef]
20. Shen, K.-M.; Biró, T.S.; Wang, E.-K. Different non-extensvie modles for heavy-ion collisions. *Phys. A Stat. Mech. Appl.* **2018**, *492*, 2353–2360. [CrossRef]
21. Kaniadakis, G. Theoretical foundations and mathematical formalism of the power-law tailed statistical distributions. *Entropy* **2013**, *15*, 3983–4010. [CrossRef]

Article

The Abundance of the Species [†]

Helmut Satz

Fakultät für Physik, Universität Bielefeld, D-33501 Bielefeld, Germany; satz@physik.uni-bielefeld.de
† This paper is dedicated to the memory of Jean Cleymans, friend and colleague for many years.

Abstract: I review the pioneering work of Jean Cleymans in establishing the statistical description of multihadron production in high energy strong interaction physics.

Keywords: high-energy collisions; hadron production; statistical models

Citation: Satz, H. The Abundance of the Species. *Physics* **2022**, *4*, 912–919. https://doi.org/10.3390/physics4030059

Received: 24 June 2022
Accepted: 4 August 2022
Published: 19 August 2022

Publisher's Note: MDPI stays neutral with regard to jurisdictional claims in published maps and institutional affiliations.

Jean Cleymans joined Physics Department of the University of Bielefeld, Germany, in the year 1975, coming to us from CERN (European Organization for Nuclear Research, Geneva, Switzerland). Physics at the newly founded University of Bielefeld was only four years old then, and to establish a new research center here required attracting bright and adventurous people from throughout the world—people open not only to a new university, but also to new physics. Jean definitely was one of these. He was instrumental in putting Bielefeld physics-wise onto the map and in his later career he did the same for Cape Town.

In his earlier years here, Jean worked a lot on lepton pair production, but around 1980 when QCD (quantum chromodynamics) entered the scene, he quickly turned to the quark gluon-plasma and its hadronization. The 1986 *Physics Report* [1] on "Quarks and gluons at high temperatures and densities", written by Jean together with two other bright young international Bielefeld accessions, Rajiv Gavai and Esko Suhonen, quickly became the standard reference to the field and remained so for many years.

I had worked for some years on the statistical description of hadron production, and so Jean and I took this up as the mechanism for the hadronization of the quark-gluon plasma. In this paper, I elaborate a little on our joint work, which indeed turned out to be the beginning of a new approach to multi-hadron production. Much of this material is covered in Chapter 11 of Ref. [2], where more details can also be found.

1. Statistical Multihadron Production

Multiparticle production in high energy collisions of strongly interacting particles has fascinated physicists for well over half a century. As predicted by Heisenberg [3], the little bang of such collisions produce with increasing energy an ever growing number of mesons and baryons of different quantum states, and from the beginning, the large numbers were a challenge to describe these reactions by collective or statistical approaches. It was tempting to go even further, to imagine that what they produced were really droplets of strongly interacting matter, thus providing a means to access the thermodynamics of strong interaction physics in the laboratory.

The main features observed in high energy collisions are the multiplicity, i.e., the number of produced particles as a function of the collision energy, the momentum spectra of the particles, their correlations, and the relative abundances of the different species. These then are also the basic quantities which any theoretical framework has to provide.

The first statistical treatment was formulated by Fermi [4]. He assumed that the collision deposits a great amount of energy in a small spatial region around the colliding particles, and that the energy of this fireball is then distributed among the various observable degrees of freedom, the emitted mesons and nucleons, according to statistical laws. Thus, the description of the production process is determined by the grand canonical phase space volume $Q(E, V)$ of a gas of non-interacting hadrons; here, E denotes the collision energy in the center of mass of the colliding particles, and V the interaction volume. This phase space volume was to be calculated with whatever constraints are imposed by conservation laws (charge, baryon number, etc.).

The main features obtained from Fermi's model are:

- A multiplicity $n(E) \sim E^{3/4}$, growing as a power of E;
- Isotropic production of secondary particles; and
- Average secondary momenta, $\langle |p| \rangle \sim E^{1/4}$, also increasing as a power of E.

Modified versions of the model [5,6] lead to slightly changed powers, but the basic features remain. The available collision energy is equidistributed among the isotropically emitted secondaries; any increase of E goes partially into making more secondaries and partially into making each constituent more energetic.

Experimental data showed that with increasing collision energy, this picture became untenable for two main reasons:

- In a nucleon-nucleon collision, the incident nucleons always retained a considerable fraction of the collision energy ("leading particle effect");
- The secondaries were not emitted isotropically; their average transverse momenta (orthogonal to the collision axis) reached a constant value, independent of the incident energy, while the average longitudinal momenta increased with E.

Certainly, this meant that not all information about the initial state was lost in the collision; the reaction retained a memory both of the conserved quantum numbers of the incident particles and of the collision axis. Fermi had already suggested that the spatial volume, as seen in the overall center-of-mass system, should be Lorentz-contracted along the collison axis. However, as long as there is no interrelation between the momenta and the coordinates of the secondaries, this does not produce anisotropic particle production.

From another point of view, the combination of the anisotropic secondary momentum distributions and the leading particle effect seemed to indicate that at high energy the incident nucleons could not fully stop each other. Instead, they seemed to "pass through" one another, losing only part of their energy in the process. This "transparency" was subsequently explained by Gribov [7] as a consequence of hadronic size and the finite speed of information transmission.

It thus became evident that high-energy collisions could not be understood in terms of the the formation of just one fireball, in the sense of a single isotropic energy deposit into a small spatial volume. The collision instead seems more like the passage of an energetic charge through a medium, leaving behind a condensation trail of smaller fireballs superimposed along the collision axis. Each of these bubbles could now, in principle, have the phase space structure envisioned by Fermi, and if one attributes the conserved baryon numbers to the fastest bubbles in each direction, the scenario would also provide the leading particle effect.

While this does bring in the desired longitudinal momentum growth, the energy of each bubble could also still increase, and this in turn results in an energy dependence (albeit weaker) for the average transverse momenta. The basic puzzle of the field, or, looked at in a more positive way, the most important hint provided by nature, was the constancy of the average transverse momenta, p_t, of the secondaries. Making this even more tantalizing was the observation that while different species of secondaries led to different (energy-independent) transverse *momentum* distributions, the transverse *energies*, m_t^i, appeared to follow one universal pattern, with

$$\frac{dN_i}{dm_t^i} \sim \exp\left(-\lambda m_t^i\right), \quad m_t^i = \sqrt{(p_t^i)^2 + m_i^2}, \tag{1}$$

describing the functional form of the distribution of all species i of different masses, m_i, in terms of one universal parameter λ.

The first explanation of this universality was proposed by Hagedorn [8,9], based on the resonance structure governing the interaction of the different hadron species. Experiment had shown that multiparticle production with increasing energy did not just produce a shower of many pions, kaons and nucleons. Instead, it led to the production of more and different excited resonant states which decayed strongly into less excited states and finally into the ground state hadrons. Starting in the nineteen-sixties, an ever growing number of such hadronic resonances were discovered, and today the standard compilation [10] lists hundreds of them. It thus was necessary to obtain a scheme to determine how many states of each mass are produced in the collision, and Hagedorn's statistical bootstrap model provided that in terms of a self-similar composition law, claiming that resonances consist of resonances and so on, with a universal composition equation. The number of states, $\rho(m)$, of a resonance of mass m, its "degeneracy", is then given as the number of different composition patterns. This led to an exponential growth of $\rho(m)$,

$$\rho(m) \sim m^a e^{bm}, \tag{2}$$

while the power a depends on the details of the partition problem, the coefficient b was expressed in terms of fundamental features of strong interaction physics, such as the hadronic size, the range of the strong force, or the Regge resonance pattern.

At this stage then, the experiment had shown several basic deviations from Fermi's original fireball picture. The assumption of a completely random production process failed: the system retained some information of the initial state, secondary particle distributions were different in directions along and orthogonal to the collision axis, and there were leading particles carrying a baryon number. Moreover, the emitted pions, kaons and nucleons, had gone through some intermediate interactive stage, with resonance formation and decay as the dominant process. The first of these features, anisotropy and leading particles, could be accounted for through a superposition of fireballs, using initial state dynamical information as input.

To solve the resonance problem, Hagedorn invoked a result first obtained by Beth and Uhlenbeck [11] and subsequently generalized by Dashen, Ma and Bernstein [12]. They had argued that if the interaction of a gas of constituents is indeed dominated by resonance formation, then one can replace the interacting system of elementary particles with a non-interacting system of all possible resonances. The relevant phase space for the states of multiparticle production would thus be that of an ideal resonance gas, with an exponentially growing resonance mass spectrum, and contained in an interaction volume V_0.

The grand-canonical partition function $Z(T, V_0)$ of such a resonance gas diverges for

$$T > T_H = \frac{1}{b}, \tag{3}$$

so T_H becomes an upper bound on the temperature. Increasing the energy of such a system does not increase the momentum of the secondaries and hence its temperature; instead, it leads to more species of more massive hadrons. As a result, the momentum spectra now have the form (1); the experimentally observed universal transverse mass pattern with a parameter λ is thus accounted for as the universal limiting temperature of an ideal resonance gas. This indeed agreed quite well with the observed transverse energy spectra. The energy-dependent and unbounded longitudinal momenta, on the other hand, arise from the superposition of an energy-dependent number of such Hagedorn-type fireballs. Conceptually, this provides the basis for the superposition of fireballs moving at different rapidities, as a phenomenological picture of high energy multihadron production.

2. The Abundance of the Species

For an ideal resonance gas, the hadronization temperature, T_H, determines not only the momentum spectra, but also the relative abundances of the different species. At a fixed temperature, a heavy meson is less likely to be present than a lighter one. In the case of heavy ion collisions, the overall baryon density in the clusters enters as still another factor, modifying the relation between baryon and antibaryon abundances. This is generally taken into account through a baryochemical potential, μ_B; the overall relative abundance of species i in an ideal resonance gas is thus described in terms of two parameters, the temperature T_H and the baryochemical potential u_B. The corresponding multiplicity of species i then given by

$$N_i = d_i \frac{VT m_i^2}{2\pi^2} K_2(m_i/T_H) \exp\left(B\mu_B/T_H\right) \tag{4}$$

where d_i specifies the degeneracy of species i, K_2 is the modified Bessel function of the second kind, V denotes the overall volume, and $B = 0, \pm 1, \pm 2$ is the baryon number of the species. In other words, the ratio of the multiplicities of species i to j is predicted as the ratio of the corresponding phase space weights,

$$\frac{N_i}{N_j} = \left(\frac{d_i}{d_j}\right)\left(\frac{m_i}{m_j}\right)^2 \frac{K_2(m_i/T_H)}{K_2(m_j/T_H)} \exp\left[(B_i\mu_i - B_j\mu_j)/T_H\right] \tag{5}$$

Hence, if high energy collision results are specified just by the pure phase space of a resonance gas, they should also lead to corresponding ideal resonance gas ratios of the form (5). Let us consider this in more detail.

The statistical hadronization model assumes that hadronization in high energy collisions is a universal process proceeding through the formation of multiple colorless massive clusters or fireballs of finite spacial extension at fixed temperature T_H and baryochemical potential μ_B. These clusters are assumed to decay into hadrons according to a purely statistical law: every multi-hadron state of the fireball phase space defined by its mass, volume and charges is equally probable. The mass distribution and the distribution of charges (electric, baryonic and strange) among the clusters and their (fluctuating) number are, however, in principle determined in the prior dynamical stage of the process, which determines how fireballs are emitted along the collision axis.

Hence, one would seem to need this dynamical information in order to make definite quantitative predictions to be compared with data. Nevertheless, for Lorentz-invariant quantities such as multiplicities, one can introduce a simplifying assumption and thereby obtain a simple analytical expression in terms of T_H and μ_B. The key point is to assume that the distribution of masses and charges among clusters is again purely statistical [13–15], so that, as far as the calculation of multiplicities is concerned, the set of many clusters becomes equivalent, on the average, to one large cluster (an *equivalent global cluster*) whose volume is the sum of the individual proper cluster volumes and whose charge is the sum of cluster charges (and thus the conserved charge of the initial colliding system). In such a framework, the global cluster can be hadronized on the basis of statistical equilibrium.

The primary multiplicity of a hadron species i due to fireball decay in the Boltzmann limit of phase space is then given by Equation (4), where V now is the production volume (the sum of all fireballs). In abundance ratios, V cancels out, so it is a parameter needed only for absolute multiplicities. Let N_i denotes the observed multiplicity; since all heavier resonances will still decay into lighter ones, so that the actually observed multiplicities are obtained from (4) by

$$N_i = N_i^{\mathrm{primary}} + \sum_k N_k \mathrm{Br}(k \to i), \tag{6}$$

summing over the various branching ratios Br as measured. The final pion multiplicity, for example, is in fact several times larger than the primary one.

The other species parameter in Equation (5), apart from the mass, is the degeneracy d_i. It had been noted above that self-similar resonance composition leads to a degeneracy $\exp\{bm\}$ increasing exponentially in mass (see Equation (2)). Such an increase in turn leads to a critical point $T_H = 1/b$, limiting hadron thermodynamics to $T \leq T_H$. A more empirical alternative is the *physical* resonance gas, in which the resonance spectrum is taken to consist of the actually measured and tabulated resonances, with only their spin and isospin degeneracy taken into account. This approach is apparently limited in resonance mass since little is known about states above 2.5 to 3.0 GeV. With an upper limit in resonance mass, the partition function is analytic and hence there is no critical point; the energy density and all higher derivatives remain finite for all values to the temperature, and also those above a Hagedorn T_H. As far as the measurable abundances are concerned, the missing high mass states do not play a significant role: including all excited states up to 2.0 GeV covers almost all of the feed-down sources for pions, for example, since the higher mass states are strongly suppressed, for $m \gg T_H$. Studies comparing species abundances with cuts at 1.5 or 2.0 GeV thus show very little difference.

If the basic assumption of statistical hadronization—the equivalence of interacting hadron gas and ideal resonance gas—is indeed correct, the abundances of the species can thus be used to determine the hadronization temperature. In particular, the measured ratio of specific hadron abundances provides by Equations (4) and (6) an equation in terms of T and μ_B and thus leads to a line in the T–μ_B plane. Consider the ratio of two hadron species abundances, h_1 and h_2,

$$R_{12} = \frac{N_1(T,\mu_B)}{N_2(T,\mu_B)} = a \pm \Delta a. \tag{7}$$

The crossing region of ratio R_{12} and another ratio R_{34} then specifies the thermal parameter values at hadronization; see Figure 1.

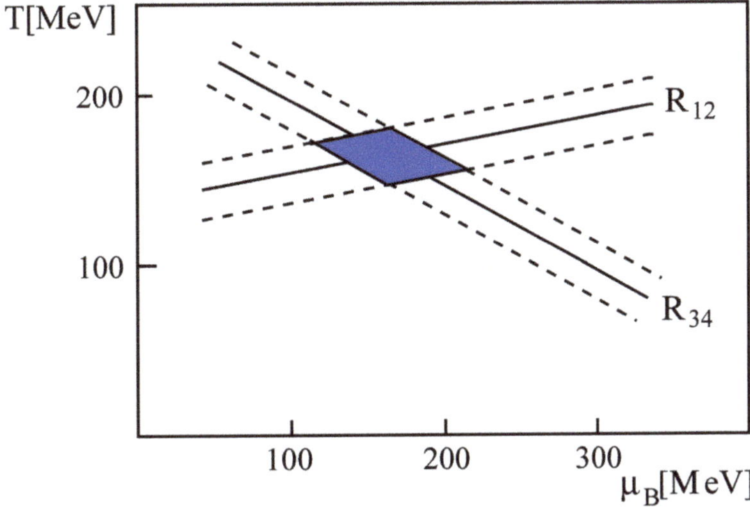

Figure 1. Thermal parameter values (shaded area) in the T–μ_B plane obtained from two different abundance ratios, R_{12} and R_{34}; the error bands correspond to the measurement errors of the ratios. See text for details.

The ratios of further species abundances give different different lines and, if statistical hadronization is correct, all lines should cross at a common point, thus specifying the temperature and the baryon density of the medium at hadronization. This was proposed and first carried out in 1992 in a study by Jean and myself [16]. The essential point is that a variety of hadron species are measured, providing a comparatively large number of ratios, and these should all lead to the same hadronization temperature, if the model is correct. Given many ratios, the two parameters T_H and μ_B are thus over-determined.

Our initial study involved heavy baryons, Λ, Ξ and their anti-particles, the vector mesons ρ, ω and ϕ, and positive and negative kaons and pions—up to ten or more hadron species. The results were quite amazing: five ratios of heavy baryons and Λ/K led to a hadronization temperature of some 150–200 MeV, in good agreement to theoretical results both from lattice QCD and from statistical bootstrap considerations. This temperature was found to be slightly too high for the ratio (Λ/all hadrons), and definitely too high for that of kaons to all hadrons. This was an early indication of what became known as strangeness suppression in hadronic collisions, and which led to the introduction of a third parameter [17], the relative strangeness abundance, γ_s, which was found to have values around 0.6 for the data available at that time.

To some surprise, this approach with the three parameters—T_H, μ_B, and γ_s has turned out to be correct far beyond all expectations. Over the past years, the resulting predictions were tested in a variety of collision configurations, from e^+e^- annihilation [13,18–20] over $pp/p\bar{p}$ [21,22] to nucleus-nucleus collisions [16,23–27]. With some caveats to be elaborated, they were found to provide a most remarkable account for what is observed, both of species abundances and, where applicable, of transverse momentum spectra [13]. Moreover, the temperature obtained for high energy experiments turned out to be quite universal, always lying around 160–180 MeV, i.e., in a range, which partitioning arguments as well as studies of critical phenomena in QCD had pre- and postdicted.

To illustrate, we compare in Table 1 the results of pp data to those from other initial state configurations (e^+e^- and nucleus-nucleus, AA) at high collision energies [28–33]. To start from a comparable basis, the data set in all three reaction channels has been restricted as far as possible to the same 12 long-lived hadron species; the rates for short-lived and hence broader resonances are in general more difficult to measure. One can see that all channels indeed appear to converge to a hadronization temperature value of about 160–170 MeV. We also note that with increasing energy density for the collision, the strangeness suppression factor approaches unity, as predicted by a description in terms of hadronization volumes [34].

Table 1. Fit results for a set of 12 long-lived particles in high energy pp, $AuAu$ and e^+e^- collisions [22].

Collision	pp	e^+e^-	$AuAu$
CMS Energy [GeV]	200	91.25	200
Temperature [MeV]	169.8 ± 4.2	164.7 ± 0.9	168.5 ± 4.0
Average relative deviation data vs. fit [%]	12.5	9.4	11.7
Strangeness Suppression, γ_s	0.57 ± 0.03	0.66 ± 0.01	0.93 ± 0.04

Combining the results for e^+e^- annihilation, from elementary hadron-hadron interactions and from nucleus-nucleus collisions at different high energies leads to one of the truly striking observations in high energy strong interaction physics: the existence of a universal hadronization temperature T_H. An overall view of this result is given in Figure 2.

On the theoretical side, this suggests that there must be an underlying universal production mechanism, i.e., that the hadronization temperature is indeed determined by or closely related to the confinement/deconfinement transition of strongly interacting matter. On the experimental side, it leads to the remarkable prediction that the relative hadron abundances produced in high energy collisions become with increasing energy independent of the collision energy. One can thus predict such ratios with considerable confidence for the higher energy experiments at the CERN-LHC (Large Hadron Collider) [35].

Nevertheless, one important feature distinguishes elementary from nuclear interactions. It is seen in Table 1 that the strangeness suppression observed in elementary reactions has essentially disappeared in nucleus–nucleus collisions. This had been already mentioned above—the suppression factor γ_s is now compatible with unity, and whatever deviations remain, can be understood as arising from "corona" interactions, i.e., collisions at the edge of the nuclei, which do not really experience the nuclear medium [36]. The fully thermal behavior of strange hadrons in the hot medium produced in central nuclear collisions is indeed a first indication for collective features present in such interactions. For its further understanding, it is expected that elementary pp collisions at the very much higher energies of the CERN-LHC, with the much higher absolute strange hadron production rates, will also show a γ_s approaching unity [35].

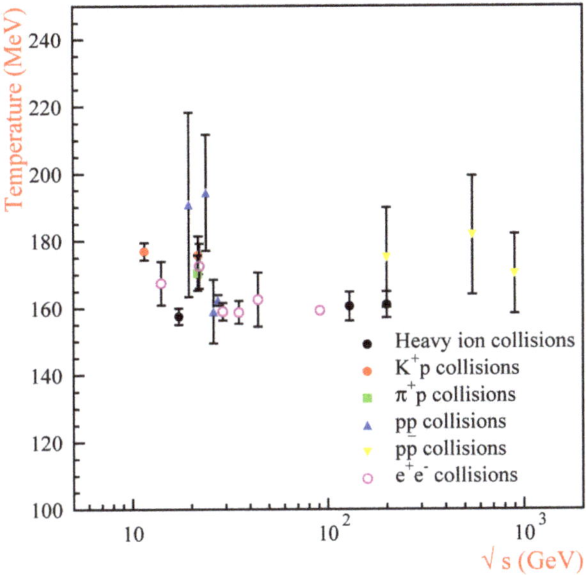

Figure 2. Hadronization temperatures obtained for various initial collision configurations at different (high) energies.

3. Conclusions

This success of the statistical hadronization model is undoubtedly one of the milestones in high energy multihadron production, and the pioneering contributions of Jean Cleymans have made it into an essential tool for the analysis of such studies. Jean continued his work in this direction and provided many further essential results [37].

Funding: This research received no external funding.

Data Availability Statement: Not applicable.

Conflicts of Interest: The authors declare no conflict of interest.

References

1. Cleymans, J.; Gavai, R.V.; Suhanen, E. Quarks and gluons at high temperatures and densities. *Phys. Rep.* **1986**, *130*, 217–292. [CrossRef]
2. Satz, H. *Extreme States of Matter in Strong Interaction Physics*; Springer: Berlin/Heidelberg, Germany, 2012. [CrossRef]
3. Heisenberg, W. Zur Theorie der explosionsartigen Schauer in der kosmischen Strahlung. II. *Z. Phys.* **1939**, *113*, 61–86 [CrossRef]
4. Fermi, E. High energy nuclear events. *Progr. Theor. Phys.* **1950**, *5*, 570–583. [CrossRef]
5. Srivastava, P.P.; Sudarshan, G. Multiple production of pions in nuclear collisions. *Phys. Rev.* **1958**, *110*, 765–766. [CrossRef]
6. Satz, H. The high-energy limit of the statistical model. *Il Nuovo Cim.* **1965**, *37*, 1407–1421. [CrossRef]
7. Gribov, V.N. Glauber Corrections and the Interaction between High-Energy Hadrons and Nuclei. *Sov. Phys. JETP* **1969**, *29*, 483–487. Available online: http://jetp.ras.ru/cgi-bin/e/index/e/29/3/p483?a=list (accessed on 7 August 2022).
8. Hagedorn, R. Staistical hermodynamiccs of strong interactions at high energies. *Nuovo Cim. Suppl.* **1965**, *3*, 147–186; Reprinted in *Quark-Gluon Plasma: Theoretical Foundations*; Kapusta, J., Müller, B., Rafelski, J., Eds.; Elsevier B.V.: Amsterdam, The Netherlands, 2003; pp. 24–63. Available online: https://cds.cern.ch/record/346206/ (accessed on 7 August 2022).
9. Hagedorn, R. Hadronic matter near the boiling point. *Il Nuovo Cim. A* **1968**, *56*, 1027–1057. [CrossRef]
10. Yao, W.-M.; et al. [Particle Data Group]. Review of Particle Physics. *J. Phys. G Nucl. Part. Phys.* **2006**, *33*, 1. [CrossRef]
11. Beth, E.; Uhlenbeck, G.E. The quantum theory of the non-ideal gas. II. Behaviour at low temperatures. *Physica* **1937**, *4*, 915–924. [CrossRef]
12. Dashen, R.; Ma, S.-K.; Bernstein, H.J. S-matrix formulation of statistical mechanics. *Phys. Rev.* **1969**, *187*, 345–370. [CrossRef]
13. Becattini, F.; Passaleva, G. Statistical hadronization model and transverse momentum spectra of hadrons in high energy collisions. *Eur. Phys. J. C* **2002**, *23*, 551–582. [CrossRef]

14. Braun-Munzinger, P.; Redlich, K.; Stachel, J. Particle production in heavy ion collisions. In *Quark-Gluon Plasma 3*; Hwa, R.C., Wang, X.-N, Eds.; World Scientific: Singapore, 2004; pp. 491–599. [CrossRef]
15. Becattini F.; Fries, R.J. The QCD confinement transition: Hadron formation. In *Landolt-Börnstein—Group 1. Volume 23: Relativistic Heavy Ion Physics*; Stock., R., Ed.; Springer: Berlin/Heidelberg, Germany, 2010. [CrossRef]
16. Cleymans, J.; Satz, H. Thermal hadron production in high energy heavy ion collisions. *Z. Phys. C* **1993**, *57*, 135–147. [CrossRef]
17. Letessier, J.; Rafelski, J.; Tounsi, A. Gluon production, cooling, and entropy in nuclear collisions. *Phys. Rev. C* **1994**, *64*, 406–409. [CrossRef]
18. Becattini, F. A thermodynamical approach to hadron production in e^+e^- collisions. *Z. Phys. C* **1996**, *69*, 485–492. [CrossRef]
19. Becattini, F.; Castorina, P.; Manninen, J.; Satz, H. The thermal production of strange and non-strange hadrons in e^+e^- collisions. *Eur. Phys. C* **2008**, *56*, 493–519. [CrossRef]
20. Andronic, A.; Beutler, F.; Braun-Munzinger, P.; Redlich, K.; Stachel, J. Thermal description of hadron production in e^+e^- collisions revisited. *Phys. Lett. B* **2009**, *675*, 312–318. [CrossRef]
21. Becattini F.; Heinz, U. Thermal hadron production in pp and p̄p collisions. *Z. Phys. C* **1997**, *76*, 269–286. [CrossRef]
22. Becattini, F.; Castorina, P.; Milov, A.; Satz, H. A comparative analysis of statistical hadron production. *Eur. Phys. J. C* **2010**, *66*, 377–386. [CrossRef]
23. Cleymans, J.; Satz, H.; Suhonen, E.; Von Oertzen, D.W. Strangeness production in heavy ion collisions at finite baryon number density. *Phys. Lett. B* **1990**, *242*, 111–114. [CrossRef]
24. Redlich, K.; Cleymans, J.; Satz, H.; Suhonen, E. Hadronisation of quark-gluon plasma. *Nucl. Phys. A* **1994**, *566*, 391. [CrossRef]
25. Braun-Munzinger, P.; Stachel, J.; Wessel, J.P.; Xu, N. Thermal equilibration and expansion in nucleus-nucleus collisions at the AGS. *Phys. Lett. B* **1995**, *344*, 43–48. [CrossRef]
26. Becattini, F.; Gazdzicki, M.; Sollfrank, J. On chemical equilibrium in nuclear collisions. *Eur. Phys. J. C* **1998**, *5*, 143–153. [CrossRef]
27. Becattini, F.; Cleymans, J.; Keränen, A.; Suhonen, E.; Redlich, K. Features of particle multiplicities and strangeness production in central heavy ion collisions between $1.7A$ and $158A$ GeV/c. *Phys. Rev. C* **2001**, *64*, 024901.
28. Abelev, B.I.; et al. [STAR Collaboration]. Systematic measurements of identified particle spectra in pp, $d + Au$, and $Au + Au$ collisions at the STAR detector. *Phys. Rev. C* **2009**, *79*, 034909. [CrossRef]
29. Abelev, B.I.; et al. [STAR Collaboration]. Strange particle production in $p + p$ collisions at $\sqrt{s} = 200$ GeV. *Phys. Rev. C* **2007**, *75*, 064901. [CrossRef]
30. Adams, J.; et al. [STAR Collaboration]. ρ^0 production and possible modification in $Au + Au$ and $p + p$ collisions at $\sqrt{s} = 200$ GeV. *Phys. Rev. Lett.* **2004**, *92*, 092301. [CrossRef]
31. Adams, J.; et al. [STAR Collaboration]. ϕ meson production in $Au + Au$ and $p + p$ collisions at $\sqrt{s} = 200$ GeV. *Phys. Lett. B* **2005**, *612*, 181–189. [CrossRef]
32. Adams, J.; et al. [STAR Collaboration]. Scaling properties of hyperon production in $Au + Au$ collisions at $\sqrt{s} = 200$ GeV. *Phys. Rev. Lett.* **2007**, *98*, 062301. [CrossRef]
33. Abelev, B.I.; et al. [STAR Collaboration]. Strange baryon resonance production in $\sqrt{s} = 200$ GeV $p + p$ and $Au + Au$ collisions. *Phys. Rev. Lett.* **2006**, *97*, 132301. [CrossRef]
34. Castorina, P.; Satz, H. Strangeness production in AA and pp collisions. *Eur. Phys. J. A* **2016**, *52*, 200. [CrossRef]
35. Castorina, P.; Plumari, S.; Satz, H. Universal strangeness production in hadronic and nuclear collisions. *Int. J Mod. Phys. E* **2016**, *25*, 1650058. [CrossRef]
36. Becattini, F.; Gaździcki, M.; Keränen, A.; Manninen, J.; Stock, R. Chemical equilibrium study in nucleus-nucleus collisions at relativistic energies. *Phys. Rev. C* **2004**, *69*, 024905. [CrossRef]
37. Cleymans, J.; Oeschler, H.; Redlich, K.; Wheaton, S. Comparison of chemical freeze-out criteria in heavy-ion collisions. *Phys. Rev. C* **2006**, *73*, 034905. [CrossRef]

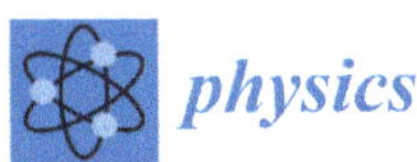

Article

Emergent Flow Signal and the Colour String Fusion

Daria Prokhorova * and Evgeny Andronov

Laboratory of Ultra-High Energy Physics, St Petersburg University, 7-9 Universitetskaya Emb.,
199034 St. Petersburg, Russia; e.v.andronov@spbu.ru
* Correspondence: d.prokhorova@spbu.ru

Abstract: In this study, we develop the colour string model of particle production, based on the multi-pomeron exchange scenario, to address the controversial origin of the flow signal measured in proton–proton inelastic interactions. Our approach takes into account the string–string interactions but does not include a hydrodynamic phase. We consider a comprehensive three-dimensional dynamics of strings that leads to the formation of strongly heterogeneous string density in an event. The latter serves as a source of particle creation. The string fusion mechanism, which is a major feature of the model, modifies the particle production and creates azimuthal anisotropy. Model parameters are fixed by comparing the model distributions with the ATLAS experiment proton–proton data at the centre-of-mass energy $\sqrt{s} = 13$ TeV. The results obtained for the two-particle angular correlation function, $C(\Delta\eta, \Delta\phi)$, with $\Delta\eta$ and $\Delta\phi$ differences in, respectively, pseudorapidities and azimuthal angles between two particles, reveal the resonance contributions and the near-side ridge. Model calculations of the two-particle cumulants, $c_2\{2\}$, and second order flow harmonic, $v_2\{2\}$, also performed using the two-subevent method, are in qualitative agreement with the data. The observed absence of the away-side ridge in the model results is interpreted as an imperfection in the definition of the time for the transverse evolution of the string system.

Keywords: proton interactions; multi-pomeron exchange; colour strings; string fusion; azimuthal flow; near-side ridge

1. Introduction

In recent decades, a significant effort has been made in the study of a unique state of matter called [1] quark–gluon plasma (QGP). The first experimental evidence of QGP formation was claimed by CERN in 2000 [2]. In 2005, this statement was quantitatively confirmed by RHIC experiments [3–6], whose studies raised questions about the properties of QGP.

Among numerous illuminating hints of a QGP signal, there was evidence [4,5] of the azimuthal anisotropy of particles produced in semi-peripheral heavy ion collisions. The measurements were done in a model-independent way [7] using transverse flow harmonics, v_n, of different orders, n: directed flow (v_1), elliptic flow (v_2), triangular flow (v_3) etc. A considerable magnitude of v_2 was observed [4], whose contribution is assumed to be a dominant one following the almond shape of the intersection region of two nuclei. The values of v_2 were comparable to the calculations [8] obtained when relativistic hydrodynamical fluid was considered in the early times of the collision. Therefore, it is the collective motion of thermally equilibrated partonic degrees of freedom that is believed to convert spatial anisotropies into momentum space under pressure gradients of the compressed medium. This rejected the view that QGP is a gas, but nevertheless, left still to establish whether this represents a perfect or viscous fluid.

Since then, plenty of flow measures have been proposed that vary in sensitivity to non-flow effects. Those are pair correlations [9], multi-particle correlations [10] measured with cumulants [11] or with subevent cumulant method [12], symmetric [13] and asymmetric [12]

Citation: Prokhorova, D.; Andronov, E. Emergent Flow Signal and the Colour String Fusion. *Physics* **2024**, *6*, 264–289. https://doi.org/10.3390/physics6010019

Received: 1 January 2024
Revised: 27 January 2024
Accepted: 2 February 2024
Published: 20 February 2024

cumulants considered in subevent method and a peculiar combination of flow-mean particle transverse momentum correlations [14].

The surprising evidence that connects these studies and the present paper is the experimental observation of collective behaviour in small systems. Namely, the unexpected ridge signal [15,16] and associated flow harmonics [17] seen in high-multiplicity proton–proton (pp) interactions at the LHC. These observations should not be explained by the fluid nature of the medium produced in such a small droplet of matter as it should not be able to reach thermal equilibrium prior to hadronization [18]. Nevertheless, there are successful attempts [19] to apply hydro description even in this case, which raises questions about the use of the same approach in all colliding systems [20].

On the one hand, from the first principles of the theory of strong interaction, the description of multi-particle production is complicated by the feature that the majority of the particles are produced in soft processes. It means that their transverse momenta, p_T, do not exceed about 1 GeV, therefore, the perturbative calculations in quantum chromodynamics (QCD) are inapplicable in this regime. This forces us to work in phenomenological approaches that can effectively describe the transition from, for instance, a few colliding hadrons to hundreds and thousands of particles produced.

One of the successful techniques is based on the concept of formation of colour strings between colliding partons which then fragment into observable hadrons [21]. This method appeared to be effective both in phenomenological calculations (see e.g., dual parton model [22], string percolation model [23], colour-glass condensate and glasma model [24]) and as the basis of many Monte-Carlo event generators, such as, e.g., EPOS [25], PYTHIA [26], HIJING [27], AMPT [28] models. In this paper, we consider the colour string picture of multi-particle production to model the transverse flow effect measured in pp collisions. It is the strings' interaction in the form of fusion [23] that plays a primary role in the model.

Our interest in this topic is inspired by the approach developed in the series of papers [29–31] in which the flow signal appears due to the quenching of particle transverse momentum in a string medium [30]. Namely, it is assumed that particles emit gluons while passing through the strings, which is similar to the energy losses of charged particles moving in the electromagnetic field in quantum electrodynamics (QED). Hence, the appearance of forbidden azimuthal angles changes the distribution of particles in the transverse plane. Here, the fusion of strings creates a heterogeneous medium density along the path of a particle.

In addition, we consider that the fusion accelerates the overlapped strings, resulting in a boost of the produced particles. It creates distinct directions of motion in the transverse plane for particles originating from the rearranged colour field. This approach originates from the pioneering paper [32] and has already been tested in more recent models [33,34]. The interplay of these two mechanisms, driven by string fusion, allows one to obtain specific joint correlations in pseudorapidity and azimuthal angle, η and ϕ, respectively, of the same origin in collision systems of different types.

In this paper, we extend our model [35,36] that provided a thorough description of particle correlations and fluctuations in η, for inelastic pp interactions at top RHIC and LHC energies. In this paper, we follow up [37] and test the model application to the problem of describing flow in pp interactions, aiming to obtain characteristic η–ϕ correlations at the pp system centre-of-mass energy $\sqrt{s} = 13$ TeV.

In Ref. [36], we emphasized that the previous model implementation lacks the short-range correlations introduced. It did not allow us to fully describe the ALICE experiment data, but we were able to catch the trend. In the present investigation, on the contrary, we aim to take advantage of this drawback of the model since getting rid of non-flow effects is the main challenge in flow studies. However, there is still a need to partially modify the model.

In the present study, the basic model features, as in Ref. [36], are the 3D (three-dimensional) dynamics of strings and string fusion. The strings consideration is slightly modified, but the main point stays intact: longitudinal contraction and shift of strings in the rapidity space, inspired by [38], and their transverse motion governed by an attractive Yukawa potential via the σ-meson exchange, following [39].

The main set of changes introduced in this study concerns the string fusion mechanism [40,41]. Namely, we abandon the consideration of the simplified cellular approach [42] to find the overlapped strings by looking after their centres. Instead, we divide the transverse plane into pixels that are much smaller than the string's size. The contribution of a pixel to particle production is determined by the degree of strings fusion that occurs inside it. We apply these changes with an eye toward future studies of nucleus-nucleus (AA) collisions.

It is worth noting separately that the mechanisms of momentum loss and string boosts applied to AA collisions provide a good description of the observed flow signal [43,44]. However, the additional introduction of the transverse motion of strings would naively result in the formation of a single colour flux tube after string fusion is taken into account. Therefore, the problem with the previous implementation of the model [36] is that a modified but uniform colour field could not serve as a source of azimuthal anisotropy.

The approach with the fine granularity of the configuration–momentum space that we propose in the present paper solves this issue, but we first test it in the description of pp collision results. Although the interpretation of pp results on azimuthal anisotropy provokes some tension in the flow community, it is technically more convenient for us to start the model development with the small colliding system, since keeping the fine cellular structure is much more challenging in AA, CPU-time wise. Furthermore, there are other improvements of the model mostly at the stage of proton assembly and event simulation implemented in this work.

To summarize, we develop the Monte-Carlo model based on the colour strings approach that addresses the problem of azimuthal particle correlations via the string–string interactions and particles' momentum loss. Thus, we anticipate that our preliminary estimates will demonstrate whether the nature of collective effects observed in pp collisions can be revealed using such a simplified but authentic model.

The paper is organized as follows. Section 2 presents the model framework, with special attention given to improvements introduced to the model compared to its previous version [36]. The description of the flow mechanism implemented in the model is included in a separate Section 3. In Section 4, we demonstrate the inference of free model parameters based on comparison with ATLAS experiment pp collision data at $\sqrt{s} = 13$ TeV [45]. Section 5 introduces the flow measures that we calculate in the model. Results obtained are presented in Section 6 and discussed in Section 7.

2. Colour String Model

The colour string model [21] is a phenomenological approach that is used to describe the soft particle production regime where the perturbative QCD calculations are not applicable. Strings that are extended colour field objects (also called tubes of gluon field) are stretched between partons participating in the collision. We consider them all to be parallel with neither kinks nor twists. It is believed that strings have a finite size in the transverse dimension, which enables them to overlap and interact. String longitudinal dynamics is described by a yo-yo solution, which means that a string periodically stretches and contracts with time. The energy of the colour field inside the tube grows as the string extends, while its massive endpoints slow down. It causes a probabilistic break-up of the colour field via the creation of quark–antiquark, q–$\bar{q}$, pairs and the breaking of strings into observed particles.

2.1. Multi-Pomeron Exchange in pp Collisions

In this paper, we consider inelastic pp interactions at $\sqrt{s} = 13$ TeV in the context of multi-pomeron exchange, with each pomeron represented by a cylindrical diagram. The uncut diagrams contribute to the elastic cross-section, while their unitarity cut creates two colour strings [22] that fragment into observed particles. Therefore, in this approach, the number of strings, n_{str}, in an event is determined by the number of exchanged pomerons, n_{pom}, as $n_{str} = 2n_{pom}$.

With the natural assumption of the Gaussian distribution of the transverse positions of partons inside the proton, one can calculate a probability of parton-parton interaction that can be interpreted as the probability of string formation [46]. The resulting distribution of the number of pomeron exchanges coincides with the Regge-theory parametrization [47] that neglects the three-pomeron vertices:

$$P(n_{pom}) = A(z)\frac{1}{zn_{pom}}\left(1 - \exp(-z)\sum_{l=0}^{n_{pom}-1}\frac{z^l}{l!}\right),$$

(1)

where $A(z)$ is a normalising coefficient, $z = \frac{2C\gamma s^\Delta}{R^2 + \alpha' \ln s}$, $C = 1.5$ is the quasi-eikonal parameter related to the small-mass diffraction dissociation of incoming hadrons, $\Delta = \alpha(0) - 1 = 0.2$ is the residue of the pomeron trajectory, $\alpha(0)$ is the intercept of the pomeron trajectory, $\gamma = 1.035\,\text{GeV}^{-2}$ and $R^2 = 3.3\,\text{GeV}^{-2}$ characterise the coupling of the pomeron trajectory with the initial hadrons, $\alpha' = 0.05\,\text{GeV}^{-2}$ is the slope of the pomeron trajectory.

Figure 1 presents the event distribution in the number of strings, n_{str}, obtained with these parameters and used in our simulation.

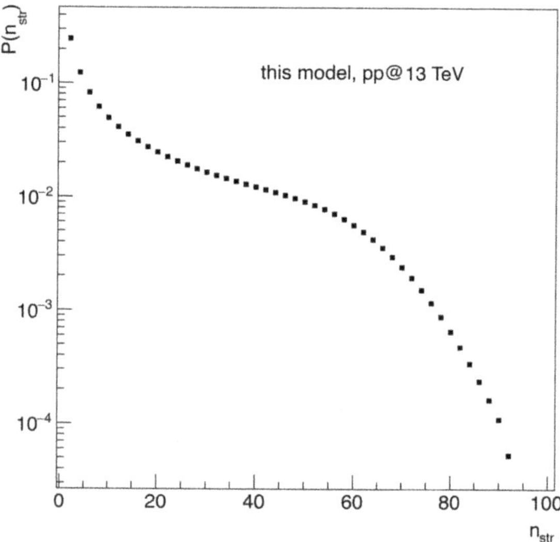

Figure 1. Model event distribution, $P(n_{str})$, of the number of strings for inelastic pp interactions at the centre-of mass energy $\sqrt{s} = 13$ TeV.

To determine the values of parameters in Equation (1), the data on multiplicity distributions and cross sections were used. We substitute into Equation (1) the parameters from Ref. [48], where it is assumed that strings that were initially formed in an event can overlap to various degrees and fuse, forming the string clusters with the modified fragmentation characteristics, which changes the obtained multiplicity. This is in contrast to another approach used in Ref. [49], where string fusion is effectively implemented already at the moment of string formation. Therefore, there are a number of free strings in Ref. [48] and a number of string clusters in Ref. [49]. The values of parameters in Equation (1) for these two methods will differ, but both the models should lead to the same charged particle multiplicity distribution, $P(N_{ch})$ (see Equation (2) below), corresponding to the data. We follow the first approach that allows us to track the 3D evolution of string density.

Thus, the proton no longer consists of a single quark and a diquark and forms not only two strings in pp interaction, as we move away from the conventional picture. Instead, we take into account the multi-parton configurations of sea quarks whose independent and simultaneous interactions result in the formation of the number of pomerons in an event.

In this approach, $P(N_{ch})$ is determined as a convolution of $P(n_{pom})$ (1) and the multiplicity distribution at fixed number of pomerons, $P_{n_{pom}}(N_{ch})$,

$$P(N_{ch}) = \sum_{n_{pom}=1}^{\infty} P(n_{pom}) P_{n_{pom}}(N_{ch}). \tag{2}$$

Due to the complexity of the analytical formulation of $P_{n_{pom}}(N_{ch})$, we treat it using a Monte-Carlo simulation. In Ref. [35], we demonstrated that for quite a simple case of non-interacting strings, the correlation observables calculated numerically and analytically are in good agreement.

2.2. Proton Assembly

Our current study is based on the assumption that all partons of a proton form strings with all partons of another proton. Therefore, we call an event the creation of n_{str} between two protons, each with the number of partons, n_{part}, which is equal to the number of strings in an event, $n_{part} = n_{str}$.

To fulfil this requirement, we prepare an extensive set of protons with all the possible even numbers of partons, $n_{part} = 2k, k = 1, \ldots, 50$. In this study, we consider the parton composition of a proton as one valence quark, one valence diquark, and $(n_{part} - 2)$ sea quarks. The algorithm provided in Appendix A is more time- and CPU-efficient than the permutations of partons between two selected protons that were implemented in Ref. [36].

2.3. Event Simulation

The first step of a simulation of an event is to sample the number of exchanged pomerons, n_{pom}, from Equation (1). Thus, we know the number of strings that should be created in an event, $n_{str} = 2n_{pom}$. By creating the event, we mean searching for two protons, with the same number of partons, $n_{part} = n_{str}$, so that all their partons can form strings under certain conditions (see Sections 2.3.1 and 2.3.2).

First, we select two random protons from the prepared set (Section 2.2) with the certain n_{part}. Second, we permute partons in one of the two protons by performing n_{part} replacements and checking whether this leads to the formation of n_{part} good pairs (see the requirements in the next two subsections). If not, we select another pair of protons.

2.3.1. String Formation: Electric Charge

In this study, we ensure the conservation of electric charge. Thus, we accept a string candidate if the electric charge of the string, Q_{str}, defined as the sum of the partons' charges that form a string, equals one of the possible integer numbers: $-1, 0, +1, +2$.

2.3.2. String Formation: Energy and Mass of Decay Products

We calculate the string energy, E_{str}, by summing the energies of partons at the ends of the string, E_{part1} and E_{part2}, as

$$E_{str} = \sqrt{m_{part1}^2 + p_{part1}^2} + \sqrt{m_{part2}^2 + p_{part2}^2}, \tag{3}$$

where p_{part} is the initial momentum of a parton and m_{part} is a dynamically defined parton mass (for details of their definitions, see Appendix A.3).

We accept the string candidate if E_{str} is sufficient to decay at least in two hadrons at rest, with masses $M_{daughter1}$ and $M_{daughter2}$, i.e., $E_{str} \geq M_{daughter1} + M_{daughter2}$.

In order to test this condition, it is necessary to identify the species of the pair of daughter particles based on the flavours of the string ends, e.g., the quark–diquark string should decay at least to a baryon and a meson. For completeness, we provide the list of minimum permitted combinations of daughter products, based on the flavours of quarks that we consider in the model: $m_{\text{nucleon}} + m_\pi$, $m_{\text{nucleon}} + m_K$, $m_{\text{nucleon}} + m_D$, $2 \cdot m_\pi$, $2 \cdot m_K$, $2 \cdot m_D$, $m_\pi + m_K$, $m_\pi + m_D$ and $m_K + m_D$. We use $m_{\text{nucleon}} = 0.9396$ GeV, $m_\pi = 0.1396$ GeV, $m_K = 0.4977$ GeV, $m_D = 1.8696$ GeV for nucleons, pions, kaons, and D-mesons, respectively.

2.4. String Transverse Dynamics

In general, the colour confinement in QCD, a non-abelian gauge theory, is viewed as the gathering of the colour field between two colour charges in the flux tube of finite transverse size [50]. However, lattice QCD demonstrates that the presence of a colour string modifies the QCD vacuum. For example, the correlator between the quark–antiquark chiral condensate, $\langle q\bar{q} \rangle$, and the Wilson loop, W, is appeared to be not a constant as a function of the transverse distance from a string [51]. Indeed, at large distances it reaches the value of unity, meaning that there is no string influence. In the meantime, its values decrease in the vicinity of the string, which indicates a partial restoration of chiral symmetry in this region of the space.

We follow the approach of Refs. [39,52], where the authors interpret these lattice calculations (left-hand side of Equation (4)), as a field created by a cloud of σ-mesons (right-hand side of Equation (4)):

$$\frac{\langle q\bar{q}(r_\perp)W \rangle}{\langle q\bar{q} \rangle \langle W \rangle} = 1 - K_0(m_\sigma \widetilde{r_\perp}), \tag{4}$$

where $\widetilde{r_\perp} = \sqrt{r_\perp^2 + s_{\text{str}}^2}$ is the regularized distance in the transverse plane, $r_\perp$ is a 2D distance between strings in a pair of strings, $s_{\text{str}} = 0.176$ fm [39] is a genuine string width, unlike the effective string width, which is a result of quantum fluctuations, K_0 is zero-th modified second-kind Bessel function corresponding to a massive scalar propagator in two dimensions and $m_\sigma = 0.6$ GeV [39] is the mass of the σ-meson that is proposed to be a mediator of the force between strings.

In this paradigm, one can consider string–string interactions in the created Yukawa potential, which results in the non-relativistic attraction between them similar to nuclear forces. Thus, the equations of motion of the string system in an event are defined by the 2D Yukawa interaction [39],

$$\ddot{\vec{r}}_i = \sum_{j \neq i} \vec{f}_{ij} = 2m_\sigma(g_N\sigma_T) \sum_{j \neq i} \frac{\vec{r}_{ij}}{\tilde{r}_{ij}} K_1(m_\sigma \tilde{r}_{ij}), \tag{5}$$

where $\vec{r}_{ij}$ and $\tilde{r}_{ij}$ correspond to $r_\perp$ and $\widetilde{r_\perp}$, respectively, from Equation (4). The i and j subscripts indicate that the quantities are constructed for the i-th and j-th strings. Here, g_N is the QCD string self-interaction coupling and σ_T is the string tension, whose dimensionless product is selected as $g_N\sigma_T = 0.2$ [52], and K_1 is the first modified second-kind Bessel function. Strings are considered to be moving as a whole according to Equations (5), without whatever kinks.

To solve this system of differential equations, the set of initial conditions is required. The Gaussian distribution of width 0.5 (model parameter) is used to sample the initial 2D transverse coordinates of string centres in an event. This simplification is to decrease the program's running time: instead of applying the Glauber approach at the partonic level and finding which partons are to form strings, we assume that some configuration of strings has already been created in an event.

Transverse strings' evolution, governed by Equation (5), can be terminated at some proper time, τ_{stop}, which affects the final string density. In Ref. [36], we showed that the largest number of overlaps between strings occurs after the time that we called τ_{deepest}. The latter is the time it takes for a string transverse configuration to attain the global minimum of the potential energy, evolving according to Equation (5). Ref. [36] considers other cases, including no transverse evolution and transverse evolution until the conventional time $\tau_{\text{stop}} = 1.5$ fm [52] that is unsuccessful in describing the data, especially the $\langle p_T \rangle$–N

correlation, where $\langle p_T \rangle$ is the event-averaged transverse momentum of a particle and N denotes the particle multiplicity. Thus, in the present study, we consider $\tau_{\text{stop}} = \tau_{\text{deepest}}$. It varies event-by-event and depends not only on the initial transverse positions of the strings, but also on the number of strings, and thus on the collision energy. It is worth noting that τ_{deepest} cannot be found for any event. This means that for some configurations of strings, the minimum of the potential energy of the system can only be reached at $\tau_{\text{stop}} > 1.5$ fm. In this case, we set $\tau_{\text{stop}} = 1.5$ fm as the typical time for string hadronization and restrict system evolution to it.

Figure 2 shows how a few events (one event a row) with different numbers of strings look before and after the transverse evolution of string density. Figure 2 (left half, left to right) demonstrates the event projections to the transverse plane, X–Y, before and after the transverse dynamics of strings. Figure 2 (right half, left to right) present the projections to the X–rapidity plane before and after the transverse evolution. The corresponding time of the transverse motion, $\tau_{\text{transv}} = \tau_{\text{deepest}}$, is indicated as relevant. One can see that the string system is highly compressed and the 2D distribution (Figure 2, second left) approaches a more spherical shape in comparison to the initial positions of the strings (Figure 2, most left). In the rapidity dimension, one obtains quite uniform distribution up to large rapidities (Figure 2, most right).

2.5. String Longitudinal Dynamics

In the model considered here, special attention is given to the simulation of partonic degrees of freedom for colliding protons (see Appendix A). It is because the initial rapidities of the string's ends, $y_{\text{init}}^{\text{part}}$, are determined by using the momenta of partons that form it as

$$y_{\text{init}}^{\text{part}} = \sinh^{-1}\left(\frac{p_{\text{part}}}{m_{\text{part}}}\right). \tag{6}$$

However, the string tension, σ_T, slows down the massive partons flying outwards with momentum p_{part} according to $dp_{\text{part}}/dt = -\sigma_T$, where t denotes the time. As a result, the rapidity of the string end decreases [38] by value

$$y_{\text{loss}}^{\text{part}} = \cosh^{-1}\left(\frac{\tau_{\text{long}}^2 \sigma_T^2}{2m_{\text{part}}^2} + 1\right), \tag{7}$$

where τ_{long} is the proper time of the string longitudinal evolution.

The aim is to relate the τ_{long} from Equation (7) with the time for the transverse evolution, τ_{transv}, to synchronise the dynamics of the string system in rapidity and X–Y dimensions. However, one has to take into account that the compressions and stretchings of a string are periodic (yo-yo solution [21]) and are followed one by another. Moreover, the movement of massive endpoints of different masses, m_{part1} and m_{part2}, in the case here is not symmetrical (we denote by 1 or 2 one of the string ends). Therefore, we define the maximum proper time for each string end, $\tau_{\text{max}}^{\text{part1,2}}$, after which the string's endpoint fully stops the deceleration (acceleration) after acceleration (deceleration) as

$$\tau_{\text{max}}^{\text{part1,2}} = \frac{m_{\text{part1,2}}}{\sigma_T}\sqrt{2(\cosh(\tilde{y}_{\text{init}}^{\text{part1,2}}) - 1)}, \tag{8}$$

where $\tilde{y}_{\text{init}}^{\text{part1,2}}$ are the initial rapidities of the string ends converted to the string rest frame.

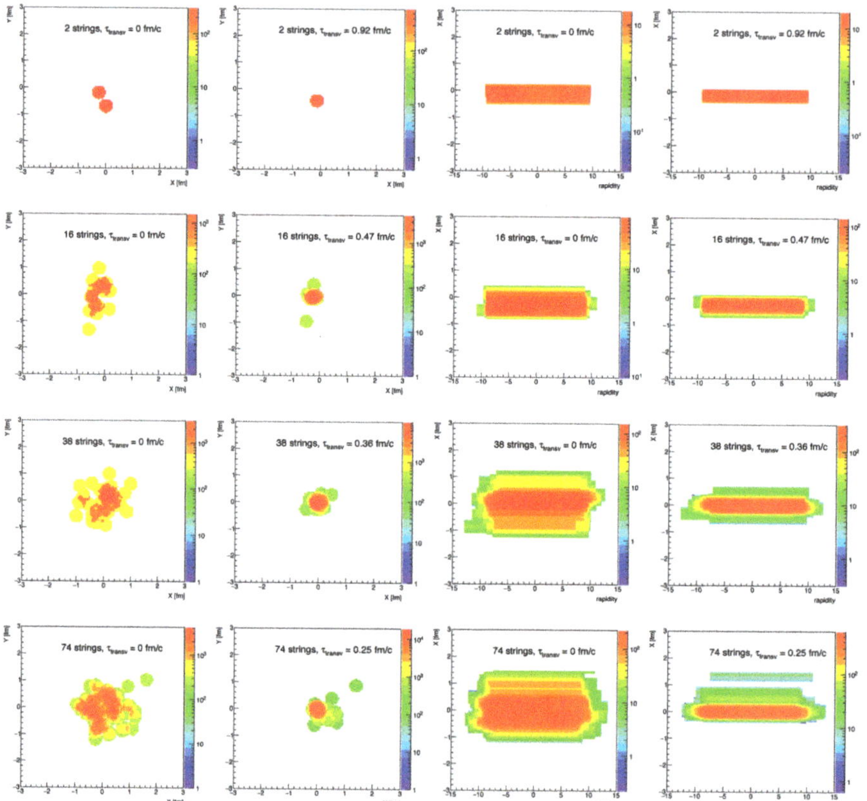

Figure 2. The simulated events (one event a row) with 2, 16, 38, and 74 strings (**top** to **bottom**) for an event projection to the transverse plane, *X–Y*, before and after the transverse evolution and to the *X*–rapidity plane before and after the transverse evolution (**left** to **right**) for the time of the transverse evolution, τ_{transv}, as indicated. The evolution runs till $\tau_{\text{transv}} = \tau_{\text{deepest}}$ (second left and most right). *Z* axis is not to scale. See text for more details.

To convert the rapidity of a string end from the laboratory frame, $y_{\text{init}}^{\text{part}}$, to the string rest frame, $\tilde{y}_{\text{init}}^{\text{part}}$, one has to know the rapidity of the string in the laboratory frame, y_{str}. It is calculated as follows. First, let us define strings' momentum, $p_{\text{str}} = p_{\text{part1}} - p_{\text{part2}}$, and strings' energy, E_{str} (3). Thus, the rapidity of a string, y_{str}, is

$$y_{\text{str}} = \frac{1}{2} \ln\left(\frac{E_{\text{str}} + p_{\text{str}}}{E_{\text{str}} - p_{\text{str}}}\right). \tag{9}$$

As soon as y_{str} is known, one can recalculate the rapidities of the string ends in the string's rest frame:

$$\tilde{y}_{\text{init}}^{\text{part1,2}} = y_{\text{init}}^{\text{part1,2}} - y_{\text{str}} \tag{10}$$

and substitute Equation (10) into Equation (8).

It is necessary to account for the periodicity of string motion. Namely, after the time $2\tau_{\text{max}}^{\text{part1,2}}$ (8), the sign of $dp_{\text{part1,2}}/dt$ flips. Therefore, it is crucial to correctly connect the τ_{deepest}, which is found from the transverse dynamics for the whole event, with $\tau_{\text{max}}^{\text{part1,2}}$, which varies for two ends of the string and for each individual string in the event.

To summarize, the time of the movement of a string end only during the last period should be used in Equation (7). The rapidities of the string ends after such loss, $y_{\text{fin}}^{\text{part1,2}}$, are found by subtraction of $y_{\text{loss}}^{\text{part1,2}}$ from $y_{\text{init}}^{\text{part1,2}}$ with the correct signs.

It is worth remarking that the longitudinal evolution changes not only the length of the strings but also their positions with respect to midrapidity. Figure 3 demonstrates how the string density of the model events from Figure 2 (already after the transverse evolution) changes after the longitudinal losses are taken into account. For clarity, we repeat Figure 2 (most right) in Figure 3 (most left). In Figure 3 (left half, left to right), one can see the event projections on the X–rapidity plane before and after the longitudinal dynamics. Figure 3 (right half, left to right) shows the Y–rapidity plane projections also before and after the longitudinal dynamics.

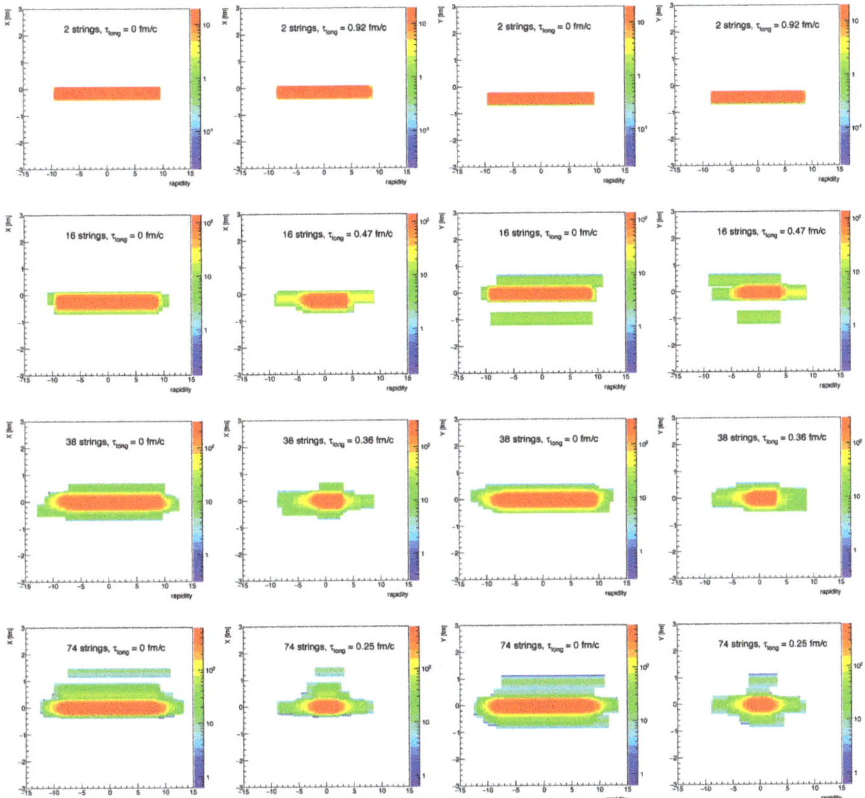

Figure 3. The simulated events (one event a row) with 2, 16, 38, and 74 strings (**top** to **bottom**) for an event projection to the X–rapidity plane before and after the longitudinal evolution and to the Y–rapidity plane before and after the longitudinal evolution (**left** to **right**) for the time of the longitudinal evolution, τ_{long}, as indicated. The evolution runs till $\tau_{\text{long}} = \tau_{\text{deepest}}$ (second left and most right). Z axis is not to scale.

It can be seen that the string system in the final state (Figure 3, second left and most right) has a significant compression in the longitudinal direction. The cloud of a high density of strings is no longer infinitely extended as it is often assumed in the string models [53]; thus, our model loses the translational invariance in rapidity. The transverse evolution of the strings, as described in Section 2.2, also leads to varying string density, which makes an event strongly heterogeneous.

2.6. String Fusion

As a result of the reductions and shifts of strings in the rapidity dimension (Section 2.5), the system of strings is a spaghetti-like structure [39], although the lengths and positions of strings vary with respect to mid-rapidity. Furthermore, due to the attractive motion of strings in the transverse plane (Section 2.4), the clusters of fully or partially overlapped strings are formed (taking into account that they have a finite transverse size). Since this overlap changes with rapidity, one obtains a non-trivial string density 3D distribution.

In this study, we follow the consideration from Ref. [23], where the colour field of fused gluonic tubes changes, which impacts particle production. This can also be interpreted as the appearance of string clusters with higher effective tension.

Apparently, there are several options for addressing the overlap and fusion of strings [42]. In our previous study [36], we employed a cellular approach that involves the fusion of strings that have the centres in the same cell in the transverse plane. In this case, the cell size is equivalent to the diameter of a string. Thus, a cluster of k overlapped strings in Ref. [36] was replaced by one string with a decreased mean multiplicity per rapidity unit, an increased mean transverse momentum of produced particles and an increased probability of producing heavier particles. The transverse position of the cluster of k fused strings can be assumed to be the mean value of their k transverse positions.

Therefore, if one considers AA collision in the approach [36] discussed just above, one can expect that after the transverse dynamics (Section 2.4) the centres of all strings appear in the same cell. This leads to the replacement of all the variety of the degrees of strings' overlaps by a single string with enhanced tension. In this scenario, it is impossible to create the anisotropy of produced particles since the information about matter density fluctuations is lost. Let us note that in the models without attractive string dynamics in the transverse plane, this is not an issue since an event picture is less dense. Thus, in the current study, we do not consider simplifying string fusion on a coarse lattice but work using fine division of the transverse plane into small enough pixels.

In the approach considered here, the transverse plane is tessellated into square bins (pixels), each of which counts the occupation numbers of strings (number of strings that cover this pixel). In order to have a fine grid, one has to select the area of the bin, S_{bin}, so that the latter is much smaller than the string transverse area, S_0. In the current implementation, we use a string diameter, d_{str}, of 0.5 fm and a pixel width, d_{bin}, of 0.05 fm.

The resembling procedure is done for the rapidity space with a slice size of 0.1. In the previous study [36], we applied a longitudinal grid separately to each string. In this paper, we use a uniform splitting for the entire longitudinal dimension.

Finally, one can calculate the number of strings that occupy each 3D bin in the mixed configuration–momentum space. Thus, in a sense, we move away from the concept of particle-producing strings towards the concept of particle-producing pixels.

Note that the program's running time is significantly impacted by the number of 3D bins, as it must iterate over all 3D cells. Hence, it is necessary to restrict the volume permitted for simulations. Actually, the collision energy determines the longitudinal (in rapidity) size of the defined space, as the beam rapidity rises with $\sqrt{s}$. In turn, the required transverse area depends on the initial distribution of strings and their final positions after the transverse dynamics. Since, in this investigation, we focus on the evolution of the system until the global minimum of its potential energy, after the motion according to Equation (5), the system becomes even more compact.

2.7. Fusion and the Kinetic Energy of Strings

Since the total energy of the system of strings in an event to be conserved, we assume that when strings overlap, a redistribution of their potential, U, and kinetic, T, energies occur. It is because the overlap of colour strings modifies the strength of the colour field inside the strings, which affects the string tension. Therefore, the partial overlap of a few strings leads to a decrease in the potential energy of each of them in this region. This modification causes an opposite change in the kinetic energy of strings, pulling them towards each other similar to that considered in Ref. [32]. The formation of a string cluster, as a consequence of the strings attraction increases the string tension in comparison to a free string. Interestingly enough, the alternative option, also mentioned in Ref. [32], includes the decrease in the effective string tension and, therefore, should lead to the string repulsion (see also Ref. [33]).

As a result, one needs to parameterize the gain of kinetic energy, $\Delta T_{i,j}$, that the i-th string in an event obtains from the overlap with the j-th string. The functional form of $\Delta T_{i,j}$ should reflect the feature that the full overlap of two strings means that their fusion is completed and those strings stop. On the other hand, $\Delta T_{i,j}$ should decrease with the distance between the strings' centres. Thus, the following dependence on $d_{i,j}$—the 2D distance between the i-th and j-th strings—is proposed:

$$\Delta T_{i,j} = \chi d_{i,j} \exp\left(\frac{-d_{i,j}^2}{4r_0^2}\right). \tag{11}$$

Here, $d_{i,j} \leq 2r_0$, where $r_0 = 0.25$ fm is the string radius, while χ is a dimensional constant measured in GeV/fm that is a free model parameter. Hence, to get a new value of the kinetic energy of a string we iterate over its neighbouring overlapping strings in every rapidity slice.

We assume that the string motion according to Equation (5) is non-relativistic; therefore, in the following, we neglect the initial momentum of a string. As a result, the i-th string gains the extra transverse momentum from the j-th string:

$$\Delta p_T^{i,j} = \sqrt{(\Delta T_{i,j} + m_{\text{str}}^i)^2 - (m_{\text{str}}^i)^2}, \tag{12}$$

where $m_{\text{str}}^i = m_{\text{part1}} + m_{\text{part2}}$ is the sum of the dynamically obtained masses of partons forming the i-th string (see Appendix A.3).

The azimuthal direction of the vector from the centre of the i-th string to the centre of the j-th string, $\phi_{i,j}$, is determined by the distances on X and Y axes between the centres of the strings, $\Delta X_{i,j}$ and $\Delta Y_{i,j}$. The projections Δp_X^i and Δp_Y^i of the total momentum that the i-th string gains from all its overlapping neighbours, Δp_T^i, are found by summing the projections of $\Delta p_T^{i,j}$ as

$$\Delta p_X^i = \sum_{\substack{j \neq i}}^{n_{\text{str}}-1} \Delta p_T^{i,j} \cos(\phi_{i,j}), \quad \Delta p_Y^i = \sum_{\substack{j \neq i}}^{n_{\text{str}}-1} \Delta p_T^{i,j} \sin(\phi_{i,j}). \tag{13}$$

Thus, the 2D vector of the string's transverse momentum, $\overrightarrow{\Delta p}_T^i$, is determined by its overlap with other strings in each rapidity slice. It is this extra momentum that is transmitted to particles produced by the string (see Section 3.1).

2.8. Particle Production

In this study, we define string hadronisation in each 3D bin. The strength of the colour field inside it determines the average number of charged particles produced and their average transverse momentum. Particles' rapidities are found from the uniform distribution in each rapidity slice.

2.8.1. Mean Multiplicity with String Fusion

Let us consider a rapidity slice of a free single string. The colour field, $\mathcal{E}_0$, inside it can be represented as a sum of colour fields inside all transverse (2D) bins that tessellate the area of this string,

$$\mathcal{E}_0 = \sum_{\text{bins}} \mathcal{E}_{\text{bin}} = \sum_{\text{bins}} \frac{S_{\text{bin}}}{S_0} \mathcal{E}_0. \tag{14}$$

Thus, the field inside a 2D bin, $\mathcal{E}_{\text{bin}}$, is proportional to the ratio of its area, S_{bin}, to the area of a string, S_0.

Let us now consider a random 3D bin in the space that is populated with k strings. The resulting colour field, $\mathcal{E}_{\text{tot}}$, inside the string, is

$$\mathcal{E}_{\text{tot}} = \sqrt{\sum_{i=1}^{k} \mathcal{E}_i^2} = \sqrt{\sum_{i=1}^{k} \left(\frac{S_{\text{bin}}}{S_0^i}\mathcal{E}_0^i\right)^2}, \tag{15}$$

where $\mathcal{E}_i$ is the field of a string in the bin.

For simplicity, the present study considers all strings to have the same transverse area, S_0^i, which does not vary along the rapidity direction, albeit the study can be more complex [54]. The colour field of a free string, $\mathcal{E}_0^i$, is assumed to be constant and is defined by the confinement properties [55]. Therefore, in the case considered here, Equation (15) can be simplified to

$$\mathcal{E}_{\text{tot}} = \sqrt{k}\frac{S_{\text{bin}}}{S_0}\mathcal{E}_0. \tag{16}$$

The average multiplicity from a free string, ν_0, is proportional to the field of a string, $\mathcal{E}_0$. Thus, by transitivity, for the k strings that overlap in a 3D bin, one obtains $\nu_{\text{tot}} \propto \mathcal{E}_{\text{tot}}$.

One can define the average multiplicity from a 3D bin with the length ε_{rap} and transverse area S_{bin} as

$$\mu_{\text{bin}} = \mu_0\varepsilon_{\text{rap}}\sqrt{k}\frac{S_{\text{bin}}}{S_0}, \tag{17}$$

with a free parameter μ_0 defining the average multiplicity of a unit of rapidity of a free string and $S_0 = \pi r_0^2$. For comparison, in our previous study [36], we calculated the mean multiplicity from the cluster of k strings (without division in the transverse plane), μ_{clust}, as $\mu_{\text{clust}} = \mu_0\varepsilon_{\text{rap}}\sqrt{k}$, following Ref. [40].

The actual multiplicity per 3D bin, N_{bin}, is sampled from the Poisson distribution with a given mean, μ_{bin} (17). Then, N_{ch} is a sum of the multiplicities of N_{bin} from all the 3D bins.

2.8.2. Mean Transverse Momentum with String Fusion

The mean transverse momentum of the particles produced by a free string, p_0, does not depend on the string's length in rapidity or its transverse area. Thus, p_0 remains unchanged despite the division of a strings into pixels. Therefore, we keep the modification of p_0 for the cluster of k fused strings, $\langle p_T \rangle_k$, from Ref. [36]. Namely,

$$\langle p_T \rangle_k = p_0 k^{\beta}, \tag{18}$$

where $\beta = 1.16[1 - (\ln\sqrt{s} - 2.52)^{-0.19}]$ is found in Refs. [49,56] from the fits to data and p_0 is a free model parameter.

Particles' transverse momentum is sampled from distribution

$$f(p_T) = \frac{\pi p_T}{2\langle p_T \rangle_k^2}\exp\left(-\frac{\pi p_T^2}{4\langle p_T \rangle_k^2}\right), \tag{19}$$

corresponding to Schwinger mechanism of quark–antiquark pair creation in the colour field of a string that leads to its break-up and the formation of final hadrons [57–59], with the mean transverse momentum, $\langle p_T \rangle_k$ (18), from the cluster.

2.8.3. Probabilities of Producing Different Particle Species

The particle species with masses m_{sp} are determined by Schwinger-like probabilities with the modified effective string tension, $\sigma_{\text{eff}} = 4p_0^2 k^{2\beta}$, according to $\propto \exp(-\pi m_{\text{sp}}^2/\sigma_{\text{eff}})$, which is consistent with Equation (19). Typically, in the models that rely on the Schwinger mechanism of particle production [60], σ_{eff} slightly differs from σ_T (see Equation (7)) to effectively take into account the mechanism of particles re-scatterings.

Pions, kaons, protons and ρ-resonances are included in the model, with the ρ-resonance decaying into two charged pions. With particle masses, the longitudinal component of the particle momentum, p_z, is calculated and, thus, the pseudorapidity is

$$\eta = \frac{1}{2}\ln\left(\frac{|\vec{p}| + p_z}{|\vec{p}| - p_z}\right), \tag{20}$$

where $|\vec{p}| = \sqrt{p_T^2 + p_z^2}$ is an absolute value of the particle momentum.

3. The Origin of Particle Flow in the Model

A purpose of introducing the fine grid in the transverse plane as described in Section 2.6 above is to account for particle azimuthal flow in our model. We follow the consideration in Refs. [29,32] on the way of introducing collective behaviour in the colour string model without a hydro phase. There are three main ingredients.

First, in this type of models, the strings in an event form some clusters [23] distributed anisotropically. Alone, these fluctuations in string density do not produce whatever flow. However, this initial state anisotropy is crucial for the two mechanisms described below as controlling the strength of the effect of those mechanisms.

Second, we consider the change in the strings' kinetic energy that occurs when the strings overlap in 3D space. As a result, the boost from a string is transferred to the particles that it is fragmented to [32]. The boost creates correlations in p_T and ϕ between particles produced from strings that interact with each other. What is essential is that string hadronisation is considered at the moment of the minimum of the potential energy of the system of strings (at τ_{deepest}). This results in the high similarity of the geometry of events with the close number of strings, which leads to the same picture of particle boosts. For instance, Figure 2 (second top, second left) shows a typical event with 16 strings. We argue that all events with this n_{str} resemble each other up to the event rotation and some statistical fluctuations. It means that the events have a non-zero flow of comparable value; therefore, the signal survives after averaging over such group of events.

Third, the particles passing through single strings or string clusters lose some part of the energy due to gluon radiation in the medium [29]. When a particle loses its entire momentum, it means that this particle failed to escape. Thus, the azimuthal directions that are forbidden appear in the event. Consequently, the particles can no longer move in a whatever direction and ϕs of those particles become narrowed and correlated. Moreover, the complicated geometry of the string medium leads to different path lengths in different azimuthal directions and to different p_T losses.

The first mechanism of those listed just above is naturally introduced into our model: transverse dynamics of strings result in the formation of string clusters of different degrees of overlap. Longitudinal dynamics make the fluctuations of string density dependent on the rapidity coordinate.

The other two mechanisms are the new features implemented in the model and are described in Sections 3.1 and 3.2 just below. Let us note that different degrees of string overlaps cause variations in the magnitude of energy loss and particle boosts bin-by-bin in the transverse plane–rapidity space. As a result, the particle production in a given event becomes highly asymmetric in azimuthal angle and pseudorapidity.

3.1. Transverse Momentum Boost

Particle's transverse momentum sampled from Equation (19) receives a Lorenz boost if the particle originates from the string piece that was accelerated due to string fusion [32].

The procedure iterates over all 3D bins and finds the strings that cover each bin. A bin is assigned the momentum projections, $\Delta p_X^{\text{bin},i}$ and $\Delta p_Y^{\text{bin},i}$, that are found as fractions of the momentum projections of the i-th string, Δp_X^i and Δp_Y^i, defined in Equation (13) as

$$\Delta p_{X,Y}^{\text{bin},i} = \Delta p_{X,Y}^i \frac{d_{\text{bin}}^2}{S_0} \frac{\varepsilon_{\text{rap}}}{\delta y^i}, \tag{21}$$

where $d_{\text{bin}} = 0.05$ fm is the bin size in X and Y in the transverse plane, $\delta y^i = |y_{\text{fin}}^{\text{part1}} - y_{\text{fin}}^{\text{part2}}|$ is the length of i-th string in rapidity which is calculated as the difference between the rapidities of the ends of the string, $y_{\text{fin}}^{\text{part1}}$ and $y_{\text{fin}}^{\text{part2}}$, and highly fluctuates.

Second, we find the part $m^{\text{bin},i}$ of the mass of the i-th string, m_{str}^i, which belongs to this bin in a similar way

$$m^{\text{bin},i} = m_{\text{str}}^i \frac{d_{\text{bin}}^2}{S_0} \frac{\varepsilon_{\text{rap}}}{\delta y^i}. \tag{22}$$

Using Equations (21) and (22), we can find the energy of i-th string contained in this 3D bin, $\Delta E^{\text{bin},i}$, as

$$\Delta E^{\text{bin},i} = \sqrt{(m^{\text{bin},i})^2 + (\Delta p_X^{\text{bin},i})^2 + (\Delta p_Y^{\text{bin},i})^2}. \tag{23}$$

If this 3D bin is covered by k strings in an event, their $\Delta p_{X,Y}^{\text{bin},i}$ from Equation (21) and $\Delta E^{\text{bin},i}$ from Equation (23) should be included in its total momentum, $p_{X,Y}^{\text{bin,total}}$, and the total energy inside it, $E^{\text{bin,total}}$, as

$$p_{X,Y}^{\text{bin,total}} = \sum_{i=1}^{k} \Delta p_{X,Y}^{\text{bin},i}, \tag{24}$$

$$E^{\text{bin,total}} = \sum_{i=1}^{k} \Delta E^{\text{bin},i}. \tag{25}$$

We perform particle production from each 3D bin in the bin rest frame as described in Section 2.8. In this frame, the particle's azimuthal angle, ϕ, is sampled from the uniform distribution from $-\pi$ to π. To obtain the laboratory reference frame, one boosts the four-momenta of the produced particles using the boost vector with $p_X^{\text{bin,total}}$, $p_Y^{\text{bin,total}}$ and $E^{\text{bin,total}}$.

Thus, we obtain the correlated transverse motion of particles produced by a 3D bin.

3.2. Transverse Momentum Loss

When a particle traverses a 2D bin with a certain density of strings (occupation number), it loses part of its initial momentum, p_{init}, due to gluon radiation, reaching the value, p_{fin}. In an analogy with the QED process of photon radiation by charged particles moving in the external electromagnetic field, it can be expressed in the following way [29]

$$p_{\text{fin}} = (p_{\text{init}}^{1/3} - \varkappa \sigma_{\text{eff}}^{2/3} l)^3, \tag{26}$$

where $\varkappa$ is a quenching coefficient that is a free model parameter. It is necessary to apply the Equation (26) iteratively since $\sigma_{\text{eff}} = 4 p_0^2 k^{2\beta}$ varies bin-by-bin based on the number of overlapped strings, k. That is why l, a 2D path that a particle accomplishes, can be approximated by $d_{\text{bin}}\sqrt{2}$ (the length of the 2D bin's diagonal) for each step. Since the density of strings fluctuates with rapidity (see Sections 2.4 and 2.5), l differs not only at different azimuthal angles, ϕ, but also within different rapidity slices, ε_{rap}.

Let us remark that, in general, the value of p_{fin} can become negative after a number of iterations of Equation (26). We interpret this as the impossibility of the particle to escape in a given azimuthal direction; thus, it is absorbed by the string environment. In this case, the energy of the particle is spent on producing another particle to replace the former one. For a new particle, the model regenerates the transverse momentum and azimuthal angle of this particle; thus, the particle gets a chance to escape in a new direction. As soon as the required number of particles to be produced from this 3D bin (from Poisson distribution with the mean from Equation (17)) is known, new particles are to be regenerated until the particle leaves the string medium.

To find the trajectory of a particle in the transverse plane, the Bresenham's line algorithm [61] is applied for each rapidity slice.

4. Inference of the Model Parameters

We determine the model parameters by comparing the model distributions with the ATLAS experiment data on inelastic proton–proton interactions at $\sqrt{s} = 13$ TeV [45] (Figure 4). We describe not only the global observables such as N_{ch}-distribution, p_T and η spectra, but also the $\langle p_T \rangle$–N_{ch} correlation function.

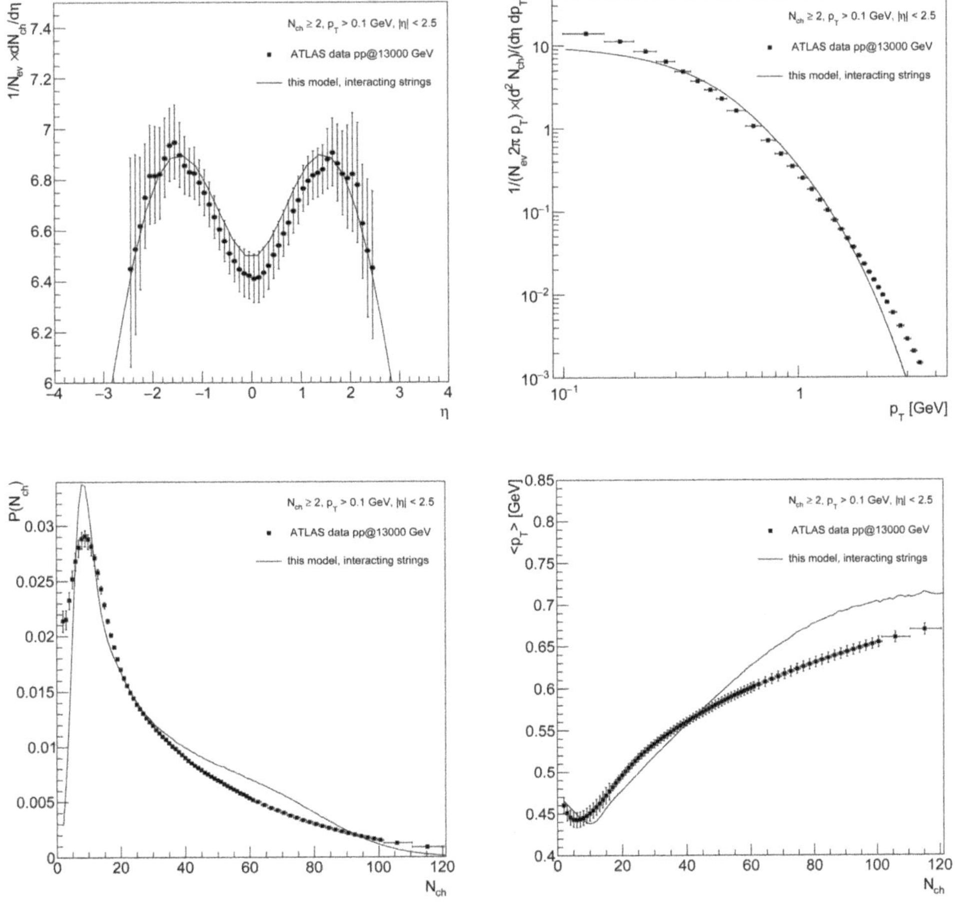

Figure 4. Model results (lines) compared to the ATLAS experiment data [45] (solid squares) for inelastic pp interactions at $\sqrt{s} = 13$ TeV: charged-particle pseudorapidity, $dN_{ch}/d\eta$, distribution (**upper left**), transverse momentum, p_T, spectrum (**upper right**), multiplicity distribution, $P(N_{ch})$, (**lower left**), and $\langle p_T \rangle$–N_{ch} correlation function (**lower right**) within the experimental $p_T > 0.1$ GeV, $|\eta| < 2.5$, and $N_{ch} \geq 2$ limits.

The $dN_{ch}/d\eta|_{\eta\approx0}$ region (Figure 4, upper left) is adjusted by finding the appropriate value of the mean particle multiplicity μ_0 (see Equation (17)) from a 3D bin of a free string. In turn, the width of the $dN_{ch}/d\eta$ distribution is controlled by the value of the string tension, σ_T (see Equation (7)).

By tuning p_0 (see Equation (18)), we settle the $\langle p_T\rangle$ at low N_{ch} (Figure 4, lower right). With the proper selection of $\varkappa$ (see Equation (26)), we are able to modify the slope of the $\langle p_T\rangle$–N_{ch} correlation function at moderate and high N_{ch}. The $\langle p_T\rangle$ is influenced by the value of χ (see Equation (11)) at high N_{ch}, when string fusion is most significant.

In the charged particle multiplicity distribution (Figure 4, lower left) the events with $N_{ch} < 2$ are removed as in the data; thus, we plot $P(N_{ch}) = P_{full}(N_{ch})/(1 - P_{full}(0) - P_{full}(1))$. Here, P_{full} denotes the charged particle multiplicity distribution that includes events without registered particles ($N_{ch} = 0$). These events are highly influenced by diffractive processes that are not considered in the model. Moreover, even with the $N_{ch} \geq 2$ selection the experimental results at low N_{ch} and low p_T are affected. Therefore, our predictions could not be directly compared in this region, but we are mostly interested in the high N_{ch} events as those events are more relevant in the studies of collective behaviour. The resulting p_T-spectrum is presented in Figure 4 (upper right).

The following values of the free model parameters were simultaneously selected to fit the ATLAS experiment data: $\mu_0 = 1.14$, $\sigma_T = 0.55\,\mathrm{GeV/fm}$, $p_0 = 0.37\,\mathrm{GeV}$, $\varkappa = 0.1$, and $\chi = 0.00001\,\mathrm{GeV/fm}$.

5. Flow Measurements and the Quantities of Interest

One can estimate the flow signal by performing the Fourier expansion of the single-particle distribution in the azimuthal angle, ϕ, [7]. It is necessary to take into account the intrinsic symmetry of particle production in every event, the latter being determined by its reaction plane, Ψ_{RP}. The plane is formed by the direction of the beam and the impact parameter and creates a preferred azimuthal orientation in an event. Thus, the relative particle azimuthal angles, ϕ–Ψ_{RP}, should be used in the calculations instead of the measured ϕ. Therefore, in a given part of phase-space, one can expand the distribution into a Fourier series as

$$E\frac{d^3N_{ch}}{d^3p} = \frac{1}{2\pi}\frac{d^2N_{ch}}{p_T\,dp_T\,dy}\left(1 + 2\sum_{n=1}^{\infty} v_n\cos(n(\phi - \Psi_{RP}))\right),\qquad(27)$$

where n is the Fourier harmonic number and v_n is the corresponding Fourier coefficient. The full set of v_ns describes the amplitudes of particle distribution asymmetry in the transverse plane averaged over particles in one event. Let us note that both v_n and Ψ_{RP} fluctuate event-by-event and, typically, only the moments of the corresponding distributions are measured experimentally.

The validity of the Fourier expansion in the real case of finite event multiplicity (especially in pp collisions) is questionable. Moreover, the reaction plane, Ψ_{RP}, cannot be directly measured, so one may substitute it by proxy [62] called the event plane, Ψ_{EP}. However, there is no unique event plane in an event; instead, one determines a set of event planes, Ψ_n, depending on n as

$$\Psi_n = \frac{1}{n}\tan^{-1}\left(\frac{\sum_i \sin(n\phi_i)}{\sum_i \cos(n\phi_i)}\right),\qquad(28)$$

where the numerator and denominator are calculated from the distributions of the particles in ϕ per event.

More recent studies have shown [63] that imprecise estimation of the event plane significantly spoils the flow signal in this approach. Therefore, more sophisticated measures to be used such as two-particle correlations that, under certain assumptions, naturally exclude the dependence on Ψ_{RP} in $\Delta\phi = \phi_1 - \phi_2$.

For example, the two-particle angular correlation function, $C(\Delta\eta, \Delta\phi)$, is defined by the differences in pseudorapidities, η, and azimuthal angles, ϕ, for particle pairs such as

$$C(\Delta\eta, \Delta\phi) = \frac{N^B_{\text{pairs}}}{N^S_{\text{pairs}}} \frac{S(\Delta\eta, \Delta\phi)}{B(\Delta\eta, \Delta\phi)}, \tag{29}$$

where S and B represent the $\Delta\eta$–$\Delta\phi$-distributions calculated for particle pairs from the same and mixed events, respectively. The S and B distributions are scaled by the numbers of pairs, N^S_{pairs} and N^B_{pairs}, respectively. The experimental analysis of this observable revealed the impressive ridge structure in both AA collisions [64,65] and inelastic pp interactions [15,16].

Another method to quantify the strength of azimuthal correlations is to compute the two-particle cumulant [9], $c_2\{2\}$,

$$c_2\{2\} = \langle\langle e^{2i(\phi_1 - \phi_2)} \rangle\rangle, \tag{30}$$

where the number in $\{\dots\}$ denotes the number of particles that were correlated, and $\langle\langle \dots \rangle\rangle$ represents the average over all particle pairs an event after been averaged over all events. Moreover, the estimate, $v_2\{2\}$, of the elliptic flow harmonic, v_2, can be found [11] as

$$v_2\{2\} = \sqrt{c_2\{2\}} \tag{31}$$

under the assumption of independent particle production (flow is the only source of correlations, providing that all non-flow effects are suppressed), which implies that $c_2\{2\}$ should be positive.

The observed change in sign from positive to negative for $c_2\{2\}$ in Ref. [66] demonstrated that contributions from non-flow correlations remain in $c_2\{2\}$. To exclude these non-flow contributions, it was proposed [12] to study cumulants in subevents separated in rapidity. Thus, by introducing a pseudorapidity gap, $\Delta\eta$, between particles with ϕ_1 and ϕ_2, one can eliminate the short-range correlations (coming from resonance decays, jets, and momentum conservation laws) and measure $c_2\{2\}_{(2\text{subs})}$ and the corresponding $v_2\{2\}_{(2\text{subs})}$.

The study [11] of multi-particle cumulants, $c_2\{m\}$, that measure the correlation between $m > 2$ particles in an event [10], is a natural extension of this method. Another generalisation is to consider higher, $n > 2$, orders of flow harmonics. However, the measurement of $c_n\{m\}$ cumulants requires larger statistics. We have not tried to go beyond $m = 2, n = 2$ as the simulation procedure is too CPU-demanding at the moment.

6. Model Results for Inelastic pp Interactions at $\sqrt{s} = 13$ TeV

Here, we repeat the procedure for event classification as in Ref. [66] based on charged-particle multiplicity in a certain acceptance. Then, the selected charged particle multiplicity, $N^{\text{sel}}_{\text{ch}}$, distributions were obtained for particles with $|\eta| < 2.5$ acceptance and for one of the following p_T intervals: $0.3 < p_T < 3.0$ GeV, $p_T > 0.2$ GeV, $p_T > 0.4$ GeV or $p_T > 0.6$ GeV. The distributions were split up into percentiles, which allows one to find to which event class based on $N^{\text{sel}}_{\text{ch}}$ a given event belongs.

In Figure 5, the model result for the two-particle angular correlation function, $C(\Delta\eta, \Delta\phi)$, calculated for particles with $|\eta| < 2.5$ and $0.3 < p_T < 3.0$ GeV, is presented for the $0 - 10\%$ event class based on the multiplicity $N^{\text{sel}}_{\text{ch}}$ of particles with $|\eta| < 2.5$ and $p_T > 0.6$ GeV.

Remarkably, the shape of $C(\Delta\eta, \Delta\phi)$ (Figure 5) contains a near-side ridge at $\Delta\phi \approx 0$ extended over the entire presented $\Delta\eta$ range. The ridge indicates the emission of particles in narrowly collimated azimuthal directions that is roughly boost-invariant at the given $\Delta\eta$ range. In the model framework, this structure is created by particles produced from the boosted string cluster that is of elongated shape in rapidity.

The prominent peak at $\Delta\phi \approx 0$, $\Delta\eta \approx 0$ and the extended structure along $\Delta\eta \approx 0$ come from the decays of the ρ-resonances [67]. However, the developed model does not reproduce the away-side ridge seen in the data [15,16,68].

To plot the results on cumulants (Figures 6 and 7) in a unified way for different event classifications, we correlate $N_{\mathrm{ch}}^{\mathrm{sel}}$ with the charged-particle multiplicity N_{ch} calculated on particles with $|\eta| < 2.5$ and $p_T > 0.4$ GeV. We find the event-mean $\langle N_{\mathrm{ch}} \rangle$ for every event class based on $N_{\mathrm{ch}}^{\mathrm{sel}}$ and plot it along the X-axis.

Figure 6 shows the model results for $c_2\{2\}$ (Figure 6, upper) and $v_2\{2\}$ (Figure 6, lower) calculated for particles with $|\eta| < 2.5$ and $0.3 < p_T < 3.0$ GeV, as a function of $\langle N_{\mathrm{ch}} \rangle$. Results are presented for event classes with a 0.5% width of the $N_{\mathrm{ch}}^{\mathrm{sel}}$ distribution calculated for different event classifications based on p_T, as indicated.

The values of $c_2\{2\}$ are positive and show an increasing trend towards high-multiplicity collisions. The associated second flow harmonic, $v_2\{2\}$, repeats this behaviour. We interpret it as follows. The string fusion plays a more significant role with the increased string density, therefore, both quenching and boosting of particles become stronger, leading to a greater flow signal.

Figure 7 presents $c_2\{2\}_{(2\mathrm{subs})}$ (Figure 7, upper) and $v_2\{2\}_{(2\mathrm{subs})}$ (Figure 7, lower) calculated in the two-subevent method for particles with $0.3 < p_T < 3.0$ GeV and $-2.5 < \eta < -0.8$ or $0.8 < \eta < 2.5$. This selection is used to suppress the non-flow correlations that may remain in the calculation of $c_2\{2\}$ ($v_2\{2\}$). One can see in Figure 7 that $c_2\{2\}_{(2\mathrm{subs})}$ and $v_2\{2\}_{(2\mathrm{subs})}$ repeat the trend of $c_2\{2\}$ and $v_2\{2\}$ from Figure 6 with the values being slightly higher. Thus, we assume that the two-subevent method indeed eliminated some of the remaining impacts of the ρ-resonance decays, even though its usage requires larger statistics of simulated events.

A similar dependence was presented by ALICE experiment (see [69]) for the two-particle correlation, $Y(\Delta\phi)$, integrated over rapidity, that grows with the event multiplicity. The results were also presented for the near-side ridge, but those differ in particle selection and acceptance.

For large $\langle N_{\mathrm{ch}} \rangle$ in Figures 6 and 7, one can see a slight splitting of the results with different $N_{\mathrm{ch}}^{\mathrm{sel}}$ definitions. The primary goal of the analysis with the different selections is to test the sensitivity of the flow to particles' transverse momenta.

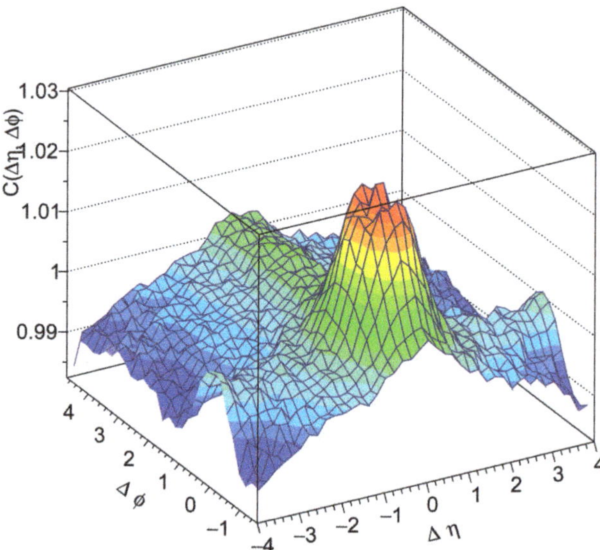

Figure 5. Model result for the two-particle correlation function, $C(\Delta\eta, \Delta\phi)$ (29), calculated for particles with $|\eta| < 2.5$ and $0.3 < p_T < 3.0$ GeV and presented for event class 0–10% based on charged particles multiplicity, $N_{\mathrm{ch}}^{\mathrm{sel}}$, under particle selection criteria $|\eta| < 2.5$ and $p_T > 0.6$ GeV for inelastic pp interactions at $\sqrt{s} = 13$ TeV.

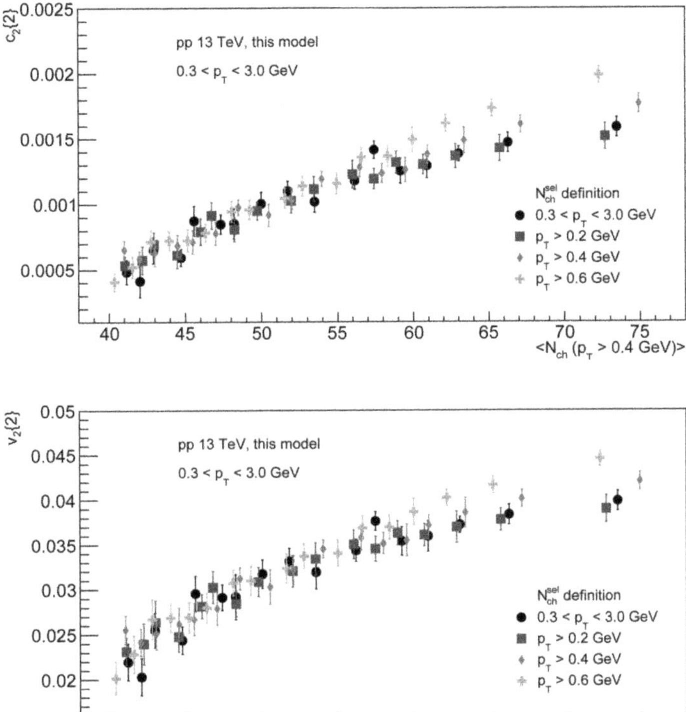

Figure 6. Model results for two-particle cumulant, $c_2\{2\}$ (30), (**upper**) and the corresponding second flow harmonic, $v_2\{2\}$ (see Equation (27)), (**lower**) calculated for charged particles with $0.3 < p_T < 3.0\,\text{GeV}$ and $|\eta| < 2.5$ as a function of $\langle N_{ch} \rangle$ estimated for different event selections of 0.5% width of N_{ch}^{sel} distribution of particles with different p_T (as indicated) and $|\eta| < 2.5$ in inelastic pp interactions at $\sqrt{s} = 13\,\text{TeV}$.

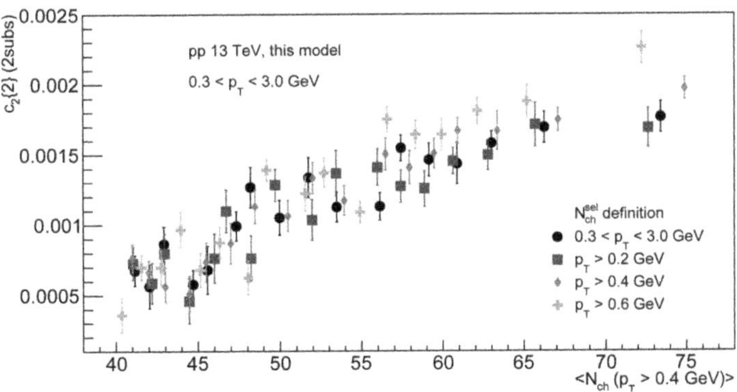

Figure 7. *Cont.*

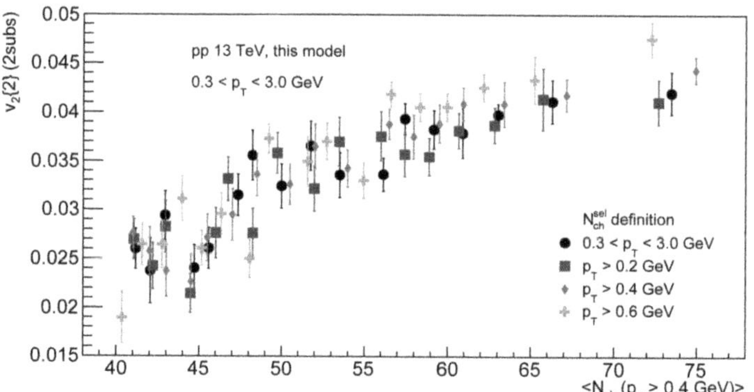

Figure 7. Model results for two-particle cumulant, $c_2\{2\}_{(2\mathrm{subs})}$, (**upper**) and the corresponding second flow harmonic, $v_2\{2\}_{(2\mathrm{subs})}$, (**lower**) calculated by two-subevent method for particles with $0.3 < p_T < 3.0$ GeV and $-2.5 < \eta < -0.83$ or $0.83 < \eta < 2.5$ as a function of $\langle N_{\mathrm{ch}} \rangle$ estimated for different event selections of 0.5% width of $N_{\mathrm{ch}}^{\mathrm{sel}}$ distribution calculated for particles with different p_T (as indicated) and $|\eta| < 2.5$ in inelastic pp interactions at $\sqrt{s} = 13$ TeV. See text for details.

To make it more visible, we show in Figure 8 the p_T dependence of two-particle cumulants, $c_2\{2\}$ and $c_2\{2\}_{(2\mathrm{subs})}$, for 0–5% event class of $N_{\mathrm{ch}}^{\mathrm{sel}}$ distribution for particles with $0.3 < p_T < 3.0$ GeV and $|\eta| < 2.5$. To achieve a similar number of particles for all model points, we perform the calculation of cumulants in the p_T-intervals of a varying size.

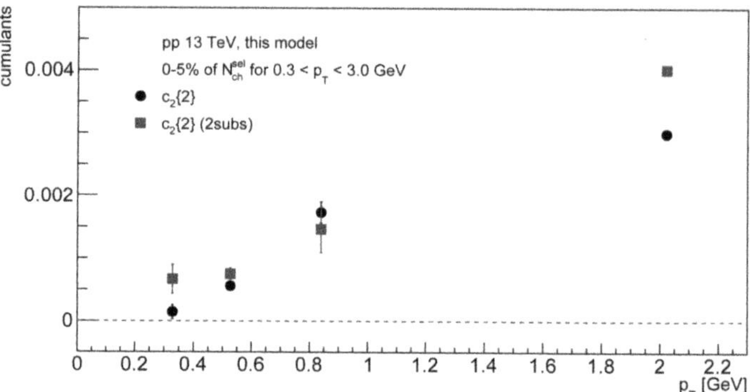

Figure 8. Model results for two-particle cumulants, $c_2\{2\}$ and $c_2\{2\}_{(2\mathrm{subs})}$, as a function of p_T of charged particles for 0–5% event class based on $N_{\mathrm{ch}}^{\mathrm{sel}}$ calculated for $0.3 < p_T < 3.0$ GeV, $|\eta| < 2.5$ in inelastic pp interactions at $\sqrt{s} = 13$ TeV. See text for details.

Figure 8 shows that $c_2\{2\}$ and $c_2\{2\}_{(2\mathrm{subs})}$ values gradually increase as the functions of the momentum of particles that are used in their calculations. This is in qualitative agreement with the ATLAS results [66] showing a similar trend.

In the model workflow, the p_T behaviour of $c_2\{2\}$ and $c_2\{2\}_{(2\mathrm{subs})}$ can be explained as follows. First of all, the quenching mechanism that was introduced produces larger anisotropy for particles with higher p_T. It is because it becomes increasingly more difficult for particles to escape the string matter while saving sufficiently large transverse momentum. In turn, string–string interactions and their acceleration in the transverse plane increase the p_T of correlated particles, leading to larger anisotropy from these boosts.

7. Discussion

In this study, we developed the workflow of the colour string model to address the challenging question of the origin of the collective flow observed in inelastic proton–proton interactions. It is interesting to draw parallels between our simplified model and the cognate event generators also based on colour string fragmentation such as in the EPOS [25] and PYTHIA [26] models.

The core concept of our model is the description of pp inelastic interaction via the multi-pomeron exchange, Equation (1), find which resembles the parallel scatterings of partons that occur in the EPOS4 event generator. The way to find momenta of string endpoints combines methods of energy-momentum sharing and saturation implemented in the EPOS4 model. All this is strikingly different from the consideration of hard scattering a starting point of multi-parton interactions in the PYTHIA model. The concept that was not implemented in any existing event generator is attractive string dynamics [39] introduced to the workflow of our model. It modifies the shape of the string density in an event resulting in the collapse of particle-producing sources. This leads to the fusion of overlapped strings into an event hot spot, which is analogues to the formation of core in the EPOS4 model. On the other hand, the concept of the formation of the fused particle source with higher tension is realised in the PYTHIA generator by using the rope mechanism. Finally, in the EPOS4 generator, evolution of the core is considered in hydro-regime, which results in a transverse flow signal. In the presented model, particle boosts (similar to the effect of string shoving mechanism in PYTHIA) and their momentum losses in string medium are the sources of azimuthal anisotropy of produced particles.

Moving on to discussing the obtained results, let us note that an event picture prior to hadronisation looks like a highly inhomogeneous string medium. Originating from the multi-pomeron exchange, the system of particle-producing strings is disordered by the longitudinal and transverse dynamics of the system. One observes the peculiar grouping of strings in the mixed configuration–momentum space. Some of them are isolated forming the debris of pp interaction. Other strings participate in the event hot spot, partially overlapping, while the most lucky ones form a dense cluster of strings, overlapping to the highest degree. It is the presence of such a core that determines the crucial collective features of the soft particle production in pp collisions. In particular, in the model, the particle anisotropy appears due to this complex structure of event shape and the effect of particle momentum quenching.

Thus, a recognisable core-corona picture of the string system [70] is observed. This picture represents the interplay of these two characteristic event regions that governs the transverse flow in the model. Namely, in the corona region, it is only the loss of the particle's momentum that plays a role. On the other hand, the impact of the complex core structure is quite complicated to deduce. The relatively low occupancy regions of the core create correlated multi-directional particle boosts, while the hottest region determines a single dominant direction and, therefore, strongly boosts particles with $\Delta\phi \approx 0$.

For example, the model result for the angular correlation function, $C(\Delta\eta, \Delta\phi)$, reveals the familiar near-side ridge that is formed by the collimated emission of particles, boosted by the clusters of strings.

On the other hand, the away-side ridge is missing from the model results. We interpret that as due to an excessive approach of strings at the τ_{deepest} moment, which results in the formation of the over-condensed string region.

In this case, the production of particles with $\Delta\phi \approx \pi$ separation in the model framework with p_T boost should come from the peripheral, low-occupancy parts of the core. It seems that, in the current implementation, this area is too scarce to create the away-side ridge. A similar observation was made in Ref. [71], where the absence of the away-side ridge is caused by complications in defining the cutoff between the core and corona.

The model results obtained for the second-order two-particle cumulants are in qualitative agreement with the data by the ALICE and ATLAS experiments on inelastic pp interactions at $\sqrt{s} = 13$ TeV. In our future studies, higher statistics would be needed to calculate multi-particle cumulants and to access high-multiplicity events.

The flow signal obtained in the model grows with the transverse momentum of charged particles in high-multiplicity events. This reflects the flow origin from the particles formed by interacting strings.

In conclusion, let us note that in this study, it was expected that the τ_{deepest} time—the time of the string system evolution in the transverse plane—that results in the highest density of strings would provide the best description of the collective behaviour of particles in pp interactions. However, it appeared that, in terms of this model, τ_{deepest} controls a degree of core-corona separation. The obtained $\Delta\eta$–$\Delta\phi$ correlation result indicates that one would have to tune τ_{deepest} more precisely, which is a subject for further studies.

Author Contributions: Conceptualization, D.P. and E.A.; methodology, D.P. and E.A.; software, D.P. and E.A.; validation, D.P. and E.A.; formal analysis, D.P.; writing—original draft preparation, D.P.; writing—review and editing, E.A.; visualization, D.P. All authors have read and agreed to the published version of the manuscript.

Funding: The authors acknowledge Saint-Petersburg State University for a research project 95413904.

Data Availability Statement: The data presented in this study are available on request from the authors.

Acknowledgments: The authors are grateful to Grigory Feofilov, Vladimir Vechernin and Vladimir Kovalenko for fruitful discussions.

Conflicts of Interest: The authors declare no conflict of interest. The funders had no role in the design of the study; in the collection, analyses, or interpretation of data; in the writing of the manuscript; or in the decision to publish the results.

Abbreviations

The following abbreviations are used in this manuscript:

AA	nucleus-nucleus
AMPT	A Multi-Phase Transport
ATLAS	A Toroidal LHC ApparatuS
CERN	Conseil Européen pour la Recherche Nucléaire (the European Organization for Nuclear Research)
CPU	central processing unit
CTEQ	Coordinate Theoretical-Experimental Project on QCD
CT10nnlo	CTEQ version 10 next-to-next-to-leading order
EPOS	Energy conserving quantum mechanical multiple scattering approach, based on Partons (parton ladders), Off-shell remnants, and Saturation of parton ladders
HIJING	Heavy Ion Jet INteraction Generator
LHA	Les Houches Accord
LHC	Large Hadron Collider
PDF	parton density function
pp	proton–proton
QCD	quantum chromodynamics
QED	quantum electrodynamics
QGP	quark-gluon plasma
RHIC	Relativistic Heavy Ion Collider
2D, 3D	2-, 3-dimensional

Appendix A

Here, we describe the procedure for creating a proton with $n_{\text{part}} \geq 2$.

Appendix A.1. Valence Quarks and Diquark

We start with the creation of valence u and d quarks by sampling their proton momentum fractions, x_u^{val} and x_d^{val}, from their valence parton distribution functions (PDFs),

$x f_u^{\text{val}}(x)$ and $x f_d^{\text{val}}(x)$. We find $f_{u,d}^{\text{val}}(x)$ as the difference between PDFs for u (d) quark and $\bar{u}$ ($\bar{d}$) antiquark taken from CT10nnlo approximation [72] based on CT10 PDFs [73], set 1 by LHAPDF [74] at the momentum transferred, $Q^2 = 1\ \text{GeV}^2$.

We consider three possible combinations of selecting 2 out of 3 valence quarks in order to create a diquark (either uu or ud). We define diquark momentum, x_{di}, as the sum of the proton momentum fractions of the two selected valence quarks, $x_u^{\text{val}} + x_{u,d}^{\text{val}}$. The three combinations are ordered by the largest $x_{u,d}^{\text{val}}$ of the valence quark that is not included in the diquark.

We set the diquark mass, m_{di}, equal to 0.1185 GeV. For quarks, both valence and sea, we start with the current (not constituent) masses: $m_u = 0.0022$ GeV, $m_d = 0.0048$ GeV, $m_s = 0.0950$ GeV, and $m_c = 1.2750$ GeV that are dynamically changed later, see Appendix A.3.

For each parton representing a quark or a diquark and enumerated by $1 \leq i \leq n_{\text{part}}$, we calculate its energy, E_i, and its fraction of proton energy, e_i, as

$$E_i = \sqrt{m_i^2 + (x_i p_{\text{beam}})^2}, e_i = \frac{E_i}{E_{\text{proton}}}, \tag{A1}$$

where $p_{\text{beam}} = \sqrt{s/4 - m_{\text{proton}}^2}$, $m_{\text{proton}} = 0.938$ GeV, $E_{\text{proton}} = \sqrt{s}/2$.

Appendix A.2. Sea Quarks

In case of a number of pomeron exchanges greater than 1, we take into account the presence of sea quarks that participate in the interaction. We sample the proton momentum fraction for each sea quark, x_i^{sea}, where i runs over values from 1 to $(n_{\text{part}} - 2)$, using the sum of PDFs for all flavours, $\sum_{\text{fl}} x f_{\text{fl}}(x)$. For a given sea quark (sampled x_i^{sea}), we define its flavour from the relative probability that is known at any given x (i.e., from the ratio between PDFs for different flavours, $f_{\text{fl}}(x)$, at x). Gluons are not considered at this stage because they are accounted for differently (see Appendix A.3). For each sea quark, we also calculate e_i according to Equation (A1).

Appendix A.3. Energy-Momentum Conservation for a Proton

At this point, we take a step back and verify that the sums of partons' x_i and e_i are less than 1. Otherwise, a proton is to be regenerated.

As soon as the goal is achieved, the energy-momentum conservation is to be taken into account within a proton. Namely, the sums of the x_i and e_i for all n_{part} should be equal to 1 each. However, at this step, there is no guarantee these conditions are fulfilled. Therefore, we find the deficiencies, x_0 and e_0, as

$$x_0 = 1 - \sum_{i=1}^{n_{\text{part}}} x_i, \text{ and } e_0 = 1 - \sum_{i=1}^{n_{\text{part}}} e_i. \tag{A2}$$

In this study, we distribute x_0 and e_0 between all the created partons (both valence and sea), which may be interpreted as a gluon contribution. For example, in the event with $n_{\text{pom}} = 1$ we split x_0 and e_0 in half between valence quark and diquark. Thus, for the bare valence quark and diquark one gets the modified fractions of proton momentum, $x_{q/\text{di}}^{\text{dressed}}$, and proton energy, $e_{q/\text{di}}^{\text{dressed}}$, from

$$x_{q/\text{di}}^{\text{dressed}} = x_{q/\text{di}}^{\text{bare}} + 0.5 \cdot x_0, \tag{A3}$$

$$e_{q/\text{di}}^{\text{dressed}} = e_{q/\text{di}}^{\text{bare}} + 0.5 \cdot e_0. \tag{A4}$$

As soon as the gluon contribution is taken into account, one can define the initial parton momentum as

$$p_{\text{part}} = x_{\text{part}}^{\text{dressed}} \cdot p_{\text{beam}} \tag{A5}$$

and the initial parton energy as

$$E_{\text{part}} = e_{\text{part}}^{\text{dressed}} \cdot E_{\text{proton}}. \tag{A6}$$

The modifications of partons' momenta (A3) and energies (A4) change the partons' masses from "bare" (current) values, m_{di} or $m_{\text{u/d/s/c}}$, to the "dressed" ones, $m_{q/\text{di}}^{\text{dressed}}$, according to

$$m_{q/\text{di}}^{\text{dressed}} = \sqrt{E_{\text{part}}^2 - p_{\text{part}}^2}. \tag{A7}$$

This approach can be naturally extrapolated to any arbitrary number of sea quarks. Namely, we assign $x_0/3$ and $e_0/3$ to each of the valence quark and the diquark and split the remaining $1/3$ between all the sea quarks in a proton. This procedure increases the masses of quarks and diquark (for valence ones to a greater extent), which makes the distributions of the string ends' rapidities more realistic.

However, it is also possible that after the procedure described, parton's energy decreases compared to parton's momentum. Therefore, one cannot calculate parton's dressed mass, $m_{q/\text{di}}^{\text{dressed}}$. In this case, the remaining two different groupings of valence quarks into diquark should be considered (see Appendix A.1). If none of the combinations allows finding $m_{q/\text{di}}^{\text{dressed}}$, the proton is to be regenerated.

References

1. Shuryak, E.V. Theory of hadron plasma. *Sov. Phys. JETP* **1978**, *47*, 212–219. Available online: http://jetp.ras.ru/cgi-bin/e/index/e/47/2/p212?a=list (accessed on 30 January 2024).
2. Heinz, U.W.; Jacob, M. Evidence for a new state of matter: An Assessment of the results from the CERN lead beam program. *arXiv* **2000**, arXiv:nucl-th/0002042. [CrossRef]
3. Arsene, I. et al. [BRAHMS Collaboration] Quark gluon plasma and color glass condensate at RHIC? The Perspective from the BRAHMS experiment. *Nucl. Phys. A* **2005**, *757*, 1–27. [CrossRef]
4. Back, B.B. et al. [PHOBOS Collaboration] The PHOBOS perspective on discoveries at RHIC. *Nucl. Phys. A* **2005**, *757*, 28–101. [CrossRef]
5. Adams, J. et al. [STAR Collaboration] Experimental and theoretical challenges in the search for the quark gluon plasma: The STAR Collaboration's critical assessment of the evidence from RHIC collisions. *Nucl. Phys. A* **2005**, *757*, 102–183. [CrossRef]
6. Adcox, K. et al. [PHENIX Collaboration] Formation of dense partonic matter in relativistic nucleus-nucleus collisions at RHIC: Experimental evaluation by the PHENIX Collaboration. *Nucl. Phys. A* **2005**, *757*, 184–283. [CrossRef]
7. Voloshin, S.; Zhang, Y. Flow study in relativistic nuclear collisions by Fourier expansion of Azimuthal particle distributions. *Z. Phys. C* **1996**, *70*, 665–672. [CrossRef]
8. Kolb, P.F.; Huovinen, P.; Heinz, U.W.; Heiselberg, H. Elliptic flow at SPS and RHIC: From kinetic transport to hydrodynamics. *Phys. Lett. B* **2001**, *500*, 232–240. [CrossRef]
9. Wang, S.; Jiang, Y.Z.; Liu, Y.M.; Keane, D.; Beavis, D.; Chu, S.Y.; Fung, S.Y.; Vient, M.; Hartnack, C.; Stoecker, H. Measurement of collective flow in heavy ion collisions using particle pair correlations. *Phys. Rev. C* **1991**, *44*, 1091–1095. [CrossRef]
10. Adler, C. et al. [STAR Collaboration] Elliptic flow from two and four particle correlations in Au+Au collisions at $\sqrt{s_{\text{NN}}} = 130$ GeV. *Phys. Rev. C* **2002**, *66*, 034904. [CrossRef]
11. Borghini, N.; Dinh, P.M.; Ollitrault, J.Y. Flow analysis from multiparticle azimuthal correlations. *Phys. Rev. C* **2001**, *64*, 054901. [CrossRef]
12. Jia, J.; Zhou, M.; Trzupek, A. Revealing long-range multiparticle collectivity in small collision systems via subevent cumulants. *Phys. Rev. C* **2017**, *96*, 034906. [CrossRef]
13. Bilandzic, A.; Christensen, C.H.; Gulbrandsen, K.; Hansen, A.; Zhou, Y. Generic framework for anisotropic flow analyses with multiparticle azimuthal correlations. *Phys. Rev. C* **2014**, *89*, 064904. [CrossRef]
14. Bożek, P. Transverse-momentum–flow correlations in relativistic heavy-ion collisions. *Phys. Rev. C* **2016**, *93*, 044908. [CrossRef]
15. Khachatryan, V. et al. [CMS Collaboration] Observation of long-range near-side angular correlations in proton-proton collisions at the LHC. *J. High Energy Phys.* **2010**, *09*, 091. [CrossRef]
16. Aad, G. et al. [ATLAS Collaboration] Observation of long-range elliptic azimuthal anisotropies in $\sqrt{s} = 13$ and 2.76 TeV *pp* collisions with the ATLAS detector. *Phys. Rev. Lett.* **2016**, *116*, 172301. [CrossRef] [PubMed]
17. Aaboud, M. et al. [ATLAS Collaboration] Measurements of long-range azimuthal anisotropies and associated Fourier coefficients for *pp* collisions at $\sqrt{s} = 5.02$ and 13 TeV and *p*+Pb collisions at $\sqrt{s_{\text{NN}}} = 5.02$ TeV with the ATLAS detector. *Phys. Rev. C* **2017**, *96*, 024908. [CrossRef]
18. Ambrus, V.E.; Schlichting, S.; Werthmann, C. Establishing the range of applicability of hydrodynamics in high-energy collisions. *Phys. Rev. Lett.* **2023**, *130*, 152301. [CrossRef]

19. Weller, R.D.; Romatschke, P. One fluid to rule them all: Viscous hydrodynamic description of event-by-event central p+p, p+Pb and Pb+Pb collisions at $\sqrt{s}$ = 5.02 TeV. *Phys. Lett. B* **2017**, *774*, 351–356. [CrossRef]
20. Zhou, Y.; Zhao, W.; Murase, K.; Song, H. One fluid might not rule them all. *Nucl. Phys. A* **2021**, *1005*, 121908. [CrossRef]
21. Artru, X. Classical string phenomenology. How strings work. *Phys. Rep.* **1983**, *97*, 147–171. [CrossRef]
22. Capella, A.; Sukhatme, U.; Tan, C.-I.; Tran Thanh Van, J. Dual parton model. *Phys. Rep.* **1994**, *236*, 225–329. [CrossRef]
23. Braun, M.; Pajares, C. A Probabilistic model of interacting strings. *Nucl. Phys. B* **1993**, *390*, 542–558. [CrossRef]
24. Gelis, F. Color glass gondensate and glasma. *Int. J. Mod. Phys. A* **2013**, *28*, 1330001. [CrossRef]
25. Werner, K. Revealing a deep connection between factorization and saturation: New insight into modeling high-energy proton–proton and nucleus-nucleus scattering in the EPOS4 framework. *Phys. Rev. C* **2023**, *108*, 064903. [CrossRef]
26. Sjöstrand, T. The PYTHIA event generator: Past, resent and future. *Comput. Phys. Commun.* **2020**, *246*, 106910. [CrossRef]
27. Wang, X.N.; Gyulassy, M. HIJING: A Monte Carlo model for multiple jet production in pp, pA and AA collisions. *Phys. Rev. D* **1991**, *44*, 3501–3516. [CrossRef] [PubMed]
28. Zhang, B.; Ko, C.M.; Li, B.A.; Lin, Z.w. A multiphase transport model for nuclear collisions at RHIC. *Phys. Rev. C* **2000**, *61*, 067901. [CrossRef]
29. Braun, M.A.; Pajares, C. Elliptic flow from color strings. *Eur. Phys. J. C* **2011**, *71*, 1558. [CrossRef]
30. Braun, M.A.; Pajares, C.; Vechernin, V.V. Anisotropic flows from colour strings: Monte-Carlo simulations. *Nucl. Phys. A* **2013**, *906*, 14–27. [CrossRef]
31. Braun, M.A.; Pajares, C.; Vechernin, V.V. Ridge from strings. *Eur. Phys. J. A* **2015**, *51*, 44. [CrossRef]
32. Abramovsky, V.A.; Gedalin, E.V.; Gurvich, E.G.; Kancheli, O.V. Long range azimuthal correlations in multiple production processes at high energies. *JETP Lett.* **1988**, *47*, 337–339. Available online: http://jetpletters.ru/ps/1093/article_16503.shtml (accessed on 30 January 2024).
33. Altsybeev, I. Mean transverse momenta correlations in hadron-hadron collisions in MC toy model with repulsing strings. *AIP Conf. Proc.* **2016**, *1701*, 100002. [CrossRef]
34. Bierlich, C.; Gustafson, G.; Lönnblad, L. Collectivity without plasma in hadronic collisions. *Phys. Lett. B* **2018**, *779*, 58–63. [CrossRef]
35. Andronov, E.V.; Prokhorova, D.S.; Belousov, A.A. Influence of quark–gluon string interactions on particle correlations in p+p collisions. *Theor. Math. Phys.* **2023**, *216*, 1265–1277. [CrossRef]
36. Prokhorova, D.; Andronov, E.; Feofilov, G. Interacting colour strings approach in modelling of rapidity correlations. *Physics* **2023**, *5*, 636–654. [CrossRef]
37. Braun, M.A.; Pajares, C. p_t dependence of the flow coefficients for pp collisions in the color string scenario: Monte Carlo simulations. *Eur. Phys. J. A* **2018**, *54*, 185. [CrossRef]
38. Shen, C.; Schenke, B. Dynamical initial state model for relativistic heavy-ion collisions. *Phys. Rev. C* **2018**, *97*, 024907. [CrossRef]
39. Kalaydzhyan, T.; Shuryak, E. Collective interaction of QCD strings and early stages of high-multiplicity pA collisions. *Phys. Rev. C* **2014**, *90*, 014901. [CrossRef]
40. Braun, M.A.; Pajares, C.; Ranft, J. Fusion of strings versus percolation and the transition to the quark gluon plasma. *Int. J. Mod. Phys. A* **1999**, *14*, 2689–2704. [CrossRef]
41. Braun, M.A.; Del Moral, F.; Pajares, C. Percolation of strings and the first RHIC data on multiplicity and tranverse momentum distributions. *Phys. Rev. C* **2002**, *65*, 024907. [CrossRef]
42. Braun, M.A.; Kolevatov, R.S.; Pajares, C.; Vechernin, V.V. Correlations between multiplicities and average transverse momentum in the percolating color strings approach. *Eur. Phys. J. C* **2004**, *32*, 535–546. [CrossRef]
43. Braun, M.A.; Dias de Deus, J.; Hirsch, A.S.; Pajares, C.; Scharenberg, R.P.; Srivastava, B.K. De-Confinement and Clustering of Color Sources in Nuclear Collisions. *Phys. Rep.* **2015**, *599*, 1–50. [CrossRef]
44. Altsybeev, I.; Feofilov, G. Azimuthal flow in hadron collisions from quark-gluon string repulsion. *EPJ Web Conf.* **2016**, *125*, 04011. [CrossRef]
45. Aaboud, M.; et al. [ATLAS Collaboration] Charged-particle distributions at low transverse momentum in $\sqrt{s}$ = 13 TeV pp interactions measured with the ATLAS detector at the LHC. *Eur. Phys. J. C* **2016**, *76*, 502. [CrossRef] [PubMed]
46. Vechernin, V.; Lakomov, I. The dependence of the number of pomerons on the impact parameter and the long-range rapidity correlations in pp collisions. *PoS* **2012**, *Baldin ISHEPP XXI*, 072. [CrossRef]
47. Kaidalov, A.B.; Ter-Martirosyan, K.A. Pomeron as quark-gluon strings and multiple hadron production at SPS collider energies. *Phys. Lett. B* **1982**, *117*, pp. 247–251. [CrossRef]
48. Vechernin, V.V.; Belokurova, S.N. The strongly intensive observable in pp collisions at LHC energies in the string fusion model. *J. Phys. Conf. Ser.* **2020**, *1690*, 012088. [CrossRef]
49. Armesto, N.; Derkach, D.A.; Feofilov, G.A. p_t–multiplicity correlations in a multi-pomeron-exchange model with string collective effects. *Phys. Atom. Nucl.* **2008**, *71*, 2087–2095. [CrossRef]
50. Wilson, K.G. Confinement of quarks. *Phys. Rev. D* **1974**, *10*, 2445–2459. [CrossRef]
51. Iritani, T.; Cossu, G.; Hashimoto, S. Analysis of topological structure of the QCD vacuum with overlap-Dirac operator eigenmode. *PoS* **2014**, *LATTICE 2013*, 376. [CrossRef]
52. Kalaydzhyan, T.; Shuryak, E. Self-interacting QCD strings and string balls. *Phys. Rev. D* **2014**, *90*, 025031. [CrossRef]

53. Belokurova, S.; Vechernin, V. Long-range correlations between observables in a model with translational invariance in rapidity. *Symmetry* **2020**, *12*, 1107. [CrossRef]
54. Lüscher, M.; Münster, G.; Weisz, P. How thick are chromo-electric flux tubes? *Nucl. Phys. B* **1981**, *180*, 1–12. [CrossRef]
55. Baker, M.; Cea, P.; Chelnokov, V.; Cosmai, L.; Cuteri, F.; Papa, A. The confining color field in SU(3) gauge theory. *Eur. Phys. J. C* **2020**, *80*, 514. [CrossRef]
56. Kovalenko, V.; Feofilov, G.; Puchkov, A.; Valiev, F. Multipomeron model with collective effects for high-energy hadron collisions. *Universe* **2022**, *8*, 246. [CrossRef]
57. Schwinger, J. On gauge invariance and vacuum polarization. *Phys. Rev.* **1951**, *82*, 664–679. [CrossRef]
58. Gurvich, E.G. The quark anti-quark pair production mechanism in a quark jet. *Phys. Lett. B* **1979**, *87*, 386–388. [CrossRef]
59. Casher, A.; Neuberger, H.; Nussinov, S. Chromoelectric flux tube model of particle production. *Phys. Rev. D* **1979**, *20*, 179–188. [CrossRef]
60. Bodnia, E.; Derkach, D.; Feofilov, G.; Kovalenko, V.; Puchkov, A. Multi-pomeron exchange model for pp and $p\bar{p}$ collisions at ultra-high energy. *PoS* **2014**, *QFTEP 2013*, 060. [CrossRef]
61. Bresenham, J.E. Algorithm for computer control of a digital plotter. *IBM Syst. J.* **1965**, *4*, 25–30. [CrossRef]
62. Poskanzer, A.M.; Voloshin, S.A. Methods for analyzing anisotropic flow in relativistic nuclear collisions. *Phys. Rev. C* **1998**, *58*, 1671–1678. [CrossRef]
63. Luzum, M.; Ollitrault, J.Y. Eliminating experimental bias in anisotropic-flow measurements of high-energy nuclear collisions. *Phys. Rev. C* **2013**, *87*, 044907. [CrossRef]
64. Adams, J.; et al. [STAR Collaboration] Minijet deformation and charge-independent angular correlations on momentum subspace (η, ϕ) in Au-Au collisions at $\sqrt{s_{NN}}$ = 130 GeV. *Phys. Rev. C* **2006**, *73*, 064907. [CrossRef]
65. Chatrchyan, S.; et al. [CMS Collaboration] Centrality dependence of dihadron correlations and azimuthal anisotropy harmonics in PbPb collisions at $\sqrt{s_{NN}}$ = 2.76 TeV. *Eur. Phys. J. C* **2012**, *72*, 2012. [CrossRef]
66. Aaboud, M.; et al. [ATLAS Collaboration] Measurement of multi-particle azimuthal correlations in pp, p+Pb and low-multiplicity Pb+Pb collisions with the ATLAS detector. *Eur. Phys. J. C* **2017**, *77*, 428. [CrossRef] [PubMed]
67. Graczykowski, L.K.; Janik, M.A. Unfolding the effects of final-state interactions and quantum statistics in two-particle angular correlations. *Phys. Rev. C* **2021**, *104*, 054909. [CrossRef]
68. Acharya, S.; et al. [ALICE Collaboration] Long- and short-range correlations and their event-scale dependence in high-multiplicity pp collisions at $\sqrt{s}$ = 13 TeV. *JHEP* **2021**, *05*, 290. [CrossRef]
69. Parkkila, J.E. Long-range correlations in low-multiplicity pp collisions at $\sqrt{s}$ = 13 TeV. In Proceedings of the 57th Rencontres de Moriond. 2023 QCD and High Energy Interactions, La Thuile, Italy, 25 March–1 April 2023; Augé, E., Dumarchez, J., Trân Thanh Vân, J., Eds.; ARISF: Lausanne, Switzerland, 2023; pp. 247–250. . [CrossRef]
70. Kanakubo, Y.; Tachibana, Y.; Hirano, T. Interplay between core and corona from small to large systems. *EPJ Web Conf.* **2023**, *276*, 01017. [CrossRef]
71. Werner, K.; Karpenko, I.; Pierog, T. The 'ridge' in proton-proton scattering at 7 TeV. *Phys. Rev. Lett.* **2011**, *106*, 122004. [CrossRef]
72. Gao, J.; Guzzi, M.; Huston, J.; Lai, H.L.; Li, Z.; Nadolsky, P.; Pumplin, J.; Stump, D.; Yuan, C.P. CT10 next-to-next-to-leading order global analysis of QCD. *Phys. Rev. D* **2014**, *89*, 033009. [CrossRef]
73. Lai, H.L.; Guzzi, M.; Huston, J.; Li, Z.; Nadolsky, P.M.; Pumplin, J.; Yuan, C.P. New parton distributions for collider physics. *Phys. Rev. D* **2010**, *82*, 074024. [CrossRef]
74. Buckley, A.; Ferrando, J.; Lloyd, S.; Nordström, K.; Page, B.; Rüfenacht, M.; Schönherr, M.; Watt, G. LHAPDF6: Parton density access in the LHC precision era. *Eur. Phys. J. C* **2015**, *75*, 132. [CrossRef]

Review

Probing Relativistic Heavy-Ion Collisions via Photon Anisotropic Flow Ratios. A Brief Review

Rupa Chatterjee [1,2,*] and Pingal Dasgupta [3]

[1] Variable Energy Cyclotron Centre, 1/AF, Bidhan Nagar, Kolkata 700064, India
[2] Homi Bhabha National Institute, Training School Complex, Anushaktinagar, Mumbai 400094, India
[3] Department of Atomic Physics, Eötvös Loránd University, 1053 Budapest, Hungary; dasgupta.pingal@ttk.elte.hu
* Correspondence: rupa@vecc.gov.in

Abstract: The anisotropic flow of photons produced in relativistic nuclear collisions is known as a promising observable for studying the initial state and the subsequent evolution of the hot and dense medium formed in such collisions. The investigation of photon anisotropic flow coefficients, v_n, has attracted high interest over the last decade, involving both theory and experiment. The thermal emission of photons and their anisotropic flow are found to be highly sensitive to the initial state of the fireball, where even slight modifications can lead to significant variations in the final state results. In contrast, the ratio of photon anisotropic flow stands out as a robust observable, exhibiting minimal sensitivity to the initial conditions. Here, we briefly review the studies of the individual elliptic and triangular flow parameters of photons as well as their ratios and how these parameters serve as valuable probes for investigating the intricacies of the initial state and addressing the challenges posed by the direct photon puzzle.

Keywords: relativistic heavy-ion collisions; quark–gluon plasma; anisotropic flow; direct photons; thermal photons; clustered initial state

Citation: Chatterjee, R.; Dasgupta, P. Probing Relativistic Heavy-Ion Collisions via Photon Anisotropic Flow Ratios. A Brief Review. *Physics* 2024, 6, 674–689. https://doi.org/10.3390/physics6020044

Received: 6 February 2024
Revised: 20 March 2024
Accepted: 26 March 2024
Published: 4 May 2024

1. Introduction

Anisotropic flow provides some of the most compelling evidence of the existence of quark–gluon plasma (QGP) in relativistic heavy-ion collisions [1–5]. It serves as a response to the initial spatial anisotropy showing how efficiently this anisotropy is transformed into the final state momentum space anisotropy of the emitted particles [6–11].

Relativistic hydrodynamics is known as one of the most successful theoretical frameworks for explaining the observed large anisotropic flow of hadrons in relativistic heavy-ion collisions. The hydrodynamic framework treats the strongly interacting matter produced in these collisions as a 'nearly perfect fluid' characterized by its collective behavior and hydrodynamic flow properties [12–19].

Initially, it was believed that non-central collisions between two spherical nuclei or central collisions of two deformed nuclei (such as the collisions of uranium nuclei with a prolate shape) could lead to a substantial elliptic flow of charged particles. Higher order even flow coefficients were anticipated to be non-zero but significantly smaller compared to the elliptic flow of charged particles [12]. However, subsequent observations revealed that even the most central collisions of spherical nuclei could produce a non-zero elliptic flow of charged particles. This observation provided confirmation that event-by-event fluctuating initial density distributions can contribute significantly to the anisotropic flow of particles produced in heavy-ion collisions [20–22]. This also resulted in significantly large odd flow coefficients, particularly the triangular flow parameter [23,24].

The anisotropic flow coefficients, v_n, are estimated by expanding the invariant particle number distribution in the transverse momentum plane using Fourier decomposition:

$$\frac{dN}{d^2 p_T dy} = \frac{1}{2\pi} \frac{dN}{p_T dp_T dy}\left[1 + 2 \sum_{n=1}^{\infty} v_n(p_T) \cos n(\phi - \psi_n)\right]. \tag{1}$$

Here, p_T denotes the transverse momentum, y denotes the rapidity and ϕ represents the azimuthal angle of the emitted particle. ψ_n denotes the event plane angle for the $n$thflow coefficient. The event plane angle is a crucial quantity for the study of anisotropic flow as it provides a quantitative measure of the orientation of the anisotropy in each event. The anisotropic flow coefficients are extracted from experimental data and can be compared to theoretical models to understand the various properties of the QGP such as its viscosity, formation time, equation of state, etc. [5,25].

Photons produced in relativistic nuclear collisions have long been recognized as a highly sensitive and unique probe for studying the initial state and its evolution [12,26–45]. They are also known as the thermometer of the produced matter from the initial days of heavy-ion collisions. Both real as well as virtual photons are emitted from every stage of the expanding fireball, suffer negligible re-scatterings with the hot and dense medium, and provide undistorted information about the medium produced [26,46–50].

The experimentally measured inclusive photon spectrum contains a large portion of the late-time decay photons originating mostly from the two-γ decay of π^0 and η mesons. After successful subtraction of the decay background, one obtains the direct photon spectrum at different beam energies and collision centralities [25,35].

It was first shown by the PHENIX Collaboration that there is an excess of direct photon yield for $200A$ GeV AuAu collisions at RHIC (Relativistic Heavy Ion Collider) at different centrality bins over the properly normalized production of photons in proton-proton (pp) collisions [51]. The prompt photons produced in initial hard scatterings are the dominant source of direct photons for pp collisions, and it is considered that the thermal contribution is negligible for these collisions as there is no formation of QGP. Although some recent experiments suggest that the high multiplicity events in pp collisions show signatures of medium formation [52], in the context of direct photon production, the prompt photons in pp collisions were found to explain the data well without any thermal contribution. Hence, the observed excess for heavy-ion collisions over the scaled proton–proton results in the direct photon spectrum is attributed to thermal radiations produced in the interaction of the thermalized medium constituents. The excess production of photons has also been reported later at the LHC (Large Hadron Collider) for PbPb collisions at $2.76A$ TeV by the ALICE Collaboration [53].

To note is that a significant contribution to the direct photon spectrum also comes from the pre-equilibrium phase as well as the production of photons by the passing of high-energy jets through QGP; see Ref. [25] and references therein for details.

In the QGP phase, the quark–gluon Compton scatterings and quark anti-quark annihilation processes are the dominant ones to produce thermal photons. The different hadronic channels (involving mostly π and ρ mesons) take part in the photon production in the hot hadronic matter.

These thermal photons hold significant promise for characterizing the initial hot and dense state of the matter produced in heavy-ion collisions at relativistic energies using appropriate observables. To note also is that a number of large and small systems have been studied in recent times, especially at the RHIC energy, which show an evidence of thermal radiation in these collisions [54–56].

2. Anisotropic Flow of Photons

The anisotropic flow of photons serves as a valuable observable for studying thermal photons given that non-thermal contributions are not subjected to the collectivity of the produced medium [57–63]. The elliptic flow of thermal photons was estimated initially considering an ideal hydrodynamic evolution of the system in collisions of gold nuclei at RHIC and using the state-of-the-art rates [57,64,65].

The thermal photons produced in typical heavy-ion collisions dominate the direct photon spectrum in the region of $p_T < 3$–4 GeV. The majority of these thermal photons with $p_T > 1$ GeV (high-p_T region) are expected to originate from the QGP, and the hadronic contribution tends to populate the lower $p_T < 1$ GeV region of the spectrum. However, understanding their individual contribution to the total anisotropic flow is complicated due to the competing contributions from these two sources.

Theoretical model calculations also show that the anisotropic flow of photons is larger for peripheral collisions than for central collisions [57]. The relative contribution of the photons produced in the hadronic medium as well as the initial spatial eccentricity (of the overlapping zone between the two incoming nuclei), increases towards peripheral collisions. These photons have a larger anisotropic flow compared to the photons produced in the QGP medium, which results in a larger total photon anisotropic flow for peripheral collisions than for central collisions. In addition, the p_T-dependent behavior of elliptic flow and triangular flow is found to be similar on a qualitative scale. Both photon v_2 and v_3 rise with p_T up to about 2 GeV and then drop as p_T is increased further (see Figure 3 in Ref. [57]). The peak value of $v_n(p_T)$ depends on the beam energy, initial conditions, as well on the centrality of the collisions.

The first experimental measurement of photon elliptic flow from AuAu collisions at RHIC shows that direct photon v_2 is consistent with zero above $p_T > 4$ GeV [66]. This confirms that the prompt photons are not subjected to collectivity, and as a result, their contribution to photon elliptic flow is negligible. In the region, $p_T < 4$ GeV, the elliptic flow of direct photons as a function of transverse momentum was found to be significantly large for mid-central AuAu collisions at RHIC.

Additionally, it was found that the elliptic flow of direct photons from experimental analysis shows a similar p_T-dependent nature as predicted earlier by hydrodynamic model calculation. However, the theoretical results tend to underestimate the experimental data by a significant margin [66,67]. This was termed as a 'direct photon puzzle', where theoretical model calculations cannot simultaneously explain experimental data for the p_T spectra and anisotropic flow of direct photons.

Several advancements were made in the last decade in theoretical model calculations in order to understand the discrepancy between experimental data and results from model calculation [37–39,68–74]. A recent study explored the influence of a weak magnetic field on direct photon production using a realistic (3+1)D (dimensional) hydrodynamic evolution containing a tilted fireball configuration [74]. This study is found to provide a good agreement of direct photon v_2 and v_3 with the experimental data at both the RHIC and the LHC energies. Directed flow (v_1) of photons from relativistic heavy-ion collisions has also been estimated using hydrodynamic model calculation [72]. Although the charged particle v_1 was estimated from heavy ion experiments, experimental determination of photon v_1 is yet to be conducted [75].

It has been shown that the inclusion of initial state nucleon shadowing in the Monte Carlo Glauber model provides a better description of the experimental data for hadronic observables [76]. This shadowing effect is also found to increase the elliptic flow of thermal photons from heavy-ion collisions at RHIC and LHC energies significantly [62]. Event-by-event hydrodynamic model calculations have revealed that the presence of fluctuations in the initial density distribution increases the elliptic flow in comparison to a scenario with a smooth initial density distribution [68,69]. Additionally, the incorporation of shear viscosity was identified as a factor that reduces elliptic flow significantly, particularly at larger p_T values.

The inclusion of the photons from the pre-equilibrium phase has been shown to affect both the photon spectra and anisotropic flow parameters marginally [77]. Estimation of photon anisotropic flow calculation from collisions of small, deformed, and clustered nuclei has also been found to provide interesting new insight into the initial state produced in those collisions.

Nevertheless, achieving a simultaneous explanation of both the photon spectra and anisotropic flow through model calculations and experimental data, often referred to as the 'direct photon puzzle', is still a challenge.

It is also important to emphasize the pivotal role played by the initial conditions in model calculations when determining the anisotropic flow of photons [78]. Photon observables exhibit greater sensitivity to initial conditions, such as the formation time and initial temperature, in contrast to hadronic observables. A smaller initial time of QGP formation, τ_0, results in a larger initial temperature, T_0, and a relatively greater contribution from the QGP phase compared to photons produced in the hadronic matter. This results in a larger relative contribution of the QGP photons in the anisotropic flow calculation. Theoretical model calculations show that the anisotropic flow of (only) QGP photons is significantly smaller than the photon v_n from (only) hadronic matter. Consequently, this leads to a reduction in the (overall) photon anisotropic flow at higher p_T [58]. Hence, there is a hope that photon anisotropic flow could serve as a valuable observable for precisely determining the thermalized time of the hot and dense QGP formed in relativistic heavy-ion collisions. However, the existing disparity between experimental data and photon v_n calculations impedes the extraction of precise information regarding the initial formation time.

3. Ratio of Photon Anisotropic Flow

The anisotropic flow of direct photons is primarily driven by the competing contributions of thermal photons originating from the QGP and hot hadronic matter. The unique p_T-dependent nature of photon v_n is mainly determined by the QGP contribution. The expanding hot fireball, with time, produces anisotropic flow and the p_T-dependent shape, which is a result of the convoluted emission of thermal photons. The high-p_T thermal photons largely originate during the initial QGP dynamics and, therefore, have a lower anisotropy, whereas the magnitude of the photon v_n is controlled by thermal photons originating from hot hadronic matter at the later part of the expansion, which is almost an order of magnitude larger compared to the photon anisotropy from the QGP phase [57]. However, it is important to note that the impact of non-thermal contributions, particularly from prompt photons, dilutes the anisotropic flow at larger p_T-values. This dilution occurs as the non-thermal contributions introduce additional weight in the denominator of the photon anisotropic flow analysis.

In a recent study, it was shown that the ratio of photon v_2 and v_3 as a function of transverse momentum within a specific centrality bin could serve as a valuable parameter for gaining insights into the dynamics of heavy-ion collisions [79]. One can reduce uncertainties arising from non-thermal contributions in the estimation of direct photon v_n by calculating the ratio of direct photon v_2 and v_3.

An event-by-event ideal hydrodynamic framework has been used to study the evolution of matter produced in collisions involving heavy nuclei at relativistic energies [20]. The standard Woods–Saxon nuclear density distribution coupled with a Monte Carlo Glauber model is employed to determine the initial density distribution in the overlapping zone between the two colliding nuclei [59,69]. A two-dimensional Gaussian distribution function of the form

$$s(x,y) = \frac{K}{2\pi\sigma^2} \sum_{i=1} \exp\left(-\frac{(x - x_i)^2 + (y - y_i)^2}{2\sigma^2} \right). \tag{2}$$

is used to distribute initial entropy/energy density around the source points. The position of the ith nucleon is denoted by (x_i, y_i) in the transverse plane, and the parameter σ denotes the granularity or the size of the initial density fluctuation. K is an overall normalization factor tuned to reproduce key observables, including charged particle multiplicity, spectra, and anisotropic flow parameters.

For AuAu (PbPb) collisions at RHIC (LHC), an initial formation time of about 0.17 (0.14) fm/c (with c the speed of light) is considered [80]. A lattice-based equation of state is used for transition from QGP to hot hadronic matter [81].

The state-of-the-art complete leading order, as well as next-to-leading-order rates of thermal photons production from QGP, has been available for quite some time now [64,82]. In the last couple of decades, there has been significant advancement in photon production from hadronic matter as well, which also includes the meson-meson and meson-baryon bremsstrahlung; see Ref. [25] and references therein for details.

Thermal photon rates are taken from Refs. [64,65,82] to calculate the anisotropic flow parameters at different centrality bins.

The total thermal emission is estimated by integrating the emission rates, $R = EdN/d^3p\,d^4x$, over the space-time four-volume:

$$E\frac{dN}{d^3p} = \int d^4x\, R(E^*(x), T(x)).\tag{3}$$

where E and p denote the photon energy and momentum, respectively, and $T(x)$ is the local temperature. $E^*(x) = p^\mu u_\mu(x)$, where p^μ is the four-momentum of the photon, and u_μ is the local four-velocity of the flow field obtained using longitudinal boost invariant ideal hydrodynamic model evolution. The Greek letter indices take the values 0 for time and 1, 2 and 3 for space (x) components.

In event-by-event hydrodynamic model calculation, the event plane angle (ψ_n in Equation (1)) is often replaced by the participant plane angle (ψ_n^{PP}), which is obtained from the initial state participant distribution [20].

The initial eccentricities are estimated using the relation

$$\epsilon_n = -\frac{\int dxdy\, r^2 \cos\left[n\left(\phi - \psi_n^{\mathrm{PP}}\right)\right]\varepsilon(x,y,\tau_0)}{\int dxdy\, r^2\varepsilon(x,y,\tau_0)}.\tag{4}$$

Here, ε is the initial energy density and $r^2 = x_1^2 + x_2^2$.

The participant plane angle ψ_n^{PP} is estimated as

$$\psi_n^{\mathrm{PP}} = \frac{1}{n}\arctan\frac{\int dxdy\, r^2 \sin(n\phi)\varepsilon(x,y,\tau_0)}{\int dxdy\, r^2 \cos(n\phi)\varepsilon(x,y,\tau_0)} + \pi/n.\tag{5}$$

The elliptic and triangular flows of thermal photons are calculated with respect to the participant plane angle. The anisotropic flow of direct photons can be estimated as follows:

$$v_n = \frac{v_n^{\mathrm{therm}} \times dN^{\mathrm{therm}}}{dN^{\mathrm{therm}} + dN^{\mathrm{non-therm}}}.\tag{6}$$

where v_n^{therm} is the anisotropic flow of thermal photons and dN^{therm} and $dN^{\mathrm{non-therm}}$ are the yields of thermal and non-thermal contributions, respectively. The $dN^{\mathrm{non-therm}}$ appears only in the denominator of Equation (6). Thus, the ratio of anisotropic flow parameters reflects only the thermal contribution by minimizing the non-thermal part. In this review, we calculate the ratio after taking an ensemble (event) average of individual anisotropic flow parameters (i.e., v_n) of thermal photons as follows:

$$\frac{\langle v_n \rangle_{\mathrm{event}}}{\langle v_m \rangle_{\mathrm{event}}} = \frac{\sum_{i=1}^{N_{\mathrm{event}}} v_n^{\mathrm{therm}(i)} dN^{\mathrm{therm}(i)}}{\sum_{j=1}^{N_{\mathrm{event}}} v_m^{\mathrm{therm}(j)} dN^{\mathrm{therm}(j)}}.\tag{7}$$

The theoretical ratio of elliptic and triangular flow parameters of thermal photons from $200A$ GeV AuAu collisions is shown in Figure 1 as a function of transverse momentum. The ratio of v_2 to and v_3 obtained from the experimental direct photon data by the PHENIX experiment at RHIC is shown in the same plot for comparison (see Figure A1 in Appendix A for the estimate of the experimental error on the ratio).

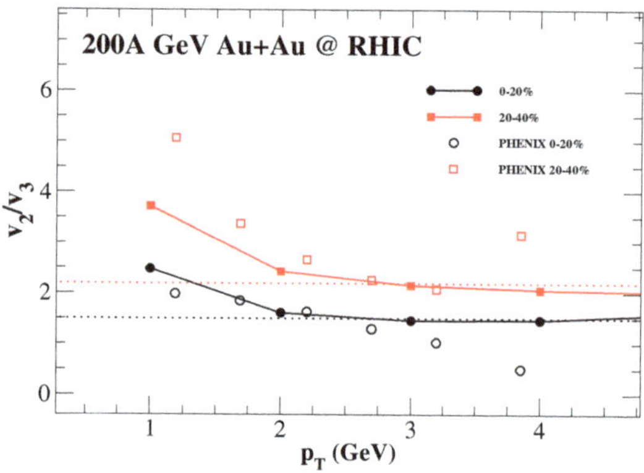

Figure 1. The calculated ratio of v_2 to v_3 for thermal photons from $200A$ GeV AuAu collisions for centrality bins 0–20% and 20–40% [79] compared to the PHENIX Collaboration experimental data at RHIC [83] (see text for details). The dotted lines parallel to p_T-axis show the approximate value of v_2/v_3 at which the ratio starts to saturate for each centrality bin.

The p_T-dependent behavior of the ratio is found to be different from the individual anisotropic flow parameters. The ratio is larger for peripheral collisions than for central collisions, as the photon triangular flow shows less sensitivity to collision centrality compared to the elliptic flow parameter. The ratio shows stronger p_T-dependence in the region $p_T \leq 2$ GeV compared to the larger-p_T region where it remains almost invariant with respect to p_T.

It was shown in Ref. [59] that the correlation between photon v_2 and v_3 is negligible for individual events. However, the event-averaged anisotropic flow parameters show similar p_T-dependent behavior in the region $p_T > 2$ GeV. As a result, negligible variation with p_T can be observed in that region for the ratio. In the lower-p_T region, the photon elliptic flow shows stronger sensitivity to p_T than the triangular flow parameter, and the ratio rises towards smaller p_T values.

Although hydrodynamic model calculations tend to significantly underestimate individual v_2 and v_3 data, an interesting observation emerges in the comparison of the v_2/v_3 ratios. Both experimental data and model calculations reveal a remarkable proximity in the 2–4 GeV p_T region. This specific p_T range is believed to be dominated by the QGP radiation in the direct photon spectrum.

It has been shown in Ref. [84], that the p_T-integrated ratio, v_2/v_3, of thermal photon anisotropic flow as a function of collision centrality shows much stronger sensitivity to the shear viscosity of the QGP medium compared to the same ratio estimated for charged particles. It was shown that both the p_T differential as well as the p_T-integrated ratio of anisotropic flow can serve as a viscometer for the QGP phase.

The v_2/v_3 ratio of thermal photons at LHC energy has been shown for three different centrality bins of PbPb collisions at the LHC energy in Figure 2. The ratio is found to be marginally dependent on transverse momentum in a relatively larger-p_T bin at the LHC energy compared to that ratio at RHIC. The lifetime, as well as temperature of the system produced at LHC, is expected to be larger than at RHIC, which resulted in significantly more production of thermal photons at LHC in the region $p_T > 2$ GeV. This might lead to a larger-p_T range over which the ratio remains flat at LHC compared to RHIC. The photon v_3 data at LHC would be valuable to confirm this behavior.

The v_2/v_3 of thermal photons from smaller systems also show similar qualitative nature as observed for AuAu and PbPb collisions; see Figure 3. The ratio is found to be slightly smaller for smaller systems as the triangular flow parameter for smaller systems is found to be relatively larger due to the more pronounced presence of initial state fluctuations.

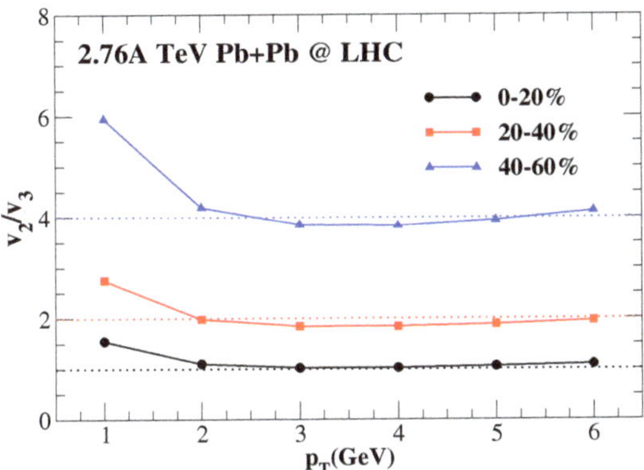

Figure 2. The predicted ratio of v_2 to v_3 of thermal photons thermal photon v_2 and v_3 from 2.76A TeV PbPb collisions for three different centrality bins [59]. The dotted lines parallel to p_T-axis show the approximate value of v_2/v_3 at which the ratio starts to saturate for each centrality bin.

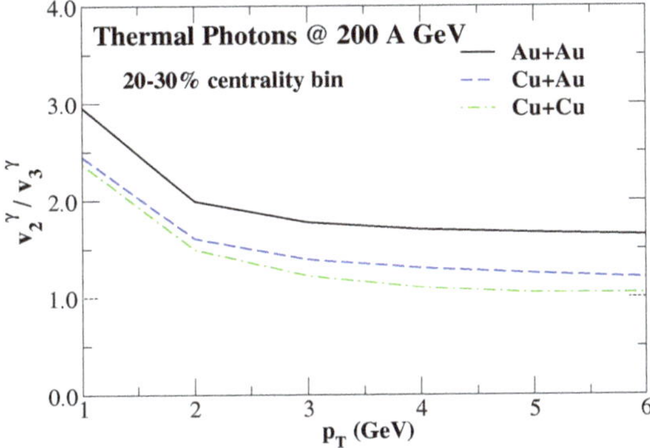

Figure 3. The predicted ratio of v_2 to v_3 of thermal photons from 20–30% centrality collisions of AuAu, CuAu, and CuCu collisions at 200A GeV [79].

The ratio is also found to be a robust quantity compared to the individual anisotropic flow parameters when the initial parameters of the model calculation are varied slightly. The initial formation time, τ_0, of QGP is not known precisely, and the τ_0 value ranging from 0.17 fm/c to 0.60 fm/c is considered mostly in different hydrodynamic model calculations to estimate the charged particle as well as photon production. The consideration of a fixed τ_0 for central as well as peripheral collisions is also an assumption to simplify the model calculation for a typical set of heavy-ion collisions. One should expect a larger value of plasma formation time for peripheral collisions, as the produced initial temperatures as well as the initial energy densities are expected to be relatively smaller for the more glancing collisions.

Although the individual anisotropic flow parameters show a strong sensitivity to the initial formation time [58], particularly at larger p_T values, due to the larger relative contribution in photon production from the hadronic phase, their p_T-dependent ratio does not change significantly in the same p_T region (see Figure 4 when the formation time τ_0 is increased from 0.17 fm/c to 0.60 fm/c at RHIC).

Additionally, a smaller value of the freeze-out temperature would result in a much larger contribution to v_n from the hadronic phase and subsequently a significantly larger total v_n for thermal photons [57]. However, it was observed that the ratio is a little sensitive to the value of kinetic freeze-out temperature. Even a significant drop in the value of T_f from 160 MeV to 120 MeV changes the ratio only marginally, as shown in Figure 5.

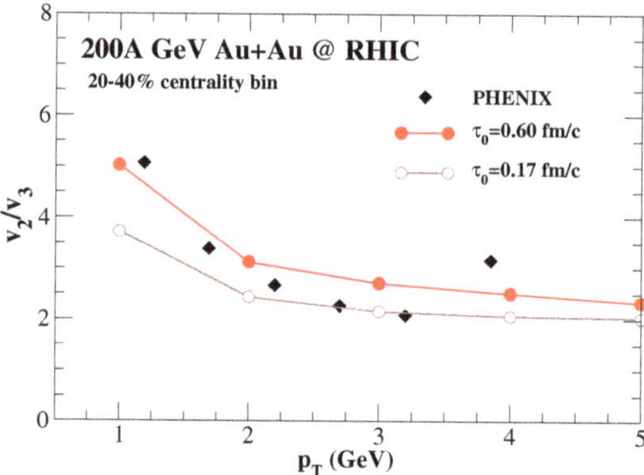

Figure 4. The predicted ratio of v_2 to v_3 of thermal photons with the quark–gluon plasma formation time, τ_0, values 0.60 fm/c and 0.17 fm/c from AuAu collisions.

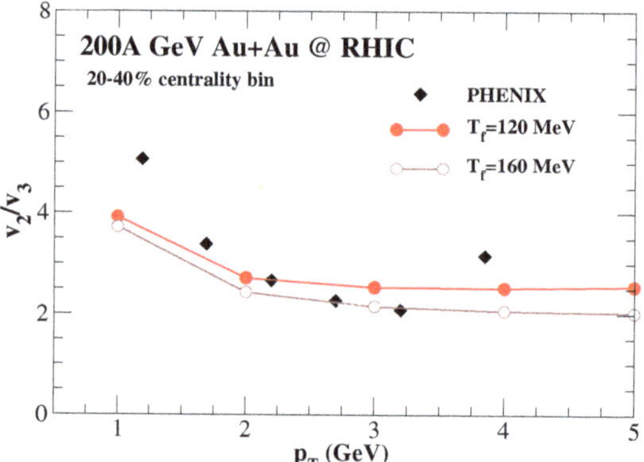

Figure 5. The predicted ratio of v_2 to v_3 of thermal photons from $200A$ GeV AuAu collisions at two different kinetic freeze-out temperatures of 120 MeV and 160 MeV [79].

To note is that the directed flow of photons shows a different p_T-dependent behavior than the elliptic or triangular flow parameters. The thermal photon $v_1(p_T)$, as calculated from hydrodynamic model calculations, is found to be negative for $p_T < 2$ GeV and becomes positive as p_T increases. The directed flow shows stronger sensitivity to the QGP phase compared to elliptic and triangular flow of photons; see Figures 3 and 6 in Ref. [72]. Thus, the ratio of photon v_1 with the elliptic (or triangular) flow parameter is more sensitive to hydrodynamic parameters (such as freeze-out temperature) than the v_2/v_3 ratio is. Therefore, v_1/v_n could be a potential parameter to understand more about photon observables and address the direct photon puzzle in relativistic nuclear collisions.

The final momentum anisotropies are considered to be a response of the initial geometry, and the correlation between ϵ_n and v_n shows how efficiently the initial spatial anisotropy is converted into final v_n [85]. Theoretical model calculations have shown that the correlation between ϵ_n and v_n is stronger for hadrons than for thermal photons. The correlation has also been found to be stronger for ϵ_2 and v_2 compared to ϵ_3 and v_3 both for photons and hadrons [59]. The linear correlation coefficient, $c(\epsilon_n, v_n)$, between ϵ_n and v_n is estimated using the relation,

$$c(\epsilon_n, v_n) = \left\langle \frac{(\epsilon_n - \langle \epsilon_n \rangle_{\text{evt}})(v_n - \langle v_n \rangle_{\text{evt}})}{\sigma_{\epsilon_n} \sigma_{v_n}} \right\rangle_{\text{evt}}. \tag{8}$$

The averages $\langle\ldots\rangle_{\text{event}}$ are taken over a sufficiently large number of random events. σ_{ϵ_n} and σ_{v_n} are the standard deviations of the initial spatial and final momentum anisotropies of thermal photons, respectively. In addition, the event averages of the initial spatial anisotropies and the anisotropic flow parameters are calculated by taking weight factors of impact parameter and thermal photon yields, respectively, for the corresponding events.

The $c(\epsilon_2, v_2)$ is found to be larger in the region $p_T \leq 3$ GeV for both RHIC and LHC (Figure 6). The correlation coefficient is found to be slightly larger at LHC than at RHIC. Probably the build-up of larger transverse flow velocity in the region $p_T \leq 3$ GeV might have resulted in a larger correlation strength in that region. Photons with relatively larger p_T are mostly from the initial hot and dense state with smaller transverse flow velocity and show a weaker correlation between ϵ_2 and v_2.

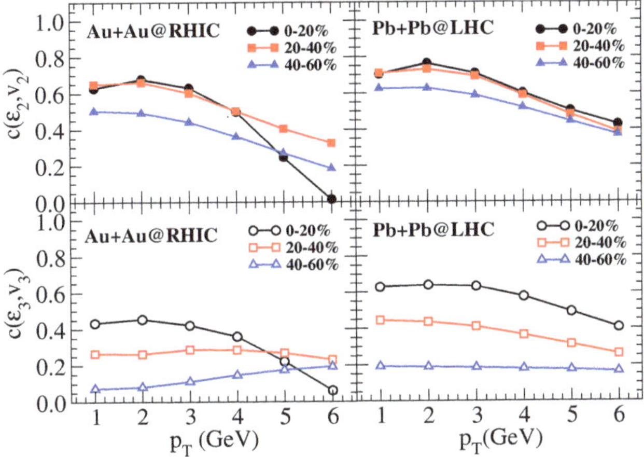

Figure 6. The predicted correlation between initial spatial anisotropies, ϵ_n, and final momentum anisotropies, v_n, of thermal photons at $200A$ GeV and at $2.76A$ TeV heavy-ion collisions for different centrality bins using relativistic ideal hydrodynamic model calculations. [59].

The photon elliptic flow parameter shows much stronger sensitivity to the collision centrality compared to the triangular flow parameter. However, the correlation coefficient for both the elliptic and triangular flow parameters (with corresponding initial spatial eccentricities) show similar p_T-dependent behavior.

To note is that the high-p_T part of the photon spectrum as well as the anisotropic flow parameters is more sensitive to the change in initial formation time and freeze-out temperature. Thus, the correlation strength is also expected to be strongly dependent on the initial parameters of the model calculation, especially at larger p_T values.

Interestingly, the larger-p_T part of the ratio of elliptic and triangular flow parameters is found to be less sensitive to the value of p_T as shown in Figures 1–5.

Thus, one can say that the correlation between v_2 and v_3 is stronger in the initial few fm time periods for high-p_T photons. One can see from Figure 7 that the ratio of total v_2/v_3 for high-p_T thermal photons saturates early and resembles the ratio of QGP contributions. In the later stage, with the strong build-up of transverse flow velocity, the relative change in elliptic flow is more than the triangular flow parameter resulting in a larger value of the ratio; see Figure 7.

Temporal evolution of the individual anisotropic flow parameters, as well as the correlation coefficient $c(\epsilon_n, v_n)$ at different p_T values, is expected to provide a more insight into the p_T-dependent behavior of the ratio of photon anisotropic flow parameters.

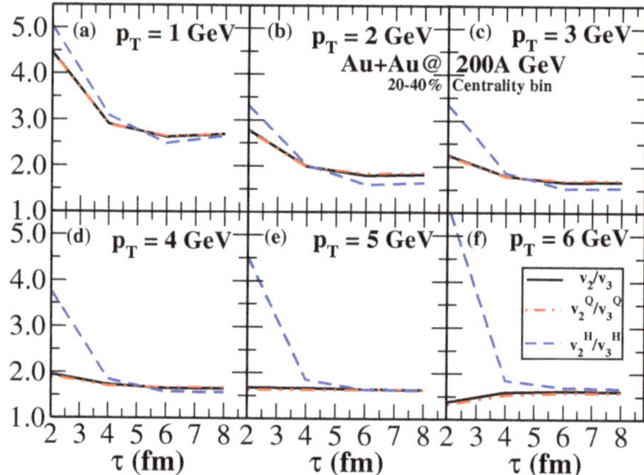

Figure 7. Time evolution of normalized total photon v_2/v_3 at different p_T [79].

4. Ratio of Anisotropic Flow in the Presence of Clustered Structure

In recent time, measurements with small collision systems such as pAu, dAu, and ^{3}HeAu have shown evidence of medium formation with the significantly large anisotropic flow of hadrons [86–88] and the scaling behavior of direct photon production over the scaled prompt photon production from pp collisions [89]. Recent studies suggest that the presence of triangular alpha-clustered structures in light nuclei might result in significant spatial anisotropies in the initial state when collided with heavy nuclei at relativistic energies [90–97]. These initial spatial anisotropies, due to the cluster structure, can be studied efficiently from final state momentum anisotropies.

Significant qualitative as well as quantitative differences between observables have been shown when alpha-clustered structures are included in light nuclei (like ^{12}C, ^{16}O, . . .) compared to the collisions of light nuclei with uniform density distributions. It is assumed that the typical time scale of relativistic nuclear collisions is too small for any slower nuclear excitation to take place, and thus, the initial clustered structure of the incoming light nuclei remains unchanged in these collisions.

Electromagnetic radiation can be a useful probe to study clustered structures in light nuclei due to their strong sensitivity to the initial state [98,99].

It has been shown that different orientations of triangular alpha-clustered carbon can give rise to different initial geometry on the transverse plane in collision with gold nuclei and subsequently different values of the anisotropic flow parameters even for most central collisions [99]. On the other hand, thermal photon production is found to be independent of the orientation of the clustered light nuclei.

A more realistic event-by-event ideal hydrodynamic model calculation of α-clustered carbons with gold at $200A$ GeV shows that the triangular flow of photons is significantly larger than the elliptic flow parameter [98]. This is contrary to the anisotropic flow results in heavy-ion collisions (such as AuAu, PbPb, . . .), where the elliptic flow is always larger than the triangular flow parameter. The event-averaged initial triangular eccentricities for most central collisions are found to be significantly larger than the elliptical eccentricities, which is reflected in the final flow results; see Figure 1 in Ref. [98].

The ratio of photon anisotropic flow from the collision of triangular α-clustered carbon with gold at RHIC energy is shown in Figure 8. As the photon $v_3(p_T)$ is larger than $v_2(p_T)$ in such collisions, the ratio is found to be less than 1 in the region $p_T > 1$ GeV.

Thus, the estimation of individual photon anisotropic flow and the ratio of the anisotropic flow parameters from light α-clustered nuclei can be complementary to the results from typical symmetric heavy-ion collisions. These findings will not only shed light on the clustered structure within light nuclei but also contribute to our understanding of the direct photon puzzle. Recent proposals for collisions of oxygen nuclei at the LHC indicate the potential to experimentally validate the presence of alpha-clustered structures in light nuclei [100].

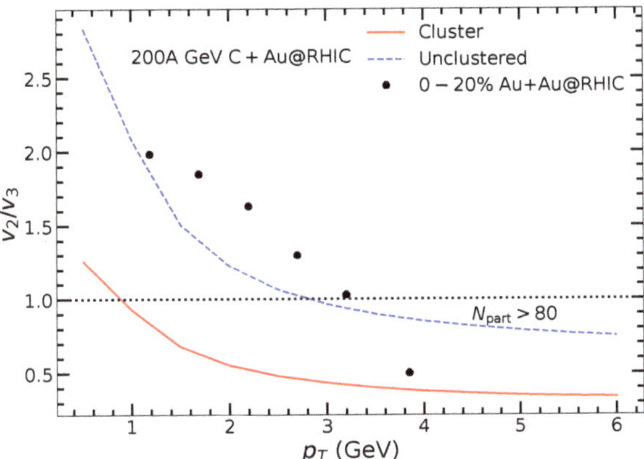

Figure 8. The predicted ratio of v_2 to v_3 of thermal photons from collisions of alpha-clustered and unclustered carbon nuclei with gold at the RHIC energy [98]. The dotted line (parallel to p_T-axis) shows the $v_2 = v_3$ case.

5. Summary and Conclusions

Direct photons originating from various stages of the evolving fireball in relativistic heavy-ion collisions dominate the differential photon spectrum in different p_T regions. Meanwhile, the differential elliptic and triangular flow parameters are predominantly influenced by thermal radiations. The presence of non-thermal photons dilutes the photon anisotropic flow, introducing an additional weight factor in the denominator of the photon v_n calculation.

The direct photon v_n measurements at RHIC and LHC energies consistently fall below the theoretical model calculations indicating a significant under-prediction of anisotropic flow.

It has been demonstrated in this review that the ratio of the elliptic and triangular flow parameters of thermal photons as a function of p_T exhibits intriguing characteristics. The ratio remains nearly independent of p_T in the region $p_T > 2$ GeV whereas it increases for smaller values of transverse momentum.

These findings straightforwardly indicate that the p_T-dependent behavior of v_2 and v_3 closely resemble each other in the $p_T > 2$ GeV region, dominated by radiation from the hot and dense plasma phase (although the prompt contribution starts to dominate the photon p_T spectrum above 4 GeV, these photons are not contributing to the anisotropic flow directly). The ratio also explains the experimental data better in the 2–4 GeV p_T-region dominated by thermal radiation. The high-p_T ($p_T > 4$ GeV) thermal photons mostly originate from the initial stage of system evolution, during which the development of transverse flow velocity is anticipated to be minimal. This likely contributes to the relatively poor explanation of the data in that p_T region. This ratio helps minimize the impact of non-thermal contributions and offers a more reliable measure of the anisotropic flow parameters associated with photon production.

By focusing on the ratio of photon v_2 to v_3 as a function of p_T within specific centrality bins along with the individual photon anisotropic flow parameters, one aims to provide a robust and insightful characterization of collision dynamics.

Funding: This research received no external funding.

Conflicts of Interest: The authors declare no conflicts of interest.

Appendix A

The estimatie of the error of the ratio of photon anisotropic flow parameters at each p_T was calculated using the given experimental values [83] of individual total uncertainties (comprising both systematic and statistical uncertainties) on v_2 and v_3. This calculation was performed as follows:

$$\frac{v_2}{v_3}\Big|_{\text{error}} = \frac{v_2}{v_3} \times \sigma\left(\frac{v_2}{v_3}\right)_{\text{rel}},$$

$$\text{where } \sigma\left(\frac{v_2}{v_3}\right)_{\text{rel}} = \sqrt{\sigma(v_2)_{\text{rel}}^2 + \sigma(v_3)_{\text{rel}}^2} \text{ and } \sigma(v_n)_{\text{rel}} = \frac{\sigma(v_n)}{v_n}, \tag{A1}$$

where σ_{rel} denotes relative uncertainity.

The estimated error of photon v_2/v_3 ratio for 0–20% and 20–40% AuAu collisions at $200A$ GeV is shown in Table A1.

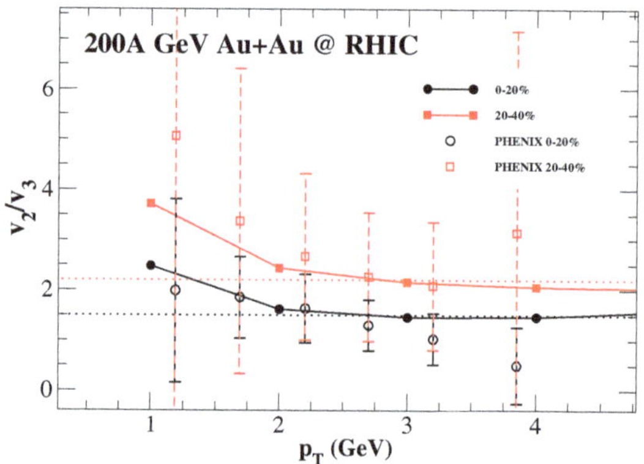

Figure A1. The model ratio of thermal photon v_2 to v_3 from $200A$ AuAu collisions for centrality bins 0–20% and 20–40% [79] compared to the PHENIX Collaboration experimental data [83], the latter at RHIC shown with the estimated errors (see text for details).

Table A1. Estimated (symmetric) error of the ratio of direct photon v_2 and v_3 in AuAu collisions at $200A$ GeV based on the experimental uncertainties of individual flow parameters.

Centrality	p_T (GeV)	v_2/v_3 (p_T)	Error Estimate
0–20% centrality	1.19	1.9832	1.8354
	1.69	1.8482	0.8103
	2.20	1.6303	0.6833
	2.70	1.2974	0.5053
	3.20	1.0289	0.5063
	3.85	0.5063	0.7587
20–40% centrality	1.19	5.0674	9.0235
	1.69	3.3807	3.0394
	2.20	2.6640	1.6579
	2.70	2.2646	1.2841
	3.20	2.0832	1.2792
	3.85	3.1558	4.0156

Due to considerably large uncertainties present in both the direct photon v_2 and v_3 data, the relative uncertainty in the ratio is also estimated to be quite large, especially for more peripheral collisions. Disregard the large error bars, the mean values of the data appear to align closely with the ratio calculated from theoretical model framework, as Figure A1 shows.

References

1. Harris, J.W.; Müller, B. The Search for the quark-gluon plasma. *Annu. Rev. Nucl. Part. Sci.* **1996**, *46*, 71–107. [CrossRef]
2. Shuryak, E. Strongly coupled quark-gluon plasma in heavy ion collisions. *Rev. Mod. Phys.* **2017**, *89*, 035001. [CrossRef]
3. Yagi, K.; Hatsuda, T.; Miake, Y. *Quark-Gluon Plasma: From Big Bang to Little Bang*; Cambridge University Press: Cambridge, UK, 2005.
4. Busza, W.; Rajagopal, K.; van der Schee, W. Heavy ion collisions: The big picture, and the big questions. *Annu. Rev. Nucl. Part. Sci.* **2018**, *68*, 339–376. [CrossRef]
5. Kolb, P.F.; Heinz, U. Hydrodynamic description of ultrarelativistic heavy-ion collisions. In *Quark–Gluon Plasma 3*; Hwa, R.C., Wang, X.-N., Eds.; World Scientific Co., Ltd.: Singapore, 2004; pp. 634–714. [CrossRef]
6. Adler, C. et al. [STAR Collaboration] Identified particle elliptic flow in Au + Au collisions at $\sqrt{s_{NN}} = 130$ GeV. *Phys. Rev. Lett.* **2001**, *87*, 182301. [CrossRef]
7. Adler, S.S. et al. [PHENIX Collaboration] Elliptic flow of identified hadrons in Au + Au collisions at $\sqrt{s_{NN}} = 200$ GeV. *Phys. Rev. Lett.* **2003**, *91*, 182301. [CrossRef]
8. Kolb, P.F.; Sollfrank, J.; Heinz, U.W. Anisotropic transverse flow and the quark hadron phase transition. *Phys. Rev. C* **2000**, *62*, 054909. [CrossRef]
9. Teaney, D.; Lauret, J.; Shuryak, E.V. A hydrodynamic description of heavy ion collisions at the SPS and RHIC. *arXiv* **2001**, arXiv:nucl-th/0110037.
10. Huovinen, P.; Kolb, P.F.; Heinz, U.W.; Ruuskanen, P.V.; Voloshin, S.A. Radial and elliptic flow at RHIC: Further predictions. *Phys. Lett. B* **2001**, *503*, 58–64. [CrossRef]
11. Heinz, U.; Snellings, R. Collective flow and viscosity in relativistic heavy-ion collisions. *Annu. Rev. Nucl. Part. Sci.* **2013**, *63*, 123–151. [CrossRef]
12. Eskola, K.J.; Honkanen, H.; Niemi, H.; Ruuskanen, P.V.; Räsänen, S.S. Predictiions for low-p_T and high-p_T hadron spectra in nearly central Pb+Pb collisions at $\sqrt{s_{NN}} = 5.5$ TeV tested at $\sqrt{s_{NN}} = 130$ and 200 GeV. . *Phys. Rev. C* **2005**, *72*, 044904. [CrossRef]
13. Huovinen, P.; Ruuskanen, P.V. Hydrodynamic models for heavy ion collisions. *Annu. Rev. Nucl. Part. Sci.* **2006**, *56*, 163–206. [CrossRef]
14. Nonaka, C.; Bass, S.A. Space-time evolution of bulk QCD matter. *Phys. Rev. C* **2007**, *75*, 014902. [CrossRef]
15. Romatschke, P.; Romatschke, U. Viscosity information from relativistic nuclear collisions: How perfect is the fluid observed at RHIC? *Phys. Rev. Lett.* **2007**, *99*, 172301. [CrossRef]
16. Gale, C.; Jeon, S.; Schenke, B. Hydrodynamic modeling of heavy-ion collisions. *Int. J. Mod. Phys. A* **2013**, *28*, 1340011. [CrossRef]
17. Schenke, B.; Jeon, S.; Gale, C. Elliptic and triangular flow in event-by-event $D = 3 + 1$ viscous hydrodynamics. *Phys. Rev. Lett.* **2011**, *106*, 042301. [CrossRef]
18. Hama, Y.; Kodama, T.; Socolowski, O., Jr. Topics on hydrodynamic model of nucleus-nucleus collisions. *Braz. J. Phys.* **2005**, *35*, 24–51. [CrossRef]
19. Andrade, R.; Grassi, F.; Hama, Y.; Kodama, T.; Socolowski, O., Jr. On the necessity to include event-by-event fluctuations in experimental evaluation of elliptical flow. *Phys. Rev. Lett.* **2006**, *97*, 202302. [CrossRef] [PubMed]
20. Holopainen, H.; Niemi, H.; Eskola, K.J. Event-by-event hydrodynamics and elliptic flow from fluctuating initial state. *Phys. Rev. C* **2011**, *83*, 034901. [CrossRef]
21. Schenke, B.; Tribedy, P.; Venugopalan, R. Fluctuating Glasma initial conditions and flow in heavy ion collisions. *Phys. Rev. Lett.* **2012**, *108*, 252301. [CrossRef]
22. Heinz, U.; Qiu, Z.; Shen, C. Fluctuating flow angles and anisotropic flow measurements. *Phys. Rev. C* **2013**, *87*, 034913. [CrossRef]
23. Alver, B.; Roland, G. Collision-geometry fluctuations and triangular flow in heavy-ion collisions. *Phys. Rev. C* **2010**, *81*, 054905. [CrossRef]
24. Qiu, Z.; Shen, C.; Heinz, U. Hydrodynamic elliptic and triangular flow in Pb–Pb collisions at $\sqrt{s_{NN}} = 2.76 A$ TeV. *Phys. Lett. B* **2012**, *707*, 151–155. [CrossRef]
25. David, G. Direct real photons in relativistic heavy ion collisions. *Rep. Prog. Phys.* **2020**, *83*, 046301. [CrossRef] [PubMed]
26. McLerran, L.D.; Toimela, T. Photon and dilepton emission from the quark-gluon plasma: Some general considerations. *Phys. Rev. D* **1985**, *31*, 545–563. [CrossRef] [PubMed]
27. Alam, J.; Sinha, B.; Raha, S. Electromagnetic probes of quark gluon plasma. *Phys. Rep.* **1996**, *273*, 243–362. [CrossRef]
28. Cassing, W.; Bratkovskaya, E.L. Hadronic and electromagnetic probes of hot and dense nuclear matter. *Phys. Rep.* **1999**, *308*, 65–233. [CrossRef]
29. Peitzmann, T.; Thoma, M.H. Direct photons from relativistic heavy-ion collisions. *Phys. Rep.* **2002**, *364*, 175–246. [CrossRef]
30. Stankus, P. Direct photon production in relativistic heavy-ion collisions. *Annu. Rev. Nucl. Part. Sci.* **2005**, *55*, 517–554. [CrossRef]
31. Shen, C.; Heinz, U.W.; Paquet, J.-F.; Gale, C. Thermal photons as a quark-gluon plasma thermometer reexamined. *Phys. Rev. C* **2014**, *89*, 044910. [CrossRef]
32. Paquet, J.-F. Probing the space-time evolution of heavy ion collisions with photons and dileptons. *Nucl. Phys. A* **2017**, *967*, 184–191. [CrossRef]
33. Gale, C. Direct photon production in relativistic heavy-ion collisions—A theory update. *PoS Proc. Sci.* **2019**, *320*, 023. [CrossRef]
34. Srivastava, D.K. Direct photons from relativistic heavy-ion collisions. *J. Phys. G Nucl. Part. Phys.* **2008**, *35*, 104026. [CrossRef]

35. Chatterjee, R.; Bhattacharya, L.; Srivastava, D.K. Electromagnetic probes. In *The Physics of the Quark-Gluon Plasma. Introductory Lectures*; Sarkar, S., Satz, H., Sinha, B., Eds.; Springer: Berlin/Heidelberg, Germany, 2010; pp. 219–264. [CrossRef]
36. Chatterjee, R. Anisotropic flow of photons in relativistic heavy ion collisions. *Pramana* **2021**, *95*, 15. [CrossRef]
37. Gale, C.; Hidaka, Y.; Jeon, S.; Lin, S.; Paquet, J.-F.; Pisarski, R.D.; Satow, D.; Skokov, V.V.; Vujanovic, G. Production and elliptic flow of dileptons and photons in a matrix model of the quark-gluon plasma. *Phys. Rev. Lett.* **2015**, *114*, 072301. [CrossRef]
38. Monnai, A. Thermal photon v_2 with slow quark chemical equilibration. *Phys. Rev. C* **2014**, *90*, 021901(R). [CrossRef]
39. McLerran, L.; Schenke, B. The Glasma, photons and the implications of anisotropy. *Nucl. Phys. A* **2014**, *929*, 71–82. [CrossRef]
40. Başar, G.; Kharzeev, D.E.; Skokov, V. Conformal anomaly as a source of soft photons in heavy ion collisions. *Phys. Rev. Lett.* **2012**, *109*, 202303. [CrossRef] [PubMed]
41. Tuchin, K. Electromagnetic radiation by quark-gluon plasma in a magnetic field. *Phys. Rev. C* **2013**, *87*, 024912. [CrossRef]
42. Zakharov, B.G. Effect of magnetic field on the photon radiation from quark-gluon plasma in heavy ion collisions. *Eur. Phys. J. C* **2016**, *76*, 609. [CrossRef]
43. Vujanovic, G.; Paquet, J.-F.; Denicol, G.S.; Luzum, M.; Schenke, B.; Jeon, S.; Gale, C. Probing the early-time dynamics of relativistic heavy-ion collisions with electromagnetic radiation. *Nucl. Phys. A* **2014**, *932*, 230–234. [CrossRef]
44. Liu, F.-M.; Liu, S.-X. Quark-gluon plasma formation time and direct photons from heavy ion collisions. *Phys. Rev. C* **2014**, *89*, 034906. [CrossRef]
45. Garcia-Montero, O.; Löher, N.; Mazeliauskas, A.; Berges, J.; Reygers, K. Probing the evolution of heavy-ion collisions using direct photon interferometry. *Phys. Rev. C* **2020**, *102*, 024915. [CrossRef]
46. Ruuskanen, P.V. Electromagnetic probes of quark-gluon plasma in relativistic heavy-ion collisions. *Nucl. Phys. A* **1992**, *544*, 169–182. [CrossRef]
47. Srivastava, D.K.; Kapusta, J.I. Photon interferometry of quark-gluon dynamics. *Phys. Lett. B* **1993**, *307*, 1–6. [CrossRef]
48. Srivastava, D.K. Intensity interferometry of thermal photons from relativistic heavy-ion collisions. *Phys. Rev. C* **2005**, *71*, 034905. [CrossRef]
49. Peressounko, D. Hanbury Brown–Twiss interferometry of direct photons in heavy ion collisions. *Phys. Rev. C* **2003**, *67*, 014905. [CrossRef]
50. Frodermann, E.; Heinz, U. Photon Hanbury-Brown–Twiss interferometry for noncentral heavy-ion collisions. *Phys. Rev. C* **2009**, *80*, 044903. [CrossRef]
51. Adare, A. et al. [PHENIX Collaboration] Centrality dependence of low-momentum direct-photon production in Au + Au collisions at $\sqrt{s_{NN}} = 200$ GeV. *Phys. Rev. C* **2015**, *91*, 064904. [CrossRef]
52. Adam, J. et al. [ALICE Collaboration] Enhanced production of multi-strange hadrons in high-multiplicity proton–proton collisions. *Nat. Phys.* **2017**, *13*, 535–539. [CrossRef]
53. Adam, J. et al. [ALICE Collaboration] Direct photon production in Pb–Pb collisions at $\sqrt{s_{NN}} = 2.76$ TeV. *Phys. Lett. B* **2016**, *754*, 235–248. [CrossRef]
54. Chatterjee, R. Electroweak physics: Summary. *PoS Proc. Sci.* **2021**, *387*, 026. [CrossRef]
55. Adare, A. et al. [PHENIX Collaboration] Low-momentum direct-photon measurement in Cu + Cu collisions at $\sqrt{s_{NN}} = 200$ GeV. *Phys. Rev. C* **2018**, *98*, 054902. [CrossRef]
56. Adare, A. et al. [PHENIX Collaboration] Beam energy and centrality dependence of direct-photon emission from ultrarelativistic heavy-ion collisions. *Phys. Rev. Lett.* **2019**, *123*, 022301. [CrossRef] [PubMed]
57. Chatterjee, R.; Frodermann, E.S.; Heinz, U.W.; Srivastava, D.K. Elliptic flow of thermal photons in relativistic nuclear collisions. *Phys. Rev. Lett.* **2006**, *96*, 202302. [CrossRef] [PubMed]
58. Chatterjee, R.; Srivastava, D.K. Formation time of QGP from thermal photon elliptic flow. *Nucl. Phys. A* **2009**, *830*, 503c–506c. [CrossRef]
59. Chatterjee, R.; Dasgupta, P.; Srivastava, D.K. Anisotropic flow of thermal photons at energies available at the BNL Relativistic Heavy Ion Collider and at the CERN Large Hadron Collider. *Phys. Rev. C* **2017**, *96*, 014911. [CrossRef]
60. Chatterjee, R.; Srivastava, D.K.; Renk, T. Triangular flow of thermal photons from an event-by-event hydrodynamic model for $2.76A$ TeV Pb + Pb collisions at the CERN Large Hadron Collider. *Phys. Rev. C* **2016**, *94*, 014903. [CrossRef]
61. Dasgupta, P.; Chatterjee, R.; Srivastava, D.K. Spectra and elliptic flow of thermal photons from full-overlap U+U collisions at energies available at the BNL Relativistic Heavy Ion Collider. *Phys. Rev. C* **2017**, *95*, 064907. [CrossRef]
62. Dasgupta, P.; Chatterjee, R.; Singh, S.K.; Alam, J. Effects of initial-state nucleon shadowing on the elliptic flow of thermal photons. *Phys. Rev. C* **2018**, *97*, 034902. [CrossRef]
63. Chatterjee, R.; Srivastava, D.K.; Heinz, U.; Gale, C. Elliptic flow of thermal dileptons in relativistic nuclear collisions. *Phys. Rev. C* **2007**, *75*, 054909. [CrossRef]
64. Arnold, P.B.; Moore, G.D.; Yaffe, L.G. Photon emission from quark gluon plasma: Complete leading order results. *J. High Energy Phys.* **2001**, *12*, 009. [CrossRef]
65. Turbide, S.; Rapp, R.; Gale, C. Hadronic production of thermal photons. *Phys. Rev. C* **2004**, *69*, 014903. [CrossRef]
66. Adare, A. et al. [PHENIX Collaboration] Observation of direct-photon collective flow in Au + Au collisions at $\sqrt{s_{NN}} = 200$ GeV. *Phys. Rev. Lett.* **2012**, *109*, 122302. [CrossRef] [PubMed]
67. Lohner, D.; ALICE Collaboration. Measurement of direct-photon elliptic flow in Pb–Pb collisions at $\sqrt{s_{NN}} = 2.76$ TeV. *J. Phys. Conf. Ser.* **2013**, *446*, 012028. [CrossRef]

68. Chatterjee, R.; Holopainen, H.; Helenius, I.; Renk, T.; Eskola, K.J. Elliptic flow of thermal photons from event-by-event hydrodynamic model. *Phys. Rev. C* **2013**, *88*, 034901. [CrossRef]

69. Chatterjee, R.; Holopainen, H.; Renk, T.; Eskola, K.J. Enhancement of thermal photon production in event-by-event hydrodynamics. *Phys. Rev. C* **2011**, *83*, 054908. [CrossRef]

70. van Hees, H.; Gale, C.; Rapp, R. Thermal photons and collective flow at energies available at the BNL Relativistic Heavy-Ion Collider. *Phys. Rev. C* **2011**, *84*, 054906. [CrossRef]

71. Linnyk, O.; Cassing, W.; Bratkovskaya, E.L. Centrality dependence of the direct photon yield and elliptic flow in heavy-ion collisions at $\sqrt{s_{NN}} = 200$ GeV. *Phys. Rev. C* **2014**, *89*, 034908. [CrossRef]

72. Dasgupta, P.; Chatterjee, R.; Srivastava, D.K. Directed flow of photons in Cu+Au collisions at RHIC. *J. Phys. G* **2020**, *47*, 085101. [CrossRef]

73. Dasgupta, P.; De, S.; Chatterjee, R.; Srivastava, D.K. Photon production from Pb + Pb collisions at $\sqrt{s_{NN}} = 5.02$ TeV at the CERN Large Hadron Collider and at $\sqrt{s_{NN}} = 39$ TeV at the proposed Future Circular Collider facility. *Phys. Rev. C* **2018**, *98*, 024911. [CrossRef]

74. Sun, J.-A.; Yan, L. The effect of weak magnetic photon emission from quark-gluon plasma. *arXiv* **2023**, arXiv:2302.07696.

75. Adamczyk, L. et al. [STAR Collaboration] Charge-dependent directed flow in Cu + Au collisions at $\sqrt{s_{NN}} = 200$ GeV. *Phys. Rev. Lett.* **2017**, *118*, 012301. [CrossRef]

76. Chatterjee, S.; Singh, S.K.; Ghosh, S.; Hasanujjaman, M.; Alam, J.; Sarkar, S. Initial condition from the shadowed Glauber model. *Phys. Lett. B* **2016**, *758*, 269–273. [CrossRef]

77. Gale, C.; Paquet, J.-F.; Schenke, B.; Shen, C. Probing early-time dynamics and quark-gluon plasma transport properties with photons and hadrons. *Nucl. Phys. A* **2021**, *1005*, 121863. [CrossRef]

78. Chatterjee, R.; Holopainen, H.; Renk, T.; Eskola, K.J. Collision centrality and τ_0 dependence of the emission of thermal photons from fluctuating initial state in ideal hydrodynamic calculation. *Phys. Rev. C* **2012**, *85*, 064910. [CrossRef]

79. Chatterjee, R.; Dasgupta, P. Ratio of photon anisotropic flow in relativistic heavy ion collisions. *Phys. Rev. C* **2021**, *104*, 064907. [CrossRef]

80. Eskola, K.J.; Kajantie, K.; Ruuskanen, P.V.; Tuominen, K. Scaling of transverse energies and multiplicities with atomic number and energy in ultrarelativistic nuclear collisions. *Nucl. Phys. B* **2000**, *570*, 379–389. [CrossRef]

81. Laine, M.; Schröder, Y. Quark mass thresholds in QCD thermodynamics. *Phys. Rev. D* **2006**, *73*, 085009. [CrossRef]

82. Ghiglieri, J.; Hong, J.; Kurkela, A.; Lu, E.; Moore, G.D.; Teaney, D. Next-to-leading order thermal photon production in a weakly coupled quark-gluon plasma. *J. High Energy Phys.* **2013**, *2013*, 10. [CrossRef]

83. Adare, A. et al. [PHENIX Collaboration] Azimuthally anisotropic emission of low-momentum direct photons in Au + Au collisions at $\sqrt{s_{NN}} = 200$ GeV. *Phys. Rev. C* **2016**, *94*, 064901. [CrossRef]

84. Shen, C.; Heinz, U.; Paquet, J.-F.; Gale, C. Thermal photon anisotropic flow serves as a quark-gluon plasma viscometer. *Nucl. Phys. A* **2014**, *932*, 184–188. [CrossRef]

85. Niemi, H.; Denicol, G.S.; Holopainen, H.; Huovinen, P. Event-by-event distributions of azimuthal asymmetries in ultrarelativistic heavy-ion collisions. *Phys. Rev. C* **2013**, *87*, 054901. [CrossRef]

86. Aidala, C. et al. [PHENIX Collaboration] Creation of quark–gluon plasma droplets with three distinct geometries. *Nat. Phys.* **2019**, *15*, 214–220. [CrossRef]

87. Aad, G. et al. [ATLAS Collaboration] Observation of long-range elliptic azimuthal anisotropies in $\sqrt{s} = 13$ and 2.76 TeV pp collisions with the ATLAS detector. *Phys. Rev. Lett.* **2016**, *116*, 172301. [CrossRef] [PubMed]

88. Chatrchyan, S. et al. [CMS Collaboration] Observation of long-range, near-side angular correlations in pPb collisions at the LHC. *Phys. Lett. B* **2013**, *718*, 795–814. [CrossRef]

89. Khachatryan, V.; PHENIX Collaboration. Direct photon production and scaling properties in large and small system collisions. *J. Phys. Conf. Ser.* **2020**, *1602*, 012015. [CrossRef]

90. Rybczyński, M.; Piotrowska, M.; Broniowski, W. Signatures of α clustering in ultrarelativistic collisions with light nuclei. *Phys. Rev. C* **2018**, *97*, 034912. [CrossRef]

91. Bożek, P.; Broniowski, W.; Ruiz Arriola, E.; Rybczyński, M. α clusters and collective flow in ultrarelativistic carbon–heavy-nucleus collisions. *Phys. Rev. C* **2014**, *90*, 064902. [CrossRef]

92. Behera, D.; Deb, S.; Singh, C.R.; Sahoo, R. Characterizing nuclear modification effects in high-energy O-O collisions at energies available at the CERN Large Hadron Collider: A transport model perspective. *Phys. Rev. C* **2024**, *109*, 014902. [CrossRef]

93. Behera, D.; Mallick, N.; Tripathy, S.; Prasad, S.; Mishra, A.N.; Sahoo, R. Predictions on global properties in O+O collisions at the Large Hadron Collider using a multi-phase transport model. *Eur. Phys. J. A* **2022**, *58*, 175. [CrossRef]

94. Li, Y.-A.; Zhang, S.; Ma, Y.-G. Signatures of α-clustering in ^{16}O by using a multiphase transport model. *Phys. Rev. C* **2020**, *102*, 054907. [CrossRef]

95. Zhang, S.; Ma, Y.G.; Chen, J.H.; He, W.B.; Zhong, C. Nuclear cluster structure effect on elliptic and triangular flows in heavy-ion collisions. *Phys. Rev. C* **2017**, *95*, 064904. [CrossRef]

96. He, J.; He, W.-B.; Ma, Y.-G.; Zhang, S. Machine-learning-based identification for initial clustering structure in relativistic heavy-ion collisions. *Phys. Rev. C* **2021**, *104*, 044902. [CrossRef]

97. Wang, Y.; Zhao, S.; Cao, B.; Xu, H.-j.; Song, H. Exploring the compactness of α cluster in the ^{16}O nuclei with relativistic ^{16}O+^{16}O collisions. *arXiv* **2024**, arXiv:2401.15723. [CrossRef].

98. Dasgupta, P.; Chatterjee, R.; Ma, G.-L. Production and anisotropic flow of thermal photons in collisions of α-clustered carbon with heavy nuclei at relativistic energies. *Phys. Rev. C* **2023**, *107*, 044908. [CrossRef]
99. Dasgupta, P.; Ma, G.-L.; Chatterjee, R.; Yan, L.; Zhang, S.; Ma, Y.-G. Thermal photons as a sensitive probe of α-cluster in C + Au collisions at the BNL Relativistic Heavy Ion Collider. *Eur. Phys. J. A* **2021**, *57*, 134. [CrossRef]
100. Brewer, J.; Mazeliauskas, A.; van der Schee, W. Opportunities of OO and pO collisions at the LHC. *arXiv* **2021**, arXiv:2103.01939. [CrossRef]

MDPI AG
Grosspeteranlage 5
4052 Basel
Switzerland
Tel.: +41 61 683 77 34

Physics Editorial Office
E-mail: physics@mdpi.com
www.mdpi.com/journal/physics